T0300133

RECLAIMING THE DESERT: TOWARDS A SUSTAINABLE ENVIRONMENT IN ARID LANDS

DEVELOPMENTS IN ARID REGIONS RESEARCH SERIES

The aim of DARE's series is to present basic and applied research related to geoengineering problems in arid lands. Areas of interest include soil science and engineering in arid lands, information technology application to geoengineering, geoenvironmental engineering, hydrogeology, applied geomechanics, novel geotechnical construction techniques, sustainable land use, and case histories.

In the series, the books are intended for use in university courses at postgraduate level and are expected to be valuable reference materials for practicing engineers.

A.M.O. Mohamed
Series Editor

BOOKS IN SERIES

1. A.M.O. Mohamed & K.I. AL Hosani, K.I. (eds.) "*Geoengnieering in Arid Lands*" 2000.
   A.A. Balkema Publishers, ISBN 90 5809 160 0, 720 pages.
2. A.M.O. Mohamed, D. Chenaf & S. El-Shahed "*DARE's Dictionary of Environmental Sciences and Engineering: English-French-Arabic*" 2003.
   A.A. Balkema Publishers, ISBN 90 5809 617 3, 556 pages.
3. A.M.O. Mohamed (ed.) "*Reclaiming the Desert: Towards a Sustainable Enviornment in Arild Lands*" 2006.
   Taylor & Francis/Balkema, ISBN 0 415 41128 9, 318 pages.

Developments in Arid Regions Research, 3

PROCEEDINGS OF THE 3RD UAE–JAPAN SYMPOSIUM ON SUSTAINABLE GCC ENVIRONMENT AND WATER RESOURCES (EWR2006), ABU DHABI, UNITED ARAB EMIRATES, JANUARY 28–30, 2006

Reclaiming the Desert: Towards a Sustainable Environment in Arid Lands

Editor

A.M.O. Mohamed
UAE University, Al Ain, United Arab Emirates

LONDON / LEIDEN / NEW YORK / PHILADELPHIA / SINGAPORE

Published by: Taylor & Francis/Balkema
P.O. Box 447, 2300 AK Leiden, The Netherlands
e-mail: Pub.NL@tandf.co.uk
www.balkema.nl, www.tandf.co.uk, www.crcpress.com

ISBN10: 0–415–41128–9 ISBN13: 978–0–415–41128–8

Printed in Great Britain

Reclaiming the Desert: Towards a Sustainable Enviornment in Arid Lands – Mohamed (ed.)
© 2006 Taylor & Francis Group, London, ISBN 0 415 41128 9

Table of Contents

Waste water treatment

EWR2006

Organized by the Research Affairs, UAE University and Japan Cooperation Center, Petroleum (JCCP), Japan.

UAE University

The *United Arab Emirates University* was established in 1976 by H.H. Sheikh Zayed Bin Sultan Al Nahayan, the father of the nation, as the flagship institution of higher education in the region and to realize his vision of knowledge based society. After almost three decades, the UAE University is standing tall as the premier national university whose mission is to meet the educational and cultural needs of the UAE society by providing programs and services of the highest quality. It contributes to the expansion of knowledge by conducting quality research and by developing and applying modern information technology and is recognized as the major research engine of the region. It plays a significant role in leading cultural, social and economic development in the country. UAEU maintains excellent teaching, an effective learning environment with an optimum student population size, quality applied research, and outstanding community services. The focus of the UAEU has also been on international recognition of excellence through accreditation and external evaluation of all its activities.

Japan Cooperation Center, Petroleum (JCCP)

Japan Cooperation Center, Petroleum (JCCP) was founded in 1981 under the auspices of the Ministry of Economy, Trade and Industry and with support from Japanese Oil and Engineering Industries. Its purpose is to work for mutual understanding between Japan and Oil-producing countries, to build relations of trust and friendship with those countries and to strengthen cooperative relationship among them. Oil-producing countries have needs in refining, marketing, and technological and human resources development. JCCP activities include: programs to accept downstream sector of the oil industry, which JCCP helps to satisfy through cooperation in overseas training and technical cooperation, etc.

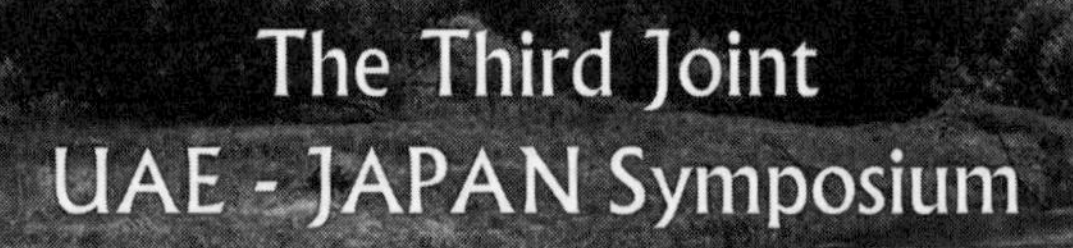

UAE University,
United Arab Emirates

Japan Cooperation
Center, Petroleum, Japan

Reclaiming the Desert: Towards a Sustainable Environment in Arid Lands – Mohamed (ed.)
© 2006 Taylor & Francis Group, London, ISBN 0 415 41128 9

About the Editor

Advisor to the Assistant Provost for Research, UAE University
Professor of Geotechnical and Geoenvironmental Engineering
Department of Civil and Environmental Engineering, UAE University
P.O. Box 17551, Al Ain, United Arab Emirates
Tel.: (9713) 7675580; Fax: (9713) 7675582;
E-mail: Mohamed.A@uaeu.ac.ae
Web: http://www.engg.uaeu.ac.ae/mohamed.a

A.M.O. Mohamed earned his M. Eng. in 1983 and Ph.D. in 1987 from the Department of Civil Engineering and Applied Mechanics, McGill University, Montreal, Canada. From 1987 to 1998 Dr. Mohamed was employed by McGill University, and was the *Associate Director* of the former Geotechnical Research Centre (GRC) and *Adjunct Professor* in the Department of Civil Engineering and Applied Mechanics. In 1998, he joined the Department of Civil and Environmental Engineering, UAE University, where he is currently *Professor of Geotechnical and Geoenvironmental Engineering*, and the *Advisor to the Assistant Provost for Research.*

Professor Mohamed is currently the *President* of the Gulf Society for Geoengineering (GSGE): http:// www.engg.uaeu.ac.ae/gsge, the *Editor-in-Chief* of Developments in Arid Regions Research (DARE) Series Published by A.A. Balkema Publishers: http://balkema.ima.nl/Scripts/cgiBalkema.exe, Regional Editor, Journal of Applied Sciences, Pakistan; Executive Board Member, Arab Healthy Water Association; http://www.mgwater.com/executive.shtml; *A member of the Geotechnical and Geoenvironmental Engineering Technical Committee* of International Society of Offshore and Polar Engineering (ISOPE); and the former *Editor-in-Chief* of the Emirates Journal for Engineering Research Published by UAE University. Prof. Mohamed also served as a *member of the International Advisory Board* for various international conferences.

Professor Mohamed's research activities have resulted in co-authoring the following three books:

(1) Yong, R.N., Mohamed, A.M.O., and Warkentin, B.P. (1992) "*Principles of Contaminant Transport in Soils*," Elsevier, The Netherlands, 327 pages; The book was *translated into Japanese in 1995.*
(2) Mohamed, A.M.O., and Antia, H.E. (1998) "*Geoenvironmental Engineering*," Elsevier, The Netherlands, 707 pages and
(3) Mohamed, A.M.O., Chenaf, D., and El-Shahed, S. (2003) "*DARE's Dictionary of Environmental Sciences and Engineering: English-French-Arabic*, A.A. Balkema Publishers, ISBN 90 5809 617 3, 556 pages.

Professor Mohamed has edited/co-edited the following four books:

(1) Yong, R.N., Hadjinicolaou, J., and Mohamed, A.M.O. (1993) "*Environmental Engineering*, Volumes 1 and 2, ASCE-CSCE
(2) Mohamed, A.M.O., and Al-Hosani, K. (2000) "*Geoengineering in Arid Lands*," A.A. Balkema, ISBN 90 5809 160 0, 720 pages.

(3) A.M.O. Mohamed (ed.) (2006) "*Arid Land Hydrogeology: In Search of Solution to a Threatened Resource*" Taylor & Francis/Balkema; and
(4) A.M.O. Mohamed (ed.) (2006) "*Reclaiming the Desert: Towards a Sustainable Environmental in Arid Lands*" Taylor & Francis/Balkema.

For publications in refereed journals and conference proceedings, Professor Mohamed published ***190 papers***, and for training of highly qualified personnel, he *supervised 28 graduate students* (M.Sc., M.Eng. and Ph.D.) in UAE and McGill Universities.

Professor Mohamed's present research activities contribute to soil properties and behaviour, ground improvement, soil-pollutant interactions, transport processes, multi-phase flow, remediation of polluted soils, monitoring of subsurface pollutants via time domain electrometry techniques, waste management, environmental impact assessment and risk management.

Professor Mohamed awarded:

(1) *The Outstanding Performance and Distinction Award in Research* in the College of Engineering, UAE University (2004),
(2) *The Outstanding Performance and Distinction Award in University and Community Services* in the College of Engineering, UAE University (2004),
(3) *The Best Interdisciplinary Research Project Award* in the College of Engineering, Research Affairs Sector, UAE University (2004),
(4) *The Decree of Merit for Outstanding Contribution to the Geoengineering Field* from International Biographical Center (IBC), Cambridge, England (2002), and
(5) *The Outstanding Performance and Distinction Award in Research* in the College of Engineering, UAE University (2001).

For Professor Mohamed's noteworthy achievements in his field, his biography has been included in Marquis Who's Who in Science and Engineering, 2002-present, Marquis Who's Who in the World, 20th Edition, 2002, and the International Biographical Center (IBC) inaugural edition of One Thousand Great Scientists, Cambridge, England (2003).

Reclaiming the Desert: Towards a Sustainable Environment in Arid Lands – Mohamed (ed.)
© 2006 Taylor & Francis Group, London, ISBN 0 415 41128 9

Preface

For the last 13 years, Japan Cooperation Center, Petroleum (JCCP), (1992-2001) as PEC, has been organizing annual symposia to promote the enhancement of the environment under desert conditions, focusing on greenery development. These symposia have been held in the Gulf Cooperation Council (GCC) countries such as Kingdom of Saudi Arabia, State of Kuwait, United Arab Emirates (UAE), Sultanate of Oman and the Kingdom of Bahrain, as well as in Japan, and have attracted considerable interest from the scientists and professionals in those countries. In bringing together these scientists, the symposia provided forums for exchanging views and information on multitudes of issues ranging from water purification, usage and water management, to desert greening.

Following the success of the first and second UAE-Japan symposia at the UAE, this third joint UAE-Japan Symposium on sustainable GCC Environment and Water Resources (January 28–30th) held in Abu Dhabi, is jointly sponsored by the Research Affairs, UAE University, and Japan Cooperation Center, Petroleum (JCCP), Japan.

The set objectives of the symposium were:

1. To bring together scientists and professionals from GCC and Japan to communicate and exchange knowledge on sustainable environment and water resources in arid environment, and
2. To provide better understanding of the environment and rehabilitation of the environment through utilization and application of innovative technologies.

Included in this book is a very good and comprehensive collection of contributions from prominent researchers and authorities, dealing with the most pressing problems in *Sustainable Practice of Environmental Scientists and Engineers in Arid Lands.* Research studies and actual experiences dealing with sulfur utilization in agriculture and construction engineering, arid land ecological issues, renewable energy sources, global warming issues in compliance with Kyoto protocol, hazardous and waste water treatment technologies are some of the many research and application topics that have been addressed.

The success of the symposium and the quality of the papers presented in this volume have been due not only to the tireless effort of the members of the organizing committee, but also to the many contributions made by the members of the Research Affairs sector, UAE university.

Although many excellent abstracts were unfortunately judged "unsuitable" for the symposium and were not included in the program, acknowledgement and appreciation should be extended to those authors for their interest in this symposium. To many paper presenters and session chairpersons, we owe our considerable gratitude for their efforts and participation. This volume proceeding of the third joint UAE-Japan symposium bears testimony to their magnificent efforts and contributions and will serve to preserve the record of sustainable GCC environmental science and engineering as represented by UAEU and JCCP.

A.M.O. Mohamed
2006

Reclaiming the Desert: Towards a Sustainable Environment in Arid Lands – Mohamed (ed.)
© 2006 Taylor & Francis Group, London, ISBN 0 415 41128 9

Summary, recommendations and suggestions for actions

A.M.O. Mohamed
Department of Civil and Environmental Engineering, UAE University, UAE

1 SUMMARY AND RECOMMENDATIONS

Papers presented at the Third Joint UAE–JAPAN Symposium on Sustainable GCC Environment and Water Resources, EWR2006, January 28–29, 2006, Abu Dhabi, UAE, was divided into two themes.

Theme A related to the environmental issues and published in "*Reclaiming the Desert: Towards a Sustainable Environment in Arid Lands*" In Developments in Arid Regions Research, DARE, Vol. 3, A.M.O. Mohamed (ed.), Taylor & Francis/Balkema, 2006

Theme B related to arid land hydrogeological issues and published in "*Arid Land Hydrogeology: In Search of a Solution to a Threatened Resource*" In Developments in Arid Regions Research, DARE, Vol. 4, A.M.O. Mohamed (ed.), Taylor & Francis/Balkema, 2006

After intensive discussions conducted by the participants of EWR 2006, the most important areas that have been selected for recommendations are:

1. Soil management;
   a. Sulfur utilization in agriculture
   b. Solar distillation
   c. Soil moisture enhancement
   d. Use of oily treated water in agriculture
   e. Soil solarization for direct thermal inactivation of soil borne-pests
   f. Utilization of adopted indigenous forage species to combat drought during vegetative and reproductive stages in the field
2. Water management challenges;
3. Groundwater recharge in arid lands;
4. Groundwater contamination and management;
5. Monitoring systems;
6. Bioreactor Landfills;
7. Treatment of contaminated soils;
8. Waste water treatment;
9. Environmental impact assessment;
10. Risk management of polluted sites; and
11. Sulfur utilization in construction industry
12. Sustainable cities
13. Global warming
14. Oil spill management

1.1 *Soil management*

The constraints on managing soils for sustainable land use in dry lands are different in nature. Some are connected with the specificity of the natural conditions including soils, others with the prevailing socioeconomic conditions. Appropriate management of soils is the prerequisite of sustainable land use in the dry lands. Certain problems still require extensive research, e.g., soil resilience, particularly in view of the diversity of dry land soils in different ecological conditions. Another area of research in the dry lands is the relationship between the processes of soil degradation and the resultant different ecological situations; to answer the question whether soil erosion is always harmful or whether it may be beneficial in certain cases depending on the nature of the soil parent material and soil formation. The above problems concerning the soils of dry lands are recommended as priorities for fundamental and adaptive research by the appropriate scientific institutions.

Alkaline soil widely spreads in arid regions. This alkaline nature of soil is a reason for the scarce vegetation in these regions, although a lack of water is also a factor. Soil pH plays a major part in the volatilization of microelements in soil, and this influences the growth and development of vegetation. Although the optimum pH range is generally considered to be 5.5–6.5, it is not rare that soil pH exceeds 8.0 in the Middle East region.

A lack of organic matters causes another problem. Organic matters in the soil play an important role in the agglomeration fertility of the soil, and the slow and continuous release of nutrition to plants. Under normal circumstances with healthy vegetation, microorganisms and animals as well as plants that inhabit the same environment contribute to a sound organic cycle. Without this cycle in place, however, mineralization generally progresses rapidly in arid regions.

a. *Sulfur utilization in agriculture*

The utilization in sulfur is a well-known method for reducing the pH level in alkaline soil. In interaction with sulfur oxidizing bacteria, sulfur is oxidized into sulfuric acid, which neutralizes alkaline soil. Generally, sulfur oxidizing bacteria are found in natural soil. However, because micro flora is scarce in the desert highly mineralized soils new technologies are needed for the utilization of the high performance species isolated and cultured in advance and then reintroduced into the soil together with the sulfur.

Moreover, sulfur is an essential nutrient for plants and is generally absorbed as sulfate by plants. Therefore, technologies to transform elemental sulfur into sulfate ions would not only contribute to the improvement of soil pH but also to the soil benefits by supplying essential sulfur nutrients.

In order to resolve the issue of a lack of organic matter, varying composts or peat moss are generally used. These methods, however, are not cost effective. For this reason, new technologies such as the use of excess sludge generated from the sewage treatment process as the organic matter should be looked at and utilized. The efficacy of sludge as an organic fertilizer has long been recognized in the world, and its use has proven successful in nurturing fertile farmland. In addition, it is beneficial to consider the reuse of the excess sludge since sewage treatment technologies are common throughout the world and the sludge can easily be obtained.

Reported studies have focused on the possible application of sulfur in plant cultivation through a combination of the other technologies such as sludge generated from waste treatment plants. Emphasis were given to the use of materials and techniques that can be procured locally, and combination of practical technologies to create a single soil amendment in view of convenient applications.

b. *Solar distillation*

Due to environmental, ecological and economical activities, particularly in an arid region (such as Middle East) and remote areas, fresh water demand mainly depends on desalination performance produced from brackish or sea water. However, such region is rich in solar energy with high intensity so that solar desalination may be an ideal solution. Solar distillation is the simplest desalination technique, compared to others such as multiple-effect distillation, multi-stage flash, reverse osmosis, electro-dialysis and biological treatment. A basin-type solar still is the most popular method of solar distillation but has undergone very little advances due to low distillate productivity and the difficulty of rapid and easy removal of salt accumulation in the basin. More studies are needed to overcome design problems and efficiency. Other techniques such as "Tubular Solar Still" should be looked at.

c. *Soil Moisture Enhancement*

Agricultural polymers are known since the 1950's and their uses were limited because of the high prices. But, recent retail prices became more affordable due to extensive research and generated private sectors on this new technology. For instance, highly crossed-polyacrylamide potassium based hydrophilic gel, was found useful to improve water absorption capability and availability of sand dunes for medium saline water.

Most of cultivated farms in Gulf Cooperation Council (GCC) countries depend on use of ground water as source for irrigation, where ground water is a limited resource and due to continuous pumping by huge quantities for irrigation and almost zero recharging, dramatic annual depletion is unavoidable. Introduction of new techniques of water conservation in the agriculture industry is a must water and energy conservation. One of the recent promising technologies is use of artificial soil conditioners, such as polymers. More research is required to evaluate the required dosage, optimum environmental conditions and environmental impact.

d. *Use of oily treated water in agriculture*

During oil production, large quantities of water are also produced. For example, Petroleum Development Oman (PDO) in 2002 produced some 330,000 m^3/day of water as a by-product of its oil output of approximately 135,000 m^3/day. The volume of this production water has been steadily increasing over time and was predicted to rise to about 650,000 m^3/day by the year 2005, as oil fields get older. Approximately 37% (123,000 m^3/day) of the currently produced water is utilized for reservoir pressure maintenance by injection mainly into northern Oman fields. About 140,000 m^3/day production water is presently disposed off into shallow (150–400 m) sub-surface formation at oil field sites mainly in south Oman.

According to the new environmental regulations in GCC countries, this practice should be stopped. The disposal of the produced water to deep formations is expensive. One of the possible alternatives to the underground disposal of produced water is land disposal by using it as irrigation water for salt tolerant crops. Special treatments are needed to clean up the produced water from its oil content. More studies are needed to assess the effect of treated oil production water on soil properties due to land disposal as well as on crop production and toxicity. Above all, risk assessment to human health and the environment is a must.

e. *Soil solarization for direct thermal inactivation of soil borne-pests*

The substantial expansion in agricultural production in United Arab Emirates (UAE) in recent years has been associated with development and spread of several weeds and other soil-borne pathogens. Methyl bromide, the sodium salt of metham and dazomet is widely used in the United Arab Emirates for soil fumigation in open fields and in green houses. However, these fumigants are considered to be harmful for all organisms. The loss of methyl bromide fumigant to the air and possibly reaching groundwater as a contaminant and its potential ozone reduction has stimulated interest in finding an alternative method for disinfestation of soil–borne pests and weeds. Soil solarization was considered as appropriate, non chemical, alternative to methyl bromide fumigation.

Soil solarization is a natural, hydrothermal process of disinfesting soil of plant pests by using passive solar heating of moist soil mulched with polyethylene sheets. This simple technique has the potentiality to increase the crop yield and improve soil chemical characters. In addition, this environmental friendly technique does not have negative effects on many of beneficial microorganisms such as arbuscular mycorrhizal fungi and nitrogen fixing bacteria.

The effectiveness of soil solarization for direct thermal inactivation of soil borne-pests is dependant on a number of physical factors including solar radiation intensity and duration, air temperature, amount of humidity at the soil surface beneath the tarps, properties of plastics used to produce the green house effect resulting in the heating of soil and properties of the soil to be treated. More research is needed to evaluate the role of these controlling parameters on the success of the technology.

f. *Utilization of adopted indigenous forage species to combat drought during vegetative and reproductive stages in the field*

The Arabian Peninsula experiences some of the most extreme climatic conditions found on the Earth. It is characterized by low erratic rainfall, high evaporation rates, extremely high temperatures during summer, and high soil salinities. A sustainable livestock industry in the Arabian Peninsula requires sustainable systems of both grazing and the production of cheap fodder plants. Currently, the main fodder crops in the area are alfalfa and Rhodes grass, which are not adapted the prevailing conditions and require vast quantities of water (up to 48,000 m^3/ha/yr), which often derived from non-renewable groundwater sources. The production of these forages has resulted in abandonment of many farms due to the increase in soil salinity. The utilization of adopted indigenous forage species to replace the exotic ones could be the solution for such problem. *Cenchrus ciliaris* (Buffel grass), indigenous to Arabian Peninsula, has been identified as one of the most tolerant forage grass to drought during vegetative and reproductive stages in the field and is currently being used as a fodder under experimental conditions in the UAE and in parts of Australia, South Africa and Pakistan under commercial conditions. In addition, this species is a highly nutritious grass and is valued for its high yields of palatable forage and intermittent grazing during droughty periods in the tropics in addition to its low cost of establishment, tolerance to drought conditions and crop pests, and its ability to withstand heavy grazing and trampling by livestock.

Developed seeds would be ready for germination following dispersal or would have a kind of dormancy that may delay their germination until the arrival of the favorable season for seedling survival. The main causes of seed dormancy are (1) a physiological inhibiting mechanism of germination in the embryo (=physiological dormancy), (2) seed coat impermeable to water (=physical dormancy), or (3) underdeveloped embryo (=morphological dormancy). Seed dormancy can influence patterns of plant distribution, recruitment dynamics and persistence in the plant community. The advantages of dormancy are to enable seeds to accumulate in the soil seed bank and protect plants from expanding their entire reproductive output at a given time. On the other hand, maintenance of seed dormancy when conditions are optimal for germination can be disadvantage, as the seeds are exposed to lethal environmental factors such as granivory and extreme temperature for long time.

Seed dormancy is generally due to several factors either internal and/or external. The internal factors include the impermeability of the teguments to water or oxygen, presence of some inhibitors in the seed coat, changes in internal hormonal balance or physiological immaturation of embryos. Several factors have been reported to overcome seed dormancy including storage under different conditions, removal of bracts or dispersal organs enclosing the seeds, presoaking the seeds in warm water, and treatment of seeds with different dormancy regulating substances.

Therefore studies are needed to assess dormancy level through the estimation of final germination percentage and germination rate as a result of (1) presence and absence of floral parts around the seeds (i.e., dispersal units or spikelets vs. naked seeds), (2) presoaking the seeds in water for different durations and different temperatures, (3) seed storage for long time periods, (4) treatment of the seeds with different dormancy regulating substances, and other related issues.

1.2 *Water management challenges*

The following major water management issues are:

1.2.1 *Water use: policy, planning and regulation*

- Reduction in quantity and quality of groundwater through over-abstraction, resulting in salinization of land, reduction in crop yields and abandonment of farms
- Lack of farm management leading to over-irrigation and drainage problems
- Unplanned development in the farming and forestry sectors
- Poor performance of the Forestry Sector due to insufficient water and poor water quality
- Little or no effort to manage the demands for water in agriculture sector
- Lack of recognition of the true economic cost of water when assigning its use
- Uncontrolled and un-regulated well drilling, leading to dry wells and wasted resources

1.2.2 *Protection, conservation and monitoring of water resources*

- Lack of a coordinated Emirate-wide water resources monitoring network and programme
- Groundwater pollution due to fertilizer use

- Lack of groundwater protection policies, e.g. no protection zones for municipal wellfields that still produce water of drinking quality, for example
- Lack of inventories on sources and demands
- Lack of qualified, technical, on-site supervision for drilling water wells
- Poor monitoring and data collection during drilling
- Insufficient water resources monitoring
- General waste of water and leakages
- Poor practices of water data and information management
- Non-availability or poor access to water resources information and data, and lack of a central, emirate-wide database to hold and analyse water resources data and information
- No well inventory, poor data collection when drilling wells

1.2.3 *Coordination of groundwater exploration and assessment*

- Need for expansion of groundwater exploration programmes, especially for deeper aquifer potential
- Lack of coordination and collaboration between existing groundwater exploration and assessment programmes.

1.2.4 *Local, regional and International cooperation and collaboration*

- Little or no technical cooperation with neighbouring Emirates and countries, especially on developments on or near to the international boundaries

1.2.5 *Strategic emergency water resources*

- No developed strategic reserve of potable quality water in case of emergency (current reserve for less than 2 days)

Common to the solution of most of the issues and problems listed above is the requirement for the establishment of a central, independent authority for Water Management in Abu Dhabi Emirate. Up until recently, responsibility for managing water sources and water use was divided between several organizations and agencies. This fragmented arrangement is unsatisfactory for effective water management and results in lack of coordination and collaboration between some of the bodies, duplication of efforts, non-assignment of responsibility for some very important aspects of water management, wasted funds and a general lack of accountability for some organizations current practices which go against the principle of sustainable water resources management and rationalization of water use.

1.3 *Groundwater recharge*

Groundwater recharge projects are considered necessary for meeting present and future demands on limited water resources. Continued growth in arid regions will require the development of a sustainable water supply, which will no doubt include improvement of our ability to store and utilize reclaimed wastewater.

Groundwater recharge using either relatively clean water or treated wastewater involves the infiltration of surplus or impaired quality water from the ground surface to underground water storage aquifers. Water is captured and stored underground in times of surplus for use in times of deficit. Underground storage avoids some of the disadvantages of conventional surface water reservoirs, such as evaporation, potential exposure to contaminants, and large capital investment become apparent.

Particularly in arid lands, treated wastewater has gained acceptance as a water supply and as an augmentation to the more conventional surface and groundwater supplies. Infiltration land treatment systems, wherein the groundwater is recharged with pretreated wastewater effluent have been suggested as a viable treatment option in arid lands.

Studies have shown that the hydraulics of water recharge basins are affected by many factors such as soil type, surface non-homogeneity, the formation of a surface clogging layer, application times for wetting/drying cycles, water quality, and climatic conditions. Clearly maintenance of a reasonable hydraulic loading capacity is of primary importance to successful soil aquifer treatment technique. Thus, a balanced approach to surface clogging layer and soil type selection for obtaining appropriate levels of water quality as well as rate of recharge must be achieved. Therefore, it is important to understand the movement of recharged water through unsaturated soils and to understand the role of unsaturated soils in the treatment process when the source of recharged water is not treated.

1.4 *Groundwater contamination and management*

Issues of concern to groundwater contamination and management are:

a. Development of innovative barrier systems for containment of hazardous and non-hazardous wastes in arid lands;
b. Development of comparative information to enable more systematic site selections, which will, in turn, make maximum usage of the natural attenuation capacity of given environmental settings to prevent groundwater contamination;
c. Environmental authorities should increase its attention to prevention of groundwater contamination and should provide programs to achieve a greater balance between prevention and remediation activities; and
d. Development of Best Management Practices (BMP) for usage of groundwater resources.

1.5 *Monitoring systems*

The use of low permeability lining materials has proven an effective method for containment of many types of waste leachates and could benefit from the use of an *in-situ* detection technique to monitor the performance of these liners. Current monitoring procedures rely mainly upon monitoring *wells*, *lysimeters*, and *leachate under-drains*. Wells are the most common means of monitoring the ground water contamination. This approach tends to be expensive and time consuming to implement. Timely detection of contaminant plume is obviously dependent on the initial layout and a number of monitoring wells. Unfortunately, wells can sample only a small volume of the aquifer. If samples collected from wells are not representative of the area or conditions for which they are intended, misleading and erroneous conclusions may result. Experience has shown that by the time a contaminant becomes detected in a monitoring well, a substantial volume of the surrounding soil and groundwater has already been polluted. In addition in arid lands, it is difficult to impossible to collect such required volume of water sample because of the high hydraulic conductivity of such soils. The risk of drilling wells and exploratory holes in unknown hazardous waste sites can be substantial. As the number of holes needed to define a problem area increases, so does the possibility of puncturing buried containers. Toxic fumes and liquids may be released. Explosions and fire may occur in extreme cases.

The limitations associated with present monitoring techniques underscore the need for an alternate approach. Undeniable, early detection and characterization of subsurface contamination can minimize its negative impact. Therefore, there is an urgent need for the development of a field diagnostic technique that allows a rapid determination of the extent of pollutants present in subsurface soils. The developed method should assist in locating a leak in the impounding boundary so that a corrective action can be taken to alleviate the problem. It should also be adaptable to a wide range of chemicals, as opposed to being ion specific.

There is an urgent need for the development of non-invasive pollutant detection systems in vadose zone for both organic and in-organic pollutants. Systems such as electrical polarizations, time domain reflectometry, and fiber optics chemical sensors are highly recommended and need further research studies.

1.6 *Solid waste management and utilization*

Due to the advancement in the understanding of landfill behavior and decomposition processes of municipal solid wastes, there has been strong thrust for upgrading existing landfill technology from storage/containment concept to process-based approach, i.e., bioreactor landfills. Bioreactor landfills allow a more active landfill management that recognizes the understanding of biological, chemical, and physical processes involved in a landfill environment. Engineered bioreactor landfill sites can provide:

a. a more controlled means by which the society can reduce the emission of global warming landfill gas;
b. immediate improvement to the surrounding local environment;
c. accelerated waste biodegradation, a means to enhance landfill gas generation rates; and
d. overall reduction in landfilling operation and maintenance costs.

Therefore, to enhance ones ability in applying such technology, there is a need to further understand the physical-chemical-biological processes that generally take place during the operation of landfills.

1.7 *Treatment of contaminated soils*

The use of microbial enhanced oil recovery concept to treat soils contaminated by hydrocarbons is a promising technology. The technology has been proven in oil industry for enhancement of oil recovery from underground reservoirs. The technology has been proven in oil industry for enhancement of oil recovery from underground reservoirs.

1.8 *Waste water treatment*

The development of petroleum and petrochemical industry in the Arabian Gulf region has been growing rapidly. In the last 50 years, exploration, drilling, extraction, refining and chemical engineering activities of oil and gas industry have all become an essential component in the economy of many of the Arabian Gulf countries. This speedy development has resulted in many changes such as landscape, economy, human development and interactions with other regions of the world as well as having impacts upon the environment and society. In fact "one of the most serious challenges facing the modern Middle East is the protection of its environment and the need to balance sustainable development with environmental security."

Water pollution is a serious concern in the UAE. This critical problem is made even more serious with the fact that water is very scarce in the region. Like most industries, oil refineries generate enormous quantities of wastewater. Such wastewater may contain several pollutants including phenols. Because of its toxicity, phenol became a wastewater quality parameter that the regulators closely look at in the effluents of heavy industry such as refineries.

Several studies have been reported in the literature on the use of biodegradation for removal of phenols from wastewater. In many of these studies, Pseudomonas putida (Pp) strain-type bacteria have been found to be effective for the degradation of phenols. The efficiency of biological methods employed for the treatment of phenol-contaminated wastewater varied with operating process variables such as pH, temperature, dissolved oxygen, and nutrients. It was further found that anaerobic degradation is typically more effective in removing phenol than degradation under aerobic conditions.

Most of the studies conducted on the removal of phenol from wastewater were limited to laboratory conditions. Transfer of laboratory results to filed conditions may not be straightforward, especially when processed wastewater is generated at condensate crude oil refineries. These refineries commonly receive crude oil from various sources, leading to fluctuations in waste characteristics. Studies to investigate the efficiency of treatment processes used in oil industry in reducing phenol concentration are needed.

1.9 *Environmental impact assessment*

Issues of concern to environmental impact assessment are:

a. A key issue that may influence the environmental impact assessment findings and subsequent interpretation is that the subsurface environment is typically characterized by very non-uniform conditions. Accordingly, consideration of such non-uniformity must be included in project evaluation and in interpretation of anticipated impacts.

b. Another issue of concern is the difficulty of taking a consistent approach for impact interpretation given the rapidly changing regulatory programs.
c. Management of soil and groundwater quality represents a new field, and considerable changes are occurring relative to the development and implementation of pertinent legislation, standards and regulations.
d. Environmental indicators and/or environmental indices can be useful tools in preparing a description of the environmental setting for a proposed project. There is a need for the development of indices for the impact of vadose zone on groundwater pollution potential.
e. Because of the need to investigate information from a number of substantive areas in the environmental impact assessment process, and because of the relative newness of the field in environmental management when compared with non traditional disciplines such as biology, chemistry and environmental engineering, there is a need for practitioners who work on the planning, conduction, and review of environmental impact studies to receive appropriate training.
f. Planning and implementation of research programs focused on answering specific impact related questions.

1.10 *Risk management of polluted sites*

Any successful approach to risk management must address a multiplicity of technical and regulatory issues. Such complexities demand the detailed attention of a coordinated, multi-disciplinary project team and integration of issues in a way perhaps unprecedented in other engineering and scientific endeavors.

Inherent uncertainties in site conditions, geochemical data, regulatory criteria, and technology performance must be recognized in formulating a risk management approach. Regardless of the level of efforts expended, uncertainty cannot be eliminated and can only be managed through an approach that recognizes this fact, strives to understand the certainties, and establishes and implements and adaptive methodology.

Regulatory attitudes exert considerable influence on the design aspects of waste containment facilities. The assessment of failure of the facility to function in a manner designed to ensure protection of public health requires one to properly appreciate what constitutes a health threat. However, it is not immediately clear that the controls needed to establish safe protection of public health are well founded. A good working knowledge of the various interactions occurring in the substrate during pollutant transport is required if one seeks to assess the fate of pollutants.

1.11 *Sulfur utilizations in construction industry*

A dramatic change has occurred in the global sulfur industry over the last decade. Sulfur consumption is dominated by a number of large-scale uses, with several complex interactions that complicate the future outlook for demand. Thus, sulfur is used to produce sulfuric acid, one of the world's largest volume chemical commodities, but this use competes with sulfuric acid recovered from ore smelting. Uses of sulfuric acid include processing phosphate rock and ore leaching in metallurgical processing (a growing market). There are a number of other uses today for sulfur, including in pigments, pesticides, and rubber vulcanization and as an agricultural nutrient.

In general, existing uses for sulfur are relatively mature, and offer limited opportunities to consume significant new supplies. Currently sulfur is in net surplus on a global basis, and with environmental regulations mandating ever greater sulfur recovery from petroleum and gas processing. The outlook is clear; there will be substantial and growing surpluses in global sulfur supply for the foreseeable future. Sulfur prices are likely to be under pressure, and producers could face substantial and growing disposal fees.

Sulfur in its elemental form is recognized as an important ingredient in several agronomic applications. These include the following:

1. an essential plant nutrient;
2. an active agent for increasing crop stress resistance;

3. environmentally benign pesticide; and
4. an efficient soil amendment aid to alleviate alkalinity.

Sulfur as an essential plant nutrient has received little scientific attention. This is explained by the facts that sulfur was obviously in sufficient supply from the atmosphere, from soil and as a by-product in mineral fertilizers. However, the use of highly concentrated fertilizers containing little or no sulfur has drastically reduced the amount of sulfur supplied to soils. Recent studies have shown that adding sulfur to soil increased crop yield, increased drought tolerance and increased nitrogen efficiency and phosphorus uptake.

All of these applications are important to the national agricultural drive in the Gulf Cooperation Council (GCC) countries. However, the fourth application, which is alkalinity amendment, is of particular significance. Soil alkalinity, in the GCC, demands the use of an acidifying agent to achieve the required neutrality. Sulfur is the major component to achieve this endeavor. Elemental sulfur is microbially oxidized to sulfuric acid, which then reacts with calcium carbonate to form gypsum. This oxidation process is highly dependent on soil moisture, temperature, microbial activity, and the size of the elemental sulfur grain. Particle size is perhaps the most critical factor from an application and product point of view.

In general, elemental sulfur granules, in their original size (250 μm), oxidize at a very slow rate. These granules are relatively large and present soil microbes with a small specific surface area for conversion. By breaking the elemental sulfur granule into smaller sizes (45 μm), the surface area is increased which in turn increases the rate of microbial oxidation and conversion of elemental sulfur to sulfate.

The phosphate fertilizer industry is the largest consumer of sulfur, primarily as a consumer of sulfuric acid used in phosphate rock processing. Sulfur itself is also a plant nutrient, mostly supplied as ammonium sulfate and potassium sulfate, and elemental sulfur is a traditional fungicide. Agriculture is thus a very large, but modestly growing outlet for sulfur.

Global sulfur demand has been relatively stagnant at about 57 million metric tons per year over the last decade. Based on new regulations limiting sulfur content in diesel and gasoline, the current small global surplus in sulfur supply is projected to reach between 6 and 12 million metric tons by 2011, or between 10 to 20 percent of demand. This projected surplus represents obvious challenges to existing producers, potentially leading to drastically reduced sulfur prices, and even the possibility of costs to producers for disposal of the surplus in some regions. On the other hand, the surplus may also represent opportunities for new uses of sulfur, driven by these very same reductions in sulfur price that can make new uses more economically feasible and attractive.

Even with relatively small surpluses, the oil and gas industry has already experienced strains on sulfur storage facilities. Sulfur is being stored on site in block, granular or palletized form, or molten, (very expensively) in rail tank cars because there is insufficient storage capacity to handle sulfur generation at refineries and gas processing plants.

One potential new market is ***sulfur solidified concrete***. This is a thermoplastic composite of mineral aggregates bound together with chemically modified sulfur. The product is more durable than Portland cement. In addition, from the environment viewpoint, there may be monetizable benefits in reducing greenhouse gas emissions that would enhance the attraction of sulfur solidified cement applications.

The main advantage of sulfur concrete is its use as a highly durable replacement for construction materials, especially Portland cement concrete, in locations within industrial plants or other locations where acid and salt environments result in premature deterioration and failure of Portland cement concrete.

There are several advantages in using sulfur concrete for construction in areas exposed to highly corrosive acids. While ultimate life or durability of sulfur concrete has not been completely established in many end use applications, enough evidence of its corrosion resistance and durability has been accumulated to show that it has several times the life of other construction materials now being used in corrosive environments. Other advantages of sulfur concrete are its fast setting time and rapid gain of high strength. Since it achieves most of its mechanical strength in less than a day, forms can be removed and the sulfur concrete placed in service without a long curing period.

The handling, mixing, and use of sulfur concrete can be accomplished with proper concern for product safety. As with any other construction material, certain measures must be taken with sulfur concrete to insure safe handling in its preparation and use. Sulfur concrete should be produced within its recommended mixing temperature range (127–141 deg. C) to minimize emissions. Adequate ventilation during construction operations and normal precautions for handling hot fluid materials (proper protective clothing, eye protection, gloves, and hard hats) should be observed.

Sulfur solidified concrete has potential use in a wide range of applications such as:

1. pre-casting activities (sewer pipe; railway ties; highway barriers; a range of agriculture products; offshore drilling platforms; construction blocks, slabs, and other building components);
2. extruding (bricks and paving stones; curbs and gutters; roof tiles); and
3. cast-in-place (bridge decks; marine installations; ship hulls and structures exposed to marine environments; drilling platforms; food processing plants; agricultural applications, including barns and effluent systems; sewage treatment plants; acid plants-drainage canals, sumps, tanks, flooring, walls and beams; fertilizer plants; foundations).

To appreciate the market size for such product in the United Arab Emirates (UAE), one briefly reviews the size and operation of cement manufacture. In UAE there are nine major cement manufacturing plants, eight producing Portland cement and one producing White Cement. The production capacity of Portland cement plants in 2000 was over 10 million tones per year (Arab Union for Cement and Building Materials, 2003) and that of the White Cement plant was about 450,000 tones per year. In 2000, these plants produced around 6,900,000 tones of cement generating over one-billon UAE Dhs (about 271 million USD) in income. Furthermore, the net profit of cement producing plants in UAE has increased from 73.3 million UAE Dhs in year 2002 to 278.5 million UAE Dhs in year 2003.

Sulfur concrete is a thermoplastic construction material, of mineral aggregates and modified sulfur as binder. Elemental sulfur has a low smelting point (120 deg. C) and appears in two solid allotropic forms, separated by considerable contraction. Hence, there is a need for modification to delay the transition. There are two main ways of modifying sulfur, both of them trying to control sulfur crystallization (unmodified sulfur crystals are brittle and subjected to failure under thermal cycles). The first one is "Plastification", where a certain chemical substance (a polymer) is added to sulfur in order to inhibit the transformation to orthorhombic structure; as a result of the chemical reaction with the substance, after cooling sulfur remains in monoclinic state, with a amorphous fraction also. Several substances have been tried for this; the most common are dicyclopentadiene, or a combination of dicyclopentadiene, cyclopentadiene and dipentene.

The second method is "controlled crystallization", where, although sulfur is allowed to reach the orthorhombic state, crystals are small and with a regular and ordered structure. To obtain this, sulfur is modified with an olefin polymer (such as RP220 or RP020 by Exxon Chemical, or Escopol) and a physical stabilizer is also added (for example, fly ash, or other fine substance).

In both ways, the issue is controlling sulfur crystallization, either chemically or physically. Depending on the ultimate use of the produced sulfur concrete, one chooses the method of treatment. Furthermore, with proper content of physical stabilizers, sulfur concrete tends to be self-extinguishing. Also it can be manufactured to be fire resistant, and/or inhibit the formation of SO_2 when heated with proper content of polymeric materials. Therefore, more research studies are needed in this area to develop the technology on large scale.

1.12 *Sustainable development*

Sustainable development has become public concern in recent years. In order to achieve sustainable development, cooperation at a global level is important to control the environmental burden caused by human activities to a degree that can be tolerated by the planet.

The three major players for achieving sustainable cities are economy, environment and society. Instead of working on the improvement of each of the three spheres separately, a result-oriented approach is to work for interlinking the economy, environment and society. Actions to improve conditions in a sustainable community take these interlinks into consideration. Sustainable environment depends on sustainable development, which in turn linked to sustainable construction practices. Engineers, Architects and Construction Managers can improve the energy and resource efficiency of commercial and residential buildings by the use of practices for water conservation, better indoor quality, use of day light, energy efficiency for which we need to impart relevant technical education and skill training. Without the emphasis of these aspects in technical education and skill training the approach towards sustainable cities is rather going to be incomplete. First starting from awareness among higher secondary school classes and later inclusion of these aspects into engineering and technical education and in skill training in vocational institutes will provide the desired results.

There is a need to introduce and include sustainable building concepts in engineering and technical education for having sustainable cities. Some introduction should also be given in general and science education, so that city people who are going to use a house or a building can ask/demand for better indoor quality, energy efficiency, etc. Further research in sustainable cities and sustainable development is needed for the better and efficient use of our resources keeping in mind healthier environment for urban population and future generations. The three areas of economy, environment and society needs to be interlinked for a purpose oriented approach. Use of local/indigenous building materials needs to be encouraged. And finally, research in local low cost environmental friendly building material to be carried out.

1.13 *Global warming*

The threat of climate change caused by the emission of anthropogenic carbon has encouraged the international community to develop agreements such as the United Nations Framework Convention on Climate Change and the Kyoto Protocol. Upon ratification of the Kyoto Protocol, Canada committed, through the 2008–2012 commitment period, to reducing its greenhouse gas (GHG) emissions to 570 MT eCO_2, from 1990 levels of 610 MT eCO_2. The solid waste sector in Canada generated 24 MT eCO_2 in 2000, 23 MT eCO_2 of which were produced by landfill gas (LFG). The transport of waste likely generated a further 740 kT eCO_2 that are not accounted for in the waste sector by the National Greenhouse Gas Inventory. It is likely that waste transport emissions will increase as fewer landfills are sited and further source separation of wastes occurs. The benefits of source reduction, recycling, LFG capture for energy recovery, composting, anaerobic digestion and incineration are parameters that need to be studied.

In its voluntary action plan for fiscal year 2010, Japan's oil refining industry established a goal of "reducing unit energy consumption in refineries by 10%." Owing to efforts that aimed to achieve the goal, it has produced results that exceed the goal. Cumulative energy conservation is estimated as approximately 3 million KL. When calculated from as far back as 1973, the year of the first oil shock, energy conservation effect amounts to a total of approximately 10 million KL. In this way, the oil industry worked to reduce CO_2 emissions through the steady implementation of energy conservation measures in refineries. At this rate, with continued efforts it should be sufficiently possible to achieve the fiscal year 2010 targets.

Japan's Kyoto Protocol Target Attainment Plan states that, in order to fulfill the commitment to achieve a 6% reduction over the value of 1990 which Japan pledged in the Kyoto Protocol, Japan must implement measures to achieve an additional emission reduction corresponding to -12% (approx. 148 million tons-CO_2 equivalent) as well as policies for their promotion in addition to those measures and policies that it are currently being implemented. The voluntary action plan of the Keidanren adopted by 34 major industries calls for a reduction in CO_2 emissions to below 0% from fiscal year 1990 to 2010. In these ways, with the steady implementation of voluntary action plans in the industrial sector, achievement of the targets looks promising. In regards the issue of CO_2 emission control in the private transportation sector, the shift to sulfur-free gasoline, diesel fuel, and other transportation fuels is expected to produce effective results in the sector.

Efforts related to reducing CO_2 emissions through the supply of environmentally compatible petroleum products are welcome. It is our hope that industrial countries will effectively introduce measures to the rationalization of energy use and the production of environmentally compatible products by oil factories throughout the world in order to achieve the current goal of reducing greenhouse gases.

1.14 *Oil spill management*

Oil spills in the marine environment raise a major concern among platform management, government authorities and the public. One of the key elements for an efficient spill control is the preparedness for an adequate and prompt response. Studies for developing framework for formulating an oil spill emergency response plan are needed.

2 PROPOSALS FOR ACTIONS

Certainly, ***individual initiatives*** are very important if the Gulf Region is to become a strong and recognizable entity in the development of Sustainable GCC environment and water resources.

These initiatives can be better ***vectorized*** by the creation of identifiable units which can act to the common advantage of the environment and water resources communities in this region. These can include the creation of:

1. The JCCP-GCC Working Group for EWR series of symposia; and
2. The JCCP-GCC Research Fund.

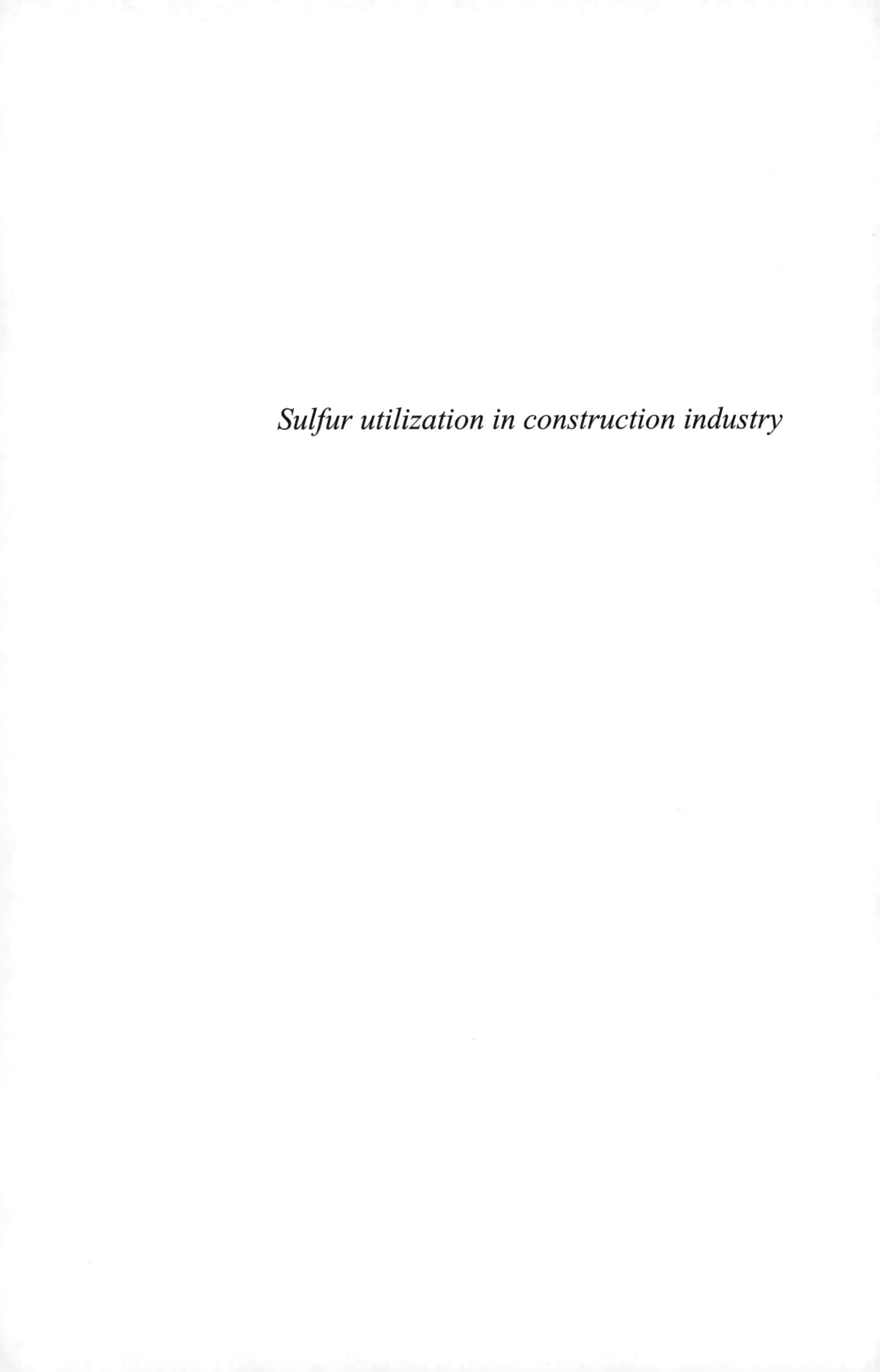

Sulfur utilization in construction industry

Reclaiming the Desert: Towards a Sustainable Environment in Arid Lands – Mohamed (ed.)
© 2006 Taylor & Francis Group, London, ISBN 0 415 41128 9

Compositional control on sulfur polymer concrete production for public works

A.M.O. Mohamed & M.M. El Gamal
Department of Civil and Environmental Engineering, UAE University, Al Ain, UAE

ABSTRACT: The aim of this study was to formulate sulfur polymer concrete having a workable consistency, good mechanical properties and high durability. In doing so novel modified sulfur was successfully prepared, which contains, in addition to the sulfur, a particularly polymeric material. Characterization as well as structural analysis of the modified sulfur was performed by C, H, S elemental analyzer, FT-IR spectra, x-ray diffraction, SEM, TGA and DSC. Comparison between sulfur concretes (using unmodified sulfur) and sulfur polymer concrete (using unmodified and modified sulfur) in addition to sand and fly ash was evaluated with respect to compressive strength, crystallization behavior, homogeneity, cumulative percentage of pore sizes, and thermal behaviors. Proper grain size distribution of the physical admixtures and suitable selection of manufacturing technological procedures have led to manufacturing novel sulfur polymer concrete with low porosity and high mechanical strength.

1 INTRODUCTION

Sulfur concrete construction materials are used in many specialized applications throughout industry and transportation. They are currently used primarily in areas where conventional materials like Portland cement concrete (PCC) failed, such as acidic and saline chemical environment. Sulfur concrete is a thermoplastic material prepared by hot mixing sulfur cements and mineral aggregates. The molten sulfur solidifies and gains strength rapidly upon cooling, as with other concrete materials.

There is evidence of even earlier uses of sulfur, and both archaeological sites and classic literature offer proofs of sulfur knowledge and utilization as a binder. Study about the use of surplus sulfur in the manufacture of construction materials was reported by Bacon and Davis (1921); they found that a mixture of 60% sand and 40% sulfur produced an acid resistant material with excellent strength. This product was further studied by Duecker (1934) who found an increase in volume on thermal cycling with loss in flexural strength. He was also able to retard both the tendency for volume increase and the resulting loss of strength on thermal cycling by modifying the sulfur with an olefin polysulfide. The testing methods for sulfur materials outlined by McKinney (1940) have been adopted and are documented in ASTM specifications for chemical resistant sulfur mortar. The acid resistant properties of materials prepared from sulfur and coke was reported by Kobbe (1924).

The work on sulfur aggregate systems was pioneered by Dale and Ludwig (1960 & 1966). This work was followed by Crow and Bates (1970) for the development of high strength sulfur basalt concrete. Construction materials such as sulfur concrete and sulfur asphalt continue to receive more attention since they are environmentally friendly and cost effective. Beginning of the 1970s, successful projects in which sulfur concrete has been used as a construction material have been carried out in different levels. The United States Department of the Interior's Bureau of Mines and The Sulfur Institute (Washington, D.C.) launched a cooperative program in 1971 to investigate and develop new uses of sulfur. At the same time, the Canada Center for Mineral and Energy Technology (CANMET) and the National Research Council (NRC) of Canada initiated a research program for the development of sulfur concrete (Malhotra 1973 & 1974 and Beaudoin 1973). Number of investigators including (Sullivan et al. 1975; Vroom 1977 and 1981; Sullivan and McBee 1976; McBee & Sullivan 1979 and 1982a, b; Funke et al. 1982; McBee et al. 1981a, b, 1983a, b and 1986, Sullivan 1986) and other published number of papers and reports investigated various aspects of sulfur and sulfur concrete. All of these activities have led to an increased awareness of the potential use of sulfur as a construction material.

While sulfur concrete materials could be prepared by hot mixing unmodified sulfur and aggregate, durability of the resulting product was a problem. Unmodified sulfur concretes failed when exposed to repeated cycles of freezing and thawing, humid conditions, or immersion in water. Studies were aimed at establishing reasons for failure of these sulfur concrete materials and determining means of preventing failures. When unmodified sulfur and aggregate are hot mixed, cast and cooled to prepare sulfur concrete products, the sulfur binder, on cooling from the liquid state, first crystallizes as monoclinic sulfur (S_β) at 114°C with a volume decrease of 7%. On further cooling to below 96°C, the S_β starts to transform to orthorhombic sulfur (S_α), which is the stable form of sulfur at ambient temperatures (Lin et al. 1995). This transformation is rapid, generally occurring in less than 24 hours. Since S_α form is more dense than S_β, high stress is induced in the material by solid sulfur shrinkage. Thus, the sulfur binder can become highly stressed and can fail prematurely.

It was further necessary to develop an economical means of modifying the sulfur so that the sulfur concrete product would have good durability and reduce the expansion/contraction of sulfur concrete during thermal cycling. Since 1984, the American Concrete Institute (ACI), through its Committee 548, polymers in concrete, and subcommittee 548D, sulfur concrete, has been active in developing guidelines for use of sulfur polymer concrete. There are two main ways for modifying sulfur, both of them trying to control sulfur crystallization chemically or physically. The first one tries to combine chemical substances to sulfur in order to inhibit the transformation to orthorhombic structure; as a result of the chemical reaction with the substance, sulfur remains in monoclinic state after cooling. Several substances have been tried for this methodology; the most common are dicyclopentadiene, or a combination of dicyclopentadiene, cyclopentadiene and dipentene (Currell 1976; Bordoloi & Pierce 1978; Beaudoin & Feldmant 1984; McBee et al. 1985 and Lin et al. 1995). The second method utilizes a modified sulfur concrete by combining sulfur with olefin hydrocarbon polymers (such as RP220 or RP020 by Exxon Chemical, or Escopol) and a physical stabilizer such as fly ash, or other fine substances (Vroom 1981 & 1992, and Nnabuife 1987).

In both ways, the issue is controlling sulfur crystallization, either chemically or physically. Depending on the ultimate use of the produced sulfur concrete, one chooses the method of treatment. Because of the abundance of the physical stabilizers in the United Arab Emirates (UAE), the second treatment method is utilized in this study whereby sulfur, polymer modifier, fly ash and sand dunes were utilized in producing sulfur polymer concrete. Each material as well as the mixture was evaluated using thermal analysis techniques, such as thermal gravimetric analysis (TGA) and differential scanning calorimetric (DSC), infrared spectra, x-ray diffraction, scanning electron microscope, and unconfined compression strength. The experimental results were evaluated in terms of the potential use of the product for public works and the optimum mixture was quantified.

2 MATERIALS

Details of materials used, specimen grain size, and procedures for preparation of modified sulfur (sulfur cement), sulfur concrete (SC) and sulfur polymer concrete (SPC) are described below.

All raw materials used for this study were obtained from UAE. The granular sulfur (99.9% purity) was obtained from Al Ruwais refinery. Bitumen with a softening point of 48.8°C, specific gravity at 20°C of 1.0289 g/cm^3 and kinematics viscosity at 135°C of 431 cSt, with a chemical analysis of C:79, H:10, S:3.3 and N:0.7%, was used. Two types of aggregates were used throughout this study; desert sand obtained from a sandy dunes quarry in Al Ain area and fly ash, India-97/591, were used. Chemical analyses were performed using Inductively Coupled Plasma – Atomic Emission Spectrometry (ICP-AES) VISTA-MPX CCD simultaneous and the results are listed in Table 1 while the grain size distributions are listed in Table 2.These aggregates should be clean, hard, tough and free of organic and other harmful substances.

Table 1. Chemical composition of aggregates using ICP analysis

Aggregate	Compound % (w/w)					
	Al_2O_3	CaO	Fe_2O_3	K_2O	MgO	SiO_2
Sand	0.47	16.35	0.676	0.13	1.158	74.4
Fly ash	32.4	0.46	4.34	0.027	0.66	60.9

Table 2. Grain size distribution of aggregates.

Sand		Fly Ash	
Sieve mesh size (mm)	Mass retained on the sieve (%)	Sieve mesh size (mm)	Mass retained on the sieve (%)
0.425	0.00	0.250	0.40
0.250	8.82	0.150	1.25
0.150	49.90	0.125	2.85
0.106	24.64	0.075	13.35
0.090	8.80	0.063	12.25
0.075	4.85	0.053	46.80
Bottom dish	3.00	0.045	18.35
		Bottom dish	4.75

3 MIX DESIGN AND SAMPLE PREPARATION

3.1 *Preparation of modified sulfur*

In oil bath, 2.5 wt % of hot bitumen, 97.5 wt % molten sulfur, and emulsifying agent were mixed and mechanically stirred at 140°C. The temperature was maintained at about 135–140°C during the mixing process, which lasted about 45–60 minutes. The progress of the reaction was monitored by the degree of homogeneity of the mixture via careful observation of the temperature and viscosity of the reacting mixture. After mixing, the prepared mixture was allowed to cool at a rate of 8–10°C/min. The final product is a sulfur containing polymer which on cooling possessed glass like properties. This modified sulfur cement is commonly called Modified Sulfur Cement (MSC). The composition and properties of the MSC were evaluated in terms of chemical analysis, FT-IR spectra, x-ray diffraction, scanning electron microscopy, and thermal processing including TGA and DSC.

3.2 *Preparation of sulfur concrete (SC) and sulfur polymer concrete (SPC)*

Sulfur concrete samples were prepared according to the procedure described in (ACI 248.2R-93). The aggregates (sand and fly ash) were heated in an oven to 170–200°C for two hours. The specified amount of sulfur was melted in a heated mixing bowl, which is placed in oil bath controlled temperature from 132–141°C. The heated aggregates were then transferred to the heated mixing bowl, and properly mixed with the molten sulfur until a homogenous viscous mixture was obtained. Cubic steel mold with dimensions $50 \times 50 \times 50$ mm was used. The molds preheated to approximately 120°C before adding the viscous sulfur concrete, as the sulfur concrete is added the material was compacted by tamping with a heated rod and simultaneously vibrated for 10 seconds on a vibrating table. The surface of each specimen was finished and left in the oven and gradually cooled at a rate of 2 degree/min. The specimens were de-molded at an age of 24 hours after placement in steel molds. Optimization of sulfur ratio in the mixture was studied; if the optimum amount of sulfur was added, the sulfur concrete will be more viscous with a good workable consistency.

Sulfur polymer concrete samples were prepared in a manner analogous to that of sulfur concrete, with exception that the modified sulfur not added directly to the molten sulfur but after the reaction between sulfur and flay ash. Temperature control is very important because the modified sulfur cement melts at 119°C, above 149°C the viscosity of sulfur concrete rapidly increases to an unworkable consistency. Also the percentage of modified sulfur plays an important role in the workability and mechanical strength of the sulfur polymer concretes. All the mix designs were analyzed by following up the compressive strength, SEM for homogeneity and porosity studies and thermal studies including TGA and DSC.

3.3 *Characterizations of modified sulfur and sulfur polymer concrete*

The determination of C, H, and S was carried out with the Finnigan Flash EA1112 CHN/S elemental analyzer. The reaction between sulfur and bitumen was studied using FTIR spectroscopy. The IR spectrum was recorded by using Nicolet FTIR Magno-IR (Model 560) system. IR spectra of bitumen taken after successive treatment with toluene, the IR spectra of bitumen modified sulfur was taken as powder mixed with a small amount of KBr powder to make the IR pellet.

X-ray powder diffraction was performed using a Philips PW/1840, with Ni filter, Cu-Kα radiation ($\lambda = 1.542$ Å) at 40 KV, 30 mA and scanning speed 0.02°/S. The diffraction peak between $2\theta = 2°$ and $2\theta = 80°$ were recorded. Measurements were made for samples to examine the interlayer activity in the pure sulfur and sulfur cement.

Sulfur, modified sulfur, sulfur concrete and sulfur polymer concrete were examined using JSM-5600 Joel microscope equipped with an energy Dispersive x-ray detector for chemical analysis. The microstructure characterization performed for sulfur concrete and sulfur polymer concrete were carried on the samples fragments, obtained from the mechanical test. Samples were fasten on a sample rock with carbon glue and coated with a 12 nm gold layer for improved SEM imaging.

Thermo Gravimetric Analyzer Perkin Elmer TGA7 was used to measure weight changes in sample materials as a function of the temperature. A furnace heats the sample while a sensitive balance monitors loss or gain of sample weight due to chemical changes, decomposition, TGA coupled with Mass Spectrometry and FT-Infrared Spectroscopy provides elemental analysis of decomposition products. Weight, temperature and furnace calibrations have been carried out for the usable range of the TGA (100–600°C) at a scan rate of 20°C/min.

Differential scanning calorimeter (Perkin Elmer DSC7) was used for measurements of heat capacity, through phase transitions on heating. 10 mg of the tested sample was heated up to 150°C, with heating rate of 5°C/min, the sample was allowed to be self cooled to room temperature for 24 hrs, then the sample was reheated up to 150°C. Mechanical strength: compressive strength of the samples was determined by Wykeham Farrance testing machine with maximum load of (200 KN).

4 RESULTS AND DISCUSSION

4.1 *Modified sulfur*

Addition of bitumen in amounts of 2.5% to sulfur initiates chemical reactions whose type depends on the bitumen content, heating temperature and the time of the reaction. For example, some competing reaction can occur, including those with bitumen incorporation into sulfur molecules or dehydrogenation with liberation of hydrogen sulfide. It should be pointed that, at $T < 95°C$ sulfur exists as a cyclooctasulfane crown with an S—S bond length of 0.206 nm, and S—S—S bond angle of 108 degree; and at $T < 119°C$ sulfur crystallizes. At 119°C (melting point of sulfur), liquid sulfur being thoroughly dispersed in bitumen forming an emulsion (role of emulsifying agent) and cyclooctasulfane turns partly into polymeric zigzag chains (bond length 0.204 nm) (Voronkov et al. 1979). The crystallization features are affected by such factors as chemical reaction of sulfur with bitumen components and its dissolution or dispersion. Sulfur at heating temperature $T < 140°C$, elementary sulfur forms polysulfide which initiate formation of a network. Such structures differ considerably in the chemical and thermal stability from unmodified sulfur. At 119–159°C, molten sulfur exists essentially as cyclooctasulfane ($\lambda -$ S). Above 159°C, eight-member ring rapidly break down into biradicals (Voronkov et al 1979). In their turn, biradicals recombine to form polymeric chains with the maximal length of up to 10^6 sulfur atoms:

$$S_8 \Leftrightarrow S^0 - S_6 - S^0$$

$$S^0 - S_6 - S^0 + S_8 \Leftrightarrow S^0 - S_6 - S^0 - S_8 \quad etc.$$

However, above 140°C, dehydrogenation of saturated bitumen components can occur; also linear polysulfide can transform into stable cyclic thiophene structures (Syroezhko et al. 2003). It is noted that modified sulfur should be produced within its recommended mixing temperature range of 135 to 141°C.

Characterization of modified sulfur performed by different tools; the chemical analyzer was used to determine the chemical composition of modified sulfur; S: 97.4 ± 0.48, C: 1.981 ± 0.074 and H: 0.1 ± 0.067 this indicates that modified sulfur contains organic compound (related to the presence of C, H). However, the reaction between bitumen and sulfur should be determined by other means. In order to investigate the nature of chemical interaction between bitumen and sulfur, the completion of bitumen sulfur reaction was strongly supported by IR measurement of the characteristic double bonds in bitumen, which is the main modifying bitumen constituent. Figure 1 shows the band at 2975 cm^{-1} which is consistent with CH stretching. Since C—H is associated with C═C double bonds, the disappearance of C—H bonds in the modified sulfur spectra suggested the consumption of aliphatic C═C bonds, which generally leads to the polymerization of sulfur with bitumen. Negative peak in 2400 cm^{-1} indicates removal of NC═O group indicating that they are sites of chemical reaction for polysulfide formation. An increase in relative intensity in spectra at 584 cm^{-1}, is corresponding to disulfide S—S bond between adjacent thiol groups.

The obtained XRD diagrams for pure sulfur and sulfur modified with a small amount of bitumen (2.5 wt%) is shown in Figure 2. XRD analysis confirms that the modification of sulfur making sulfur crystallizes in a structure different than the unmodified one. Modified sulfur gives sharp signal shifted to lower 2-theta, which in turn indicates that the newly formed structure is fine in its grain size.

Scanning electron microscopic has revealed how the bitumen controls the crystallization of sulfur. Pure sulfur crystallizes and form dense and large alpha sulfur crystals (S_α) with orthorhombic sulfur morphology as shown in Figure 3a. With addition of bitumen, the crystal growth is limited and controlled by the bitumen in such a way that all crystals are plate like of micron dimension as monoclinic sulfur crystal of beta form (S_β) as shown in Figures 3b, c. The microstructure suggested by Scanning electron microscope confirmed the presence of bitumen uniformly dispersed in sulfur, as coverage

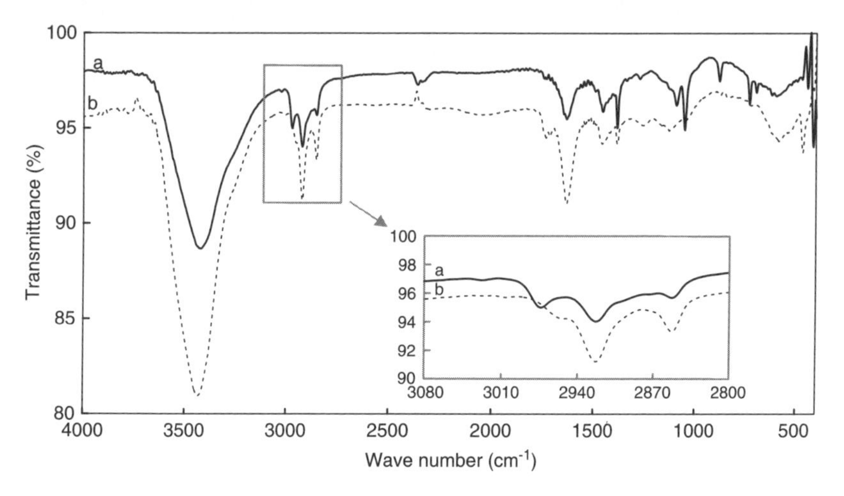

Figure 1. FTIR spectrum showing the spectra of bitumen (a), and sulfur modified with bitumen 2.5 wt % (b), viewing the disappearance of CH characteristic double bond of the bitumen after the complete reaction with sulfur.

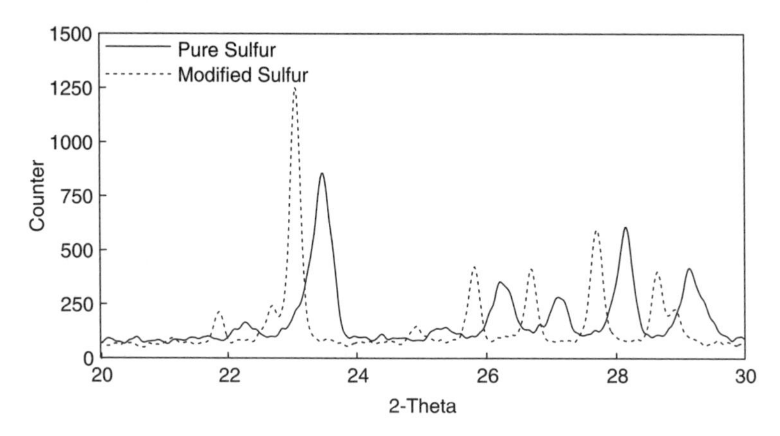

Figure 2. X-ray Diffraction patterns for pure sulfur and modified sulfur.

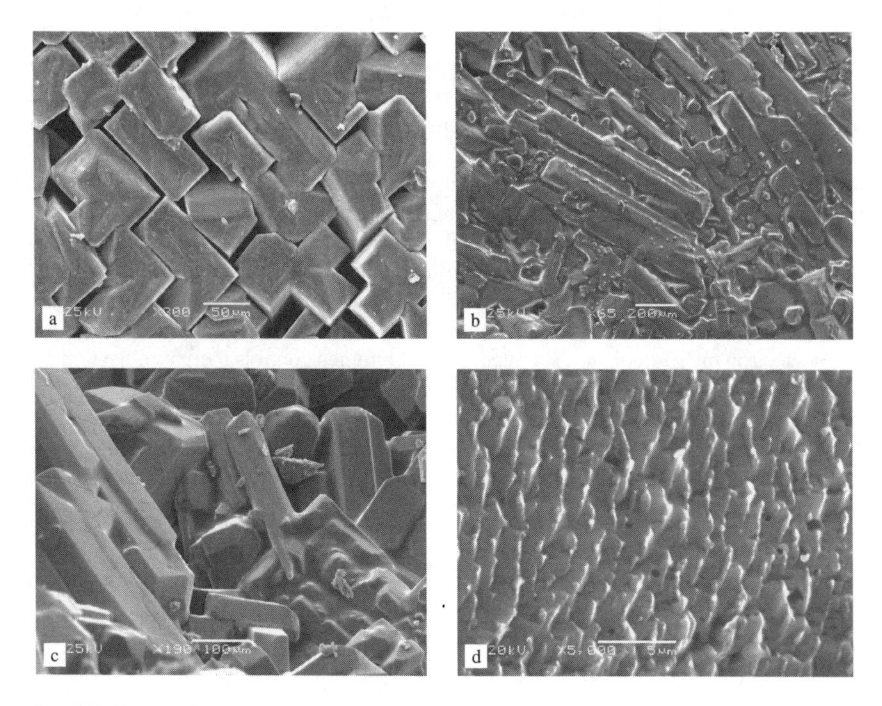

Figure 3. SEM image for; pure crystalline sulfur, (a) sulfur modified with of 2.5 wt % bitumen; surface image (b) interior image (c) and the dominant microstructure (d).

of sulfur particles and between plates and plate joints. The smallest microstructure, plate like, helps to resist cracking and to tolerate any thermal expansion. The formed microstructure is very much like sea shells as shown in Figure 3d.

The scanning electron microscopy was also used to optimize the bitumen percentage, the increasing of bitumen contents from 2.5, 3 and 3.5% by weight causing an additional significant increase in average dimensional of the micron-size microstructure of the interlocked plate like beta sulfur crystals as indicated in Table 3.

Thermal analysis techniques such as thermal gravimetric analysis (TGA) and differential scanning calorimetric (DSC) have been used effectively for determining the temperature history of sulfur and sulfur cement (Syroezhko et al. 2003). The TGA measures the weight loss of a material from a simple

Table 3. Optimization of the percentage of bitumen used for sulfur modification via beta crystal dimensions obtained from SEM.

	Bitumen (wt %)		
Crystal dimensions (μm)	2.5	3	3.5
Length	600	1000	1583
Width	147	167	279

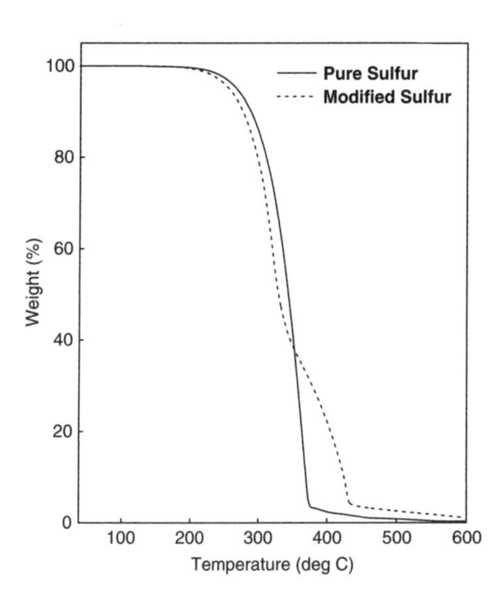

Figure 4. Thermo gravimetric analysis (TGA), for pure and modified sulfur, the heating rate was 20 deg/min.

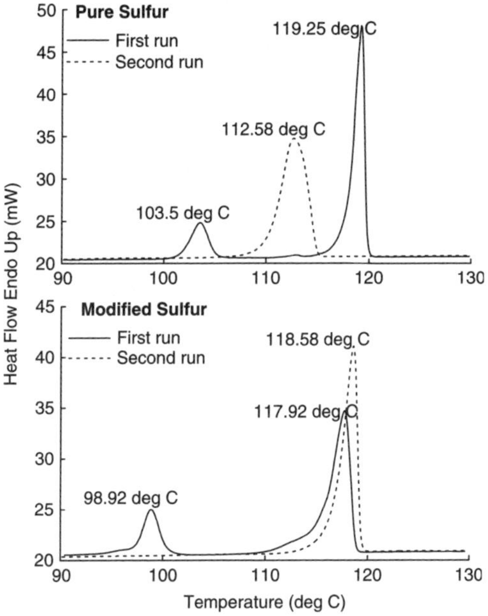

Figure 5. DSC Curves for sulfur and modified sulfur. The heating rate was 5 deg/min.

process such as drying, or from more complex chemical reactions that liberate gases and structural decomposition. Figure 4 provides comparison of the TGA curves of pure sulfur and bitumen modified sulfur. The maximal thermal effect was observed at 180–450°C for pure sulfur which was accompanied by active liberation of hydrogen sulfide, while the maximal thermal effect was observed at 200–540°C for modified sulfur. This effect reflects the increase of thermal stability of modified sulfur.

Differential scanning calorimetric (DSC) results of both sulfur and modified sulfur cement are shown in Figure 5. For pure sulfur, melting of the alpha and beta was detected in the first run, while only alpha melting was detected in the second run. For modified sulfur melting of the alpha and beta was detected in the first run, while only beta melting was detected in the second run. So, DSC thermographs show the modified sulfur remains in the monoclinic modification of beta form and does not undergo a phase transformation to the orthorhombic form (alpha form) upon solidification. The bitumen does, however, prevent the growth of macro sulfur crystals. This is important because sulfur has a relatively high linear coefficient of thermal expansion, 7.4×10^{-5}/C and a low thermal conductivity, 6.1 gcal/cm^2/sec (Δ t = 1 C), both at 40°C. When a material containing adjacent macro crystals of sulfur is subjected to changing temperatures, there will be a constant movement between these macro crystals as one example or contracts relative to its neighbor. This movement will gradually break the bonds with other cross laid crystals, causing micro fractures and eventual formation of cleavage planes (Vroom 1998). Since the modified sulfur does not go through the allotropic transformation upon solidification, it has less shrinkage and hence develops less residual stress upon cooling (Lin et al. 1995).

4.2 *Sulfur concrete and sulfur polymer concrete*

To utilize the available formulated sulfur concrete for manufacturing unmodified sulfur, it is necessary to add a finely divided viscosity increasing material such as fly ash, which in turn will increase the consistency and workability of the mixture. Fly ash is in the form of tiny hollow spheres that contains major amounts of silicon and aluminum oxides. It has been found to impart an extra durability to the final sulfur concrete. The reactive aluminosilicate and calcium aluminosilicate components of fly ash are routinely represented in their oxide nomenclatures such as silicon dioxide, aluminum oxide and calcium oxide. Fly ashes tend to contribute to concrete strength, when the aluminosilicate components react with calcium oxides to produce additional cementations materials (Mohamed 2002, 2003, Mohamed et al. 2002, 2003, 2005).

Basic mechanical properties were measured for the sulfur concrete samples in order to ensure that the developed materials met the properties found in literature for these materials. Figure 6 shows the desired optimum amount of sulfur to fly ash ratios according to the mechanical strength of the mortar. The obtained strength tended to increase as the sulfur/fly ash ratio increases up to 0.9, were all particles are coated by a thin layer of sulfur which acts as a good binder for aggregate particles that finally leads to increase the strength. However, with large sulfur addition, the compressive strength decreased, because of further increment of sulfur content increases the thickness of sulfur layers around the aggregate particles that leads to the increase of the brittleness of the formed composite material.

To have proper criteria for evaluating the difference between the performance of unmodified sulfur concrete and modified one sulfur polymer concrete samples were prepared using different modified sulfur percentages. The effect of the modified sulfur incorporated into the mortars is shown in Figure 7. The compressive strength decreased linearly with increasing amounts of modified sulfur due to the partial inhibition of the crystallization through addition of modified sulfur. These results are in agreement with that reported by Vroom (1981). Mortars with modified sulfur show higher viscosity than unmodified one. This fact has an important effect in the crystallization of sulfur. In a more viscous liquid, the growth of the crystals will be more difficult and slow causing partial reduction in compressive strength.

The microstructure characterization, performed by scanning electron microscopy (SEM), provides important data on mortar studies. Figure 8a is an image for sulfur concrete, in which some common crystallization features were observed. The microstructure shows considerable degree of packing with some pores. While sulfur polymer concrete microstructure in Figure 9a shows a sulfur binding the aggregates with filling the inner spaces in such a way that the voids are discrete and discontinues with good particles homogeneity.

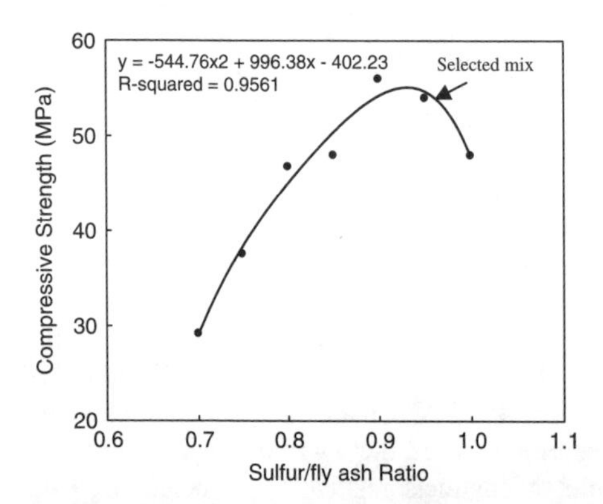

Figure 6. Effect of sulfur ratio on the compressive strength of sulfur concrete.

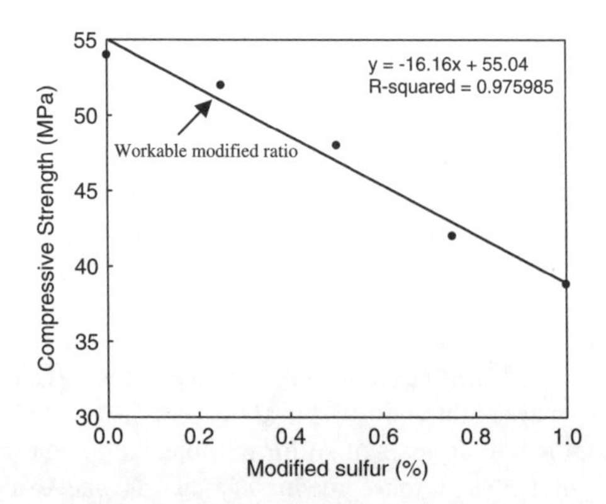

Figure 7. Effect of modified sulfur percentage on the mechanical strength of SPC.

The chemical analysis performed by energy dispersive spectroscopy (EDS) for SC and SPC are shown in Figures 8b, 9b indicates that the samples are composed mainly of sulfur and silicon containing compounds. For SC, a sharp and intense signals with low atomic sulfur percentage was found indicating high crystalline structure, while in case of SPC, a low crystalline structure was obtained from the lower of signal intensity, and the increase of atomic sulfur percentage proved the lower porosity. It was reported that porous bodies impregnated with modified sulfur, which may have surface interactive forces, are different from those of ordinary sulfur (Beaudoin and Feldmant 1984). The cumulative percentage of pore size studied by image analyzer shown in Figure 10 indicated lower porosity in sulfur polymer concrete. The presence of modified sulfur has two main advantages over than the unmodified one. The first is high homogeny, which prevents the growth of big crystals, and the second is low overall porosity.

Thermo gravimetric analysis (TGA) of SC and SPC (Figure 11) shows that the decomposition of SC is faster than that of SPC. This behavior can be considered as a further evidence for the thermal stability of SPC. According to the behavior observed from differential scanning calorimetry (DSC) for SC and SPC, shown in Figure 12, addition of modified sulfur at small percentages to the mix not only affects the crystal formation of sulfur (making it crystallizes in a structure different than the stable one), but also alter the behavior in the temperature controlled mixing period.

Crystallization of sulfur in this case is controlled by the relative percentages and distribution of space between aggregates and sulfur binder. The effect of modification seems to be an increase in degree of sulfur polymerization. Sulfur itself possesses a high tendency to polymerize; the modified sulfur would thus increase this tendency or maintain it for a longer time. In a more viscous liquid, and in which the molecules are more polymerized, the growth of the crystal will be more difficult. This

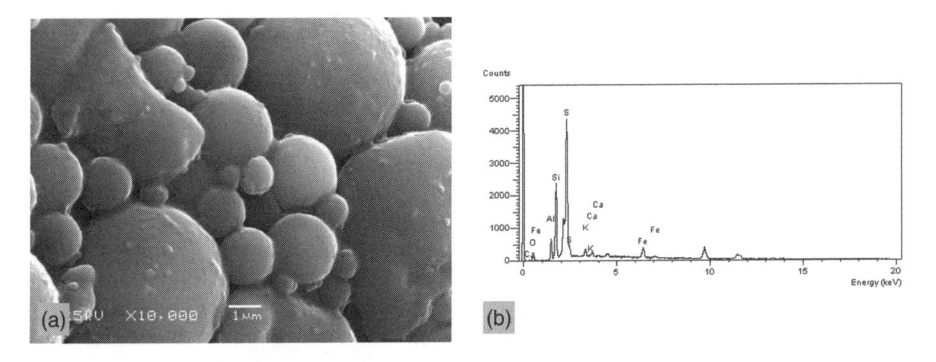

Figure 8. SEM micrograph showing a variety of particle shapes, sizes and voids for sulfur concrete (a), EDX spectrum showing the residue main chemical element for sulfur concrete (b).

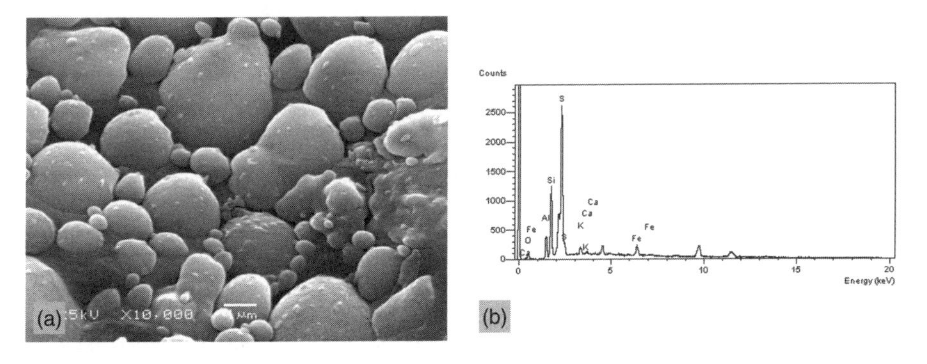

Figure 9. SEM micrograph showing a variety of particle shapes, sizes and voids for sulfur polymer concrete (a), EDX spectrum showing the residue main chemical element for sulfur polymer concrete (b).

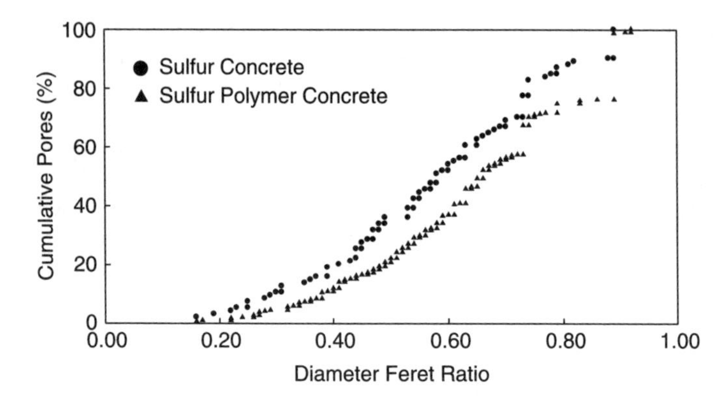

Figure 10. The cumulative percentage of pores sizes, for SC and SPC, using image analysis program.

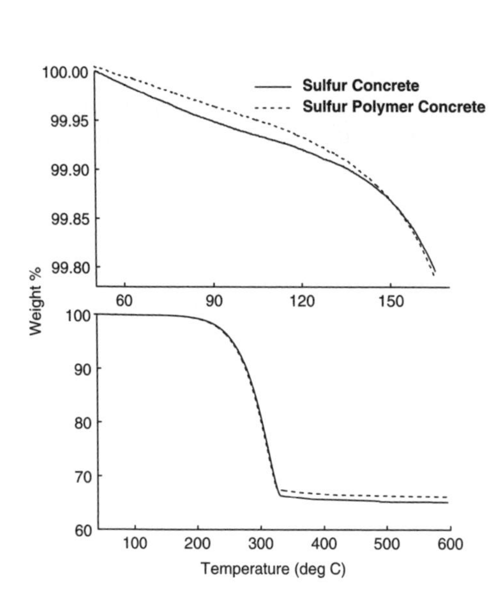

Figure 11. TGA curves for SC and SPC , the heating rate was 20 deg/min. 20deg/min.

Figure 12. DSC curves for SC and SPC. Indicating the values of area under peak, for alpha and beta sulfur crystals. The heating rate was 5 deg/min.

fact has an important effect in the crystallization of sulfur. The DSC chart for SPC shows significant reduction in the (α) and (β) sulfur crystal forms, which illustrated from the reduction of the areas under melted peaks, confirming that SPC have low order of crystalline structure compared to SC.

4 CONCLUSION

In this study, sulfur polymer concrete was manufactured with the use of modified sulfur, sulfur, and recycled aggregate materials (fly ash and sand dunes). Such material could be utilized in various public works such as retaining structure, foundations, road construction, base materials for underground piping, etc. Specific conclusions are discussed below.

1. Addition of olefin hydrocarbon polymeric material (bitumen) the sulfur has resulted in:
   a. sulfur polymerization with bitumen and increasing available sites for chemical reaction that leads to polysulfide formation;

b. forming a new structure with fine grain sizes;
c. controlling crystallization of sulfur. Pure sulfur crystallizes and form dense and large alpha sulfur crystals with orthorhombic sulfur morphology. With addition of bitumen, the crystal growth is limited and controlled by the bitumen in such a way that all crystals are plate like of micron dimension as monoclinic sulfur crystal of beta form;
d. uniform distribution of bitumen within the sulfur leading to increase resistance to crack formation and increase thermal stability;
e. modification of sulfur mineralogy. Modified sulfur remains in the monoclinic modification of beta form and does not undergo a phase transformation to the orthorhombic form (alpha form) upon solidification. Addition of bitumen does, however, prevent the growth of macro sulfur crystals;

2. Manufactured sulfur polymer concrete has resulted in:
   a. an increase of compressive strength, as the sulfur/fly ash ratio increases up to 0.9, were all particles are coated by a thin layer of sulfur. However, with large sulfur addition, the compressive strength decreased, because addition of further sulfur increases the thickness of sulfur layers around the aggregate particles leading to formation of brittle bonding;
   b. linear decrease of compressive strength with increasing amounts of modified sulfur due to the partial inhibition of the crystallization;
   c. sulfur binding of the aggregates by filling the inner spaces in such a way that the voids are discrete and discontinues with good particles homogeneity; and
   d. formation of a solid matrix with low porosity.

ACKNOWLEDGEMENTS

The authors would like to acknowledge the funding provided by the Research Affairs Sector, UAE University, for conducting this study. Also, acknowledgements are in order to the Central laboratories Unit (CLU) and Chemical and Petroleum Engineering Laboratories, UAE University, for the use of their facilities; Al Ain Cement Factory for their help in material characterization; and Al Ain Municipality for providing some of the raw materials.

REFERENCES

ACI Committee 548. 1993. Guide for mixing and placing sulfur concrete in construction (ACI 548.2R-93), *American Concrete Institute*, Farmington Hills, Mich., USA.

Bacon, R.F. & Davis, H.S. 1921. Recent advances in the American sulfur industry. Chemical and Metallurgical Engineering 24(2): 65–72.

Beaudoin, J.J. & Sereda, P.J. 1973. The freeze-thaw durability of sulfur concrete. Building Research Note, Division of Building Research, National Research Council, Ottawa, 53.

Bordoloi, B.K. & Pierce, E.M. 1978. Plastic sulfur stabilization by copolymerization of sulfur with dicyclopentadiene. Advances in Chemistry Series No. 165, American Chemical Society, Washington D.C., 31–53.

Crow, L.J. & Bates, R. C. 1970. Strength of sulfur-basalt concretes. Bureau of Mines Report No. RI 7349, U.S. Bureau of Mines, Washington, D.C., 21.

Currell, B.R. 1976. The importance of using additives in the development of new applications for sulfur. Symposium on new users for sulfur and pyrites, 105–103. Madrid.

Dale, J.M. & Ludwig, A.C. 1966. Feasibility study for using sulfur-aggregate mixtures as a structural material. Technical Report No. AFWL-TR-66-57 Southwest Research Institute, San Antonio., 40.

Dale, J.M. & Ludwig, A.C. 1968. Advanced studies of sulfur aggregate mixtures as a structural material. Technical Report No. AFWL-TR-68-21, Southwest Research Institute, San Antonio., 68.

Duecker, W.W. 1934. Admixtures improve properties of sulfur cements. Chemical and Metallurgical Engineering 41(11): 583–586.

First Annual Report of the Secretary of the Interior under the Mining and Mineral Policy Act of 1970, Public Law 91-631," U.S. Department of the Interior, 1972. Washington, D.C., 34.

Funke, R.H. Jr. & McBee, W.C. 1982. An industrial application of sulfur concretes. ASC Symposium Series No. 183, American Chemical Society, Washington, D.C., 195–208.

Kobbe, W. H. 1924. New uses for sulfur in industry. Industrial and Engineering Chemistry, 16(10): 1026–1028.
Lin, S.L. , Lai, J.S. & Chian, E.S.K. 1995. Modification of sulfur polymer cement (SPC) stabilization and solidification (S/S) process. Waste Management 15 (5/6): 441–447.
Malhotra, V.M. 1973. Mechanical properties and freeze-thaw resistance of sulfur concrete. Division Report No. IR 73-18, Energy, Mines and Resources Canada Ottwa., 30.
Malhotra, V.M. 1974. Effect of specimen size on compressive strength of sulfur concrete. Division Report No. IR 74-25, Energy, Mines and Resources Canada Ottwa.
Mc Bee, W.C., Sullivan, T.A. & Fike, H.F. 1986. Corrosion-resistant sulfur concretes. Corrosion and Chemical Resistant Masonry Materials Handbook, Noyes Publications, Park Ridge, 392–417.
Mc Bee, W.C., Sullivan, T.A. & Jong, B.W. 1981a. Modified sulfur concrete for use in concretes, flexible paving, coatings, and grouts. Bureau of Mines Report No. RI 8545, U.S. Bureau of Mines, Washington, D.C., 24.
Mc Bee, W.C., Sullivan, T.A. & Jong, B.W. 1981b. Modified sulfur concrete technology. Proceedings, SULPHUR-81 International Conference on Sulfur, Calgary, 367–388.
Mc Bee, W.C., Sullivan, T.A. & Jong, B.W. 1983. Corrosion-resistant sulfur concretes. Bureau of Mines Report No. 8758, U.S. Bureau of Mines, Washington, D.C., 28.
Mc Bee, W.C., Sullivan, T.A. & Jong, B.W. 1983. Industrial evaluation of sulfur concrete in corrosive environments. Bureau of Mines Report No. RI 8786, U.S. Bureau of Mines, Washington, D.C., 15.
Mc Bee, W.C., Sullivan, T.A. 1979. Development of specialized sulfur concretes. Bureau of Mines Report No. RI 8346, U.S. Bureau of Mines, Washington, D.C., 21.
Mc Bee, W.C. & Sullivan, T.A. 1982a. Assigned to U.S. Department of Commerce, Concrete Formulations Comprising Polymeric Reaction Products of Sulfur/Cyclopentadien, Oligmer/Dicyclopenetadiene," U.S. Patent No. 4,348,313.
Mc Bee, W.C. & Sullivan, T.A. 1982b. Assigned to U.S. Department of Commerce, Modified Sulfur Cement. U.S. Patent No. 4,311,826.
McKinney, P.V. 1940. Provisional methods for testing sulfur cements, ASTM Bulletin 96-107: 27–30.
Mohamed, A.M.O. 2002. Hydro-mechanical evaluation of soil stabilized with cement-kiln dust in arid lands. Environmental Geology, 42(8), pp. 910–921.
Mohamed, A.M.O. 2003. Geoenvironmental aspects of chemically based ground improvement techniques for pyretic mine tailings. Ground Improvement, 7(2), pp. 73–85.
Mohamed, A.M.O., Hossein, M. & Hassani, F. 2002. Hydro-mechanical evaluation of stabilized mine tailings. Environmental Geology J., 41, pp. 749–759.
Mohamed, A.M.O., Hossein, M. & Hassani, F. 2003. Role of fly ash addition on ettringite formation in lime-remediated mine tailings. J. of Cement, Concrete and Aggregates, ASTM, 25(2), pp. 49–58.
Mohamed, A.M.O., Hossein, M. & Hassani, F. 2005. Leachability potential of the newly developed ALFA technology for solidification/stabilization of mine tailings. J. of Materials in Civil Engineering, Special Edition on Advances in Physico-chemical Stabilization of Geomaterials, in press.
Nnabuife, E.C. 1987. Study of some variables affecting the properties of sulfur-reinforced sugarcane residue based boards. Indian Journal of Technology 25: 363–367.
Sullivan, T.A. & Mc Bee, W.C. & Blue, D.D. 1975. Sulfur in coatings and structural materials. Advances in Chemistry Series No. 140, American Chemical Society, Washington, D.C., 55–74.
Sullivan, T.A. & Mc Bee, W.C. 1976. Development and testing of superior sulfur concretes. BuMines Report No. RI 8160, U.S. Bureau of Mines, Washington, D.C., 30.
Sullivan, T.A. 1986. Corrosion-resistant sulfur concretes – design Manual. The Sulphur Institute, Washington D.C., 44.
Syroezhko, A.M., Begak, O.Yu., Fedorov, V.V. & Gusarova, E.N. 2003. Modification of paving asphalts with sulfur. Russian Journal of Applied chemistry 76(3): 491–496.
Voronkov, M.G., Vyazankin, N.S. & Deryagina, E.N. 1979. Reaktsii sery s organicheskimi veshchestvami (Reaction of sulfur with organic compounds), Novosibirsk: Nauka.
Vroom, A.H. 1977. Sulfur cements, process for making same and sulfur concretes made thereform," U.S. Patent No. 4,058,500.
Vroom, A.H. 1981. Sulfur Cements, Process for making same and sulfur concretes made Thereform. U.S. Patent No. 4,293,463.
Vroom, A.H. 1992. Sulfur polymer concrete and its application. In. proceeding of Seventh International Congress on polymers in concrete: 606–621 Moscow.
Vroom, A.H. 1998. Sulfur concrete goes global. Concrete International 20(1): 68–71.

Reclaiming the Desert: Towards a Sustainable Environment in Arid Lands – Mohamed (ed.)
© 2006 Taylor & Francis Group, London, ISBN 0 415 41128 9

Thermo-mechanical behavior of newly developed sulfur polymer concrete

A.M.O. Mohamed & M.M. El Gamal
Department of Civil and Environmental Engineering, UAE University, Al Ain, UAE

A.K. El Saiy
Department of Geology, UAE University, Al Ain, UAE

ABSTRACT: A concern on the utilization of sulfur polymer concrete (SPC) arises from the relatively poor durability of sulfur concrete, in response to repeated thermal cycles. Newly modified sulfur was produced and reported by Mohamed and El Gamal (2006) which consists of sulfur and bitumen (as an olefin hydrocarbon polymeric material). SPC is a thermoplastic material that is produced by mixing molten sulfur with modified sulfur and recycled aggregates such as fly as and desert sand at 135°C. This study deals with an experimental investigation to evaluate the effect of time and temperature on compressive strength of the manufactured SPC. The SPC samples were prepared and maintained at constant temperatures of 26, 40, 60, 80, and 100°C for different time periods of 1, 2, and 7 days. The experimental results were evaluated using scanning electron microscopy and image analysis software, x-ray diffraction analysis and energy dispersive x-ray to evaluate the microstructure, pore sizes, and potential formation of new cementing agents that contribute to strength development. The results indicated that the SPC developed 76% of its ultimate compressive strength after one day, 97% of its ultimate strength after three days, and no further strength gain with time. SPC compressive strength was higher than that of Portland cement concrete (PCC) by about 20% for temperatures less than 60°C and the same as that of PCC at high temperatures (i.e., 80 and 100°C).

1 INTRODUCTION

With the development of concrete technology, Sulfur Polymer Concrete (SPC) has been commonly utilized in many concrete structures around the world. Sulfur concrete has properties which, for certain applications are superior to Portland Cement Concrete (PCC). Its rapid strength development makes it a very good choice for low-temperature concreting. The strength of sulfur concrete is often greater than that of PCC and the stiffness can be modified by suitable admixtures. It also shows excellent resistance to attack by acidic and saline solutions. While SPC could be prepared by hot-mixing unmodified sulfur and aggregate, durability of the resulting product was a problem. Solid sulfur can exist in two crystal forms with different densities. Failure to stabilize the sulfur will result in a transformation to a crystalline sulfur structure with poor strength characteristics and can result in a volume change of the material. The most sever problems are caused by brittleness, poor freeze-thaw durability and excessive moisture expansions (Czarneckt & Gillott 1990).

Modification of sulfur by reaction with dicyclopentadiene (DCPD) has been investigated by many researchers (McBee et al., 1981a, b), but its practical use in commercial application has been limited because the reaction between sulfur and DCPD is exothermic and requires close control, and unstable when exposed to high temperature (greater than 140°C) especially when mixing with hot aggregates. It may react further to form an unstable sulfur product. McBee & Sullivan (1982) solved this problem through development of a process for preparing modified sulfur cement that is stable and not temperature sensitive in mixing temperature range for production of sulfur concrete. This process utilizes the control reaction of cyclopentadiene. Methods for treating sulfur for use in sulfur concretes have been reported by many researchers. They include: Leutner & Diehl (1977) for using DCPD; Gillott et al. (1980) for using crude oil and glycol; Vroom (1981) for using olefinic hydrocarbon polymer such as Escopol to prepare a modified sulfur cement; Woo (1983) for using phosphoric acid to improve freeze-thaw resistance; Nimer and Campbell (1983) for using organosilane to improve water stability; Beaudoin & Feldmant (1984) for using DCPD to improve porous durability of porous system; and Pickard (1985) for using modified sulfur cement to control expansion/contraction during thermal cycles.

When a modified sulfur cement is used as the binding agent with appropriate aggregates, the resulting sulfur concrete has shown some unique properties; high strength and fatigue resistance; excellent corrosion resistance against most acids and salts; and extremely rapid set and strength gain.

Recently Mohamed and El Gamal (2006) have produced modified sulfur using olefin hydrocarbon polymeric material for the production of SPC, which consists of modified sulfur, molten sulfur, fly as and desert sand. The physicochemical properties of the manufactured SPC were reported (Mohamed and El Gamal, 2006). However, this study deals with an experimental investigation to evaluate the effect of time and temperature on compressive strength of the manufactured SPC. The SPC samples were prepared and maintained at constant temperatures of 26, 40, 60, 80, and 100°C for different time periods of 1, 2, and 7 days. The experimental results were evaluated using scanning electron microscopy and image analysis software, x-ray diffraction analysis and energy dispersive x-ray to evaluate the microstructure, pore sizes, and potential formation of new cementing agents that contribute to strength development.

2 MATERIALS

The granular sulfur (99.9% purity) was obtained from Al Ruwais refinery, UAE. As the olefin hydrocarbon polymeric material we have use bitumen 60/70 with a softening point of 48.8°C, specific gravity at 20°C of 1.0289 g/cm^3, kinematics viscosity at 135°C of 431 cSt, and chemical analysis of; C:79, H:10, S:3.3 and N:0.7%. Two types of aggregates were used. The first is desert sand obtained from a sand dunes quarry in Al Ain area while the second is fly ash of India-97/591. Chemical

Table 1. Chemical composition of used aggregates (Mohamed and El Gamal, 2006).

	Compound % (w/w)					
Aggregate	Al_2O_3	CaO	Fe_2O_3	K_2O	MgO	SiO_2
Sand	0.47	16.35	0.676	0.13	1.158	74.4
Fly Ash	32.4	0.46	4.34	0.027	0.66	60.9

Table 2. Grain sizes distribution of aggregates (Mohamed and El Gamal, 2006).

Sand		Fly Ash	
Sieve mesh size (mm)	Mass retained on the sieve (%)	Sieve mesh size (mm)	Mass retained on the sieve (%)
0.425	0.00	0.250	0.40
0.250	8.82	0.150	1.25
0.150	49.90	0.125	2.85
0.106	24.64	0.075	13.35
0.090	8.80	0.063	12.25
0.075	4.85	0.053	46.80
Bottom dish	3.00	0.045	18.35
		Bottom dish	4.75

analyses were performed using Inductively Coupled Plasma Atomic Emission Spectrometry (ICP-AES) and grain sizes distributions are listed in Tables 1 and 2 (Mohamed and El Gamal, 2006).

3 METHODS

3.1 *Preparation of modified sulfur*

In an oil bath 2.5 wt% of hot bitumen and 97.5 wt% of molten sulfur containing emulsifying agent (iso-octylphenoxy polyethoxy ethanol) were mixed and mechanically stirred at 140°C. The range of the reaction temperature was maintained at about 135–140°C and the reaction periods were in the order of about 45–60 minutes. The progress of the reaction was monitored by the degree of homogeneity of the mixture, which was done by careful observation of the temperature of the reacting mixture, and by observation of the viscosity of the mixture. After that the mixture was allowed to cool at a rate of 8–10°C/min. The final product is a sulfur containing polymer, which on cooling possessed glass like properties. This modified sulfur cement is commonly called Modified Sulfur Cement. The microstructure of the modified sulfur cement was studied using scanning electron microscopy (SEM), and the thermal properties were studied using thermal gravimetric analysis (TGA) and differential scanning calorimetric (DSC) techniques.

3.2 *Production of SPC*

SPC consisting of elemental sulfur, modified sulfur and recycled aggregates (fly ash and desert sand) was prepared according to the procedure described in ACI 248.2R–93. The recycled aggregates were heated in an oven to 170–200°C for two hrs. The specified amount of sulfur was melted in the heated mixing bowl, which is placed in an oil bath with controlled temperature (from 132–141°C). Fly ash then transferred to the heated mixing bowl and properly mixed with the molten sulfur for about 20 minutes to insure complete reaction between sulfur and fly ash. The mix proportion is of critical importance. For example, we have maintained the sulfur/fly ash ratio of 0.9 and the modified sulfur of about 0.25 wt%. Finally sand was added with controlled ratio of sulfur/sand of 1.00. For specimen preparation, cubic steel mold with dimensions $50 \times 50 \times 50$ mm was used. The molds were pre-heated to approximately 120°C before adding the SPC. As the SPC was added, the material was compacted by tamping with a heated rod and continuous vibration for 10 seconds using vibrating table. The surface of each specimen was finished and left in the oven for gradual cooling at a rate of 2 degree/min. Specimens were de-molded at an age of 24 hours and stored in incubators at 40°C. Other series of specimens were left in the molds in ovens at different temperatures of 26, 40, 60, 80 and 100°C for different time periods of one, two and seven days to study the effect of time on curing temperatures. Specimens were then analyzed by using compressive strength, SEM set with an energy Dispersive x-ray detector (EDX) and x-ray diffraction analysis.

3.3 *Testing techniques*

Different laboratory tools were used to study the effect of temperature on sulfur, modified sulfur, and SPC. Thermal gravimetric analysis (Perkin Elmer TGA7) was used to measure weight changes in sample materials as a function of the temperature at a scan rate of 20°C/min. Differential scanning calorimeter (Perkin Elmer DSC7) was used for measurements of heat capacity through phase transitions on heating. 10 mg of the tested sample was heated up to 150°C, with heating rate of 5°C/min. The sample was allowed to be self cooled to room temperature for 24 hrs then it was reheated up to 150°C. The microstructure characterization performed for sulfur, modified sulfur and SPC was examined by SEM (JSM-5600 Joel microscope) equipped with an energy Dispersive x-ray detector (EDX). The size distribution of voids in SPC was studied using image analysis software. The shape of particles and the size of voids in the developed matrix were investigated by analyzing the Feret ratio, which is expressed as the minimum/maximum Feret diameters. The maximum distance measured between two vertical lines tangent to the ends of the void is defined as the maximum Feret diameter while, the minimum distance measured between two horizontal lines tangent to the ends of the void is defined as the minimum Feret diameter. Compressive strength of the samples was determined by Wykeham Farrance testing machine with maximum load of (200 kN), the unconfined compression tests were conducted using MST tensile testing machine. The system is equipped with universal testing software (Test Works) capable of graphic and numerical analysis of the test data with maximum load of (100 kN). x-ray diffraction was performed using Philips PW/1840, with Ni filter, Cu-Kα radiation ($\lambda = 1.542$ Å) at 40 KV, 30 mA and scanning speed 0.02°/S. The diffraction peak between $2\theta = 2°$ and $2\theta = 80°$ were recorded.

4 RESULTS AND DISCUSSION

4.1 *Properties of modified sulfur cement*

Addition of olefin hydrocarbon polymetric material (bitumen) in amounts of 2.5% to sulfur initiates chemical reactions whose type depends on the bitumen content, heating temperature and the time of the reaction. It should be pointed that, at $T < 95$°C sulfur exists as a cyclooctasulfane crown with an S-S bond length of 0.206 nm, and S-S-S bond angle of 108 degree, and at $T < 119$°C sulfur crystallizes. At 119°C (melting point of sulfur), liquid sulfur being thoroughly dispersed in bitumen forming an emulsion (role of emulsifying agent) and cyclooctasulfane turns partly into polymeric zigzag chains (bond length of 0.204 nm) (Voronkov et al., 1979). The crystallization features are affected by such factors as chemical reaction of sulfur with bitumen components and its dissolution or dispersion. Sulfur at heating temperature $T < 140$°C, elementary sulfur forms polysulfide, which initiates formation of a network. Such structures differ considerably in the chemical and thermal stability from unmodified sulfur. At 119–159°C, molten sulfur exists essentially as cyclooctasulfane (λ-S). Above 159°C, eight-member ring rapidly break down into biradicals (Voronkov et al., 1979), which are recombined to form polymeric chains with the maximal length of up to 10^6 of sulfur atoms. However above 140°C, dehydrogenation of saturated bitumen components can occur; also linear polysulfide can transform into stable cyclic thiophene structures (Syroezhko et al., 2003). Keeping in mind that modified sulfur should be produced within its recommended mixing temperature range of 135 to 141°C.

The microstructure displayed from SEM revealed the role of bitumen in controlling the crystallization of sulfur. Pure sulfur crystallizes and form dense and large alpha sulfur crystals (S_α) with orthorhombic sulfur morphology as shown in Figure 1a. With the reaction of sulfur and bitumen, the crystal growth is limited and controlled by the bitumen in such a way that all crystals are plate like of micron dimension as monoclinic sulfur crystal of beta form (S_β) as shown in Figure 1b. This plate like microstructure crystal of micron size plays a major role in relieving stresses produced from thermal expansion mismatch in concretes.

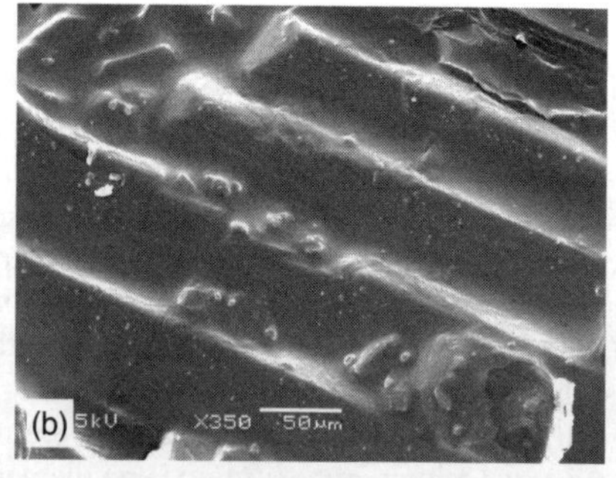

Figure 1. SEM image for (a) pure crystalline sulfur, and (b) sulfur modified with of 2.5 wt% bitumen.

Thermal analysis techniques such as TGA and DSC have been used to determine the temperature history of sulfur and sulfur cement. The TGA measures the weight loss with temperature as shown in Figure 2a which provides comparison between pure sulfur and sulfur modified with 2.5 wt% bitumen. The maximal thermal effect was observed at 180–450°C for pure sulfur, which was accompanied by active liberation of hydrogen sulfide. While for modified sulfur, the maximal thermal effect was observed at 200–540°C. This effect reflects the increase of thermal stability of modified sulfur. The DSC results of both sulfur and modified sulfur cement are shown in Figure 2b. For pure sulfur, melting of the alpha and beta was detected in the first run, while only alpha melting was detected in the second run. For modified sulfur, melting of both the alpha and beta was detected in the first run, while only beta melting was detected in the second run. This in turn indicates that the modified sulfur remains in the monoclinic modification of beta form and does not undergo a phase transformation to the orthorhombic alpha form upon solidification. It is worth noting that the existence of microstructure of the beta form will potentially resist cracking, reduce thermal expansion, minimize the development residual stress due to thermal expansion, and minimize shrinkage upon cooling (Lin et al., 1995).

4.2 *Compressive strength of SPC*

The molten sulfur, rather than the conventional Portland cement and water, acts as binder for the concrete. The raw materials used in these preparations possess some specific known characteristics. The desert sand was found to be of good quality showing a low concentration of impurities and grains with an irregular geometry. This quality of sand grains may reduce the workability of the mortar but on the other hand enables the molten sulfur to adhere more easily on the surface of the sand grains (Gemelli et al., 2004). Fly ash was found to add a finely divided viscosity increasing material, which in the form of tiny hollow spheres consisting of major amounts of silicon and aluminum oxides, due to its small particle size, shape and surface texture. It has been found to impart an extra durability to the final SPC. The rapid hardening and early strength gain change over time resulting in a very high

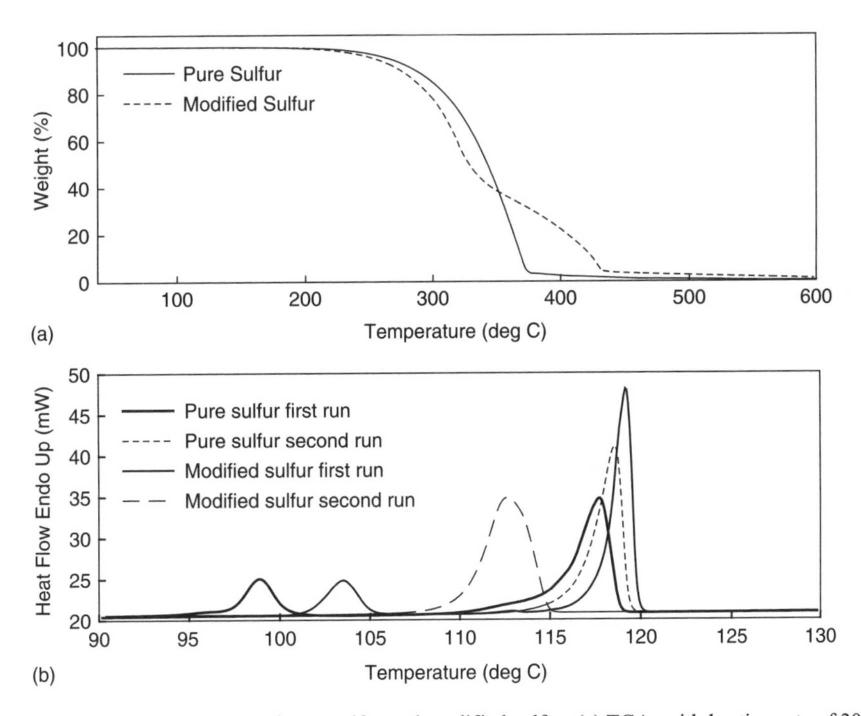

Figure 2. Thermal properties of pure sulfur and modified sulfur; (a) TGA, with heating rate of 20°C/min, (b) DSC, with heating rate of 5°C/min.

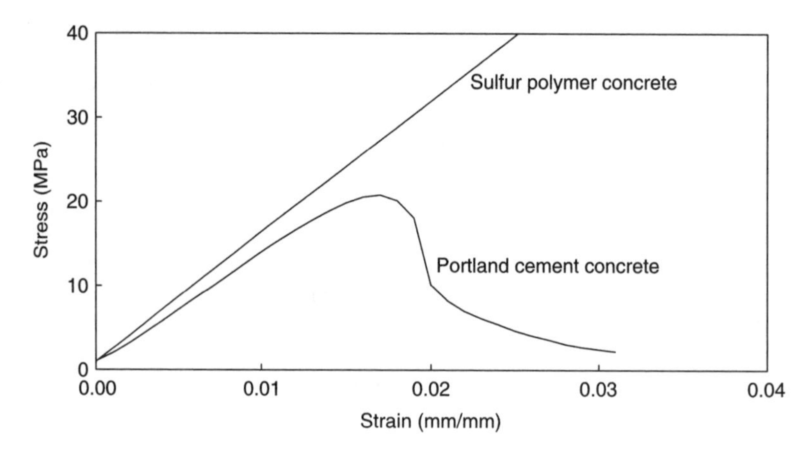

Figure 3. Stress-strain relationship for sulfur polymer concrete (SPC) and normal Portland cement concrete (PCC).

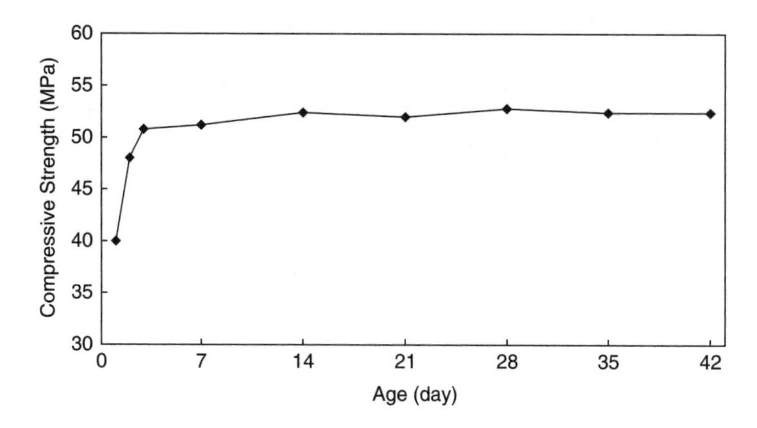

Figure 4. Variations of compressive strength with time for sulfur polymer concretes (SPC) at 40°C.

strength material with an average compression of 54 MPa and modules of elasticity of 1603 MPa. For PCC, the compressive strength is of 22 MPa and modules of elasticity of 1461 MPa.

Figure 3 shows a typical stress strain curves for SPC and PCC. The PCC compressive strength increases as strain increases, up to about 0.017 strain, reaching a maximum value of 20 MPa then, it decreases. However, for SPC, the compressive strength continues to increase with the strain in excess of 0.025 reaching a value of about 40 MPa after which the test was stopped because the stress level reaches the maximum capacity of the testing machine. The large mechanical strength for SPC is attributed to the reactive aluminosilicate and calcium aluminosilicate components that are routinely present in their oxide nomenclatures such as silicon dioxide, aluminum oxide and calcium oxide. Fly ashes tend to contribute to concrete strength, when the aluminosilicate components react with calcium oxides to produce additional cementatious materials.

4.3. *Effect of age on compressive strength*

Compressive strength was determined for all SPC samples measuring 50 × 50 × 50 mm. Specimens were cured in the oven with gradual cooling rate of 2 degree/min, were de-molded after 24 hours, and were stored in incubators at 40°C. The variation of compressive strength with time, for SPC at

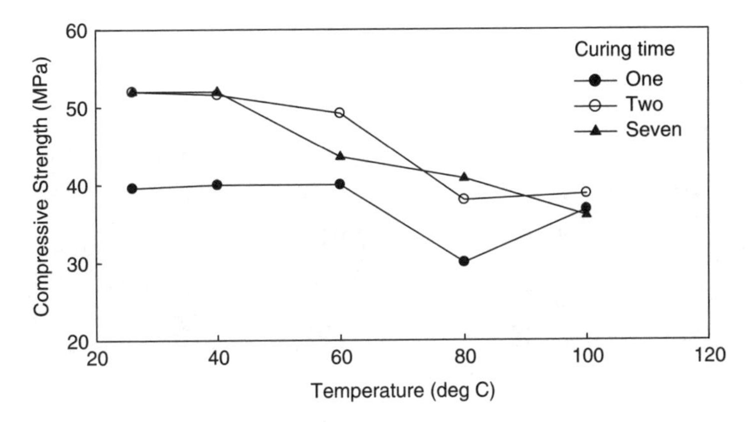

Figure 5. Variations of compressive strength with temperature and curing time for sulfur polymer concrete (SPC).

40°C, is shown in Figure 4. It is clear that SPC has developed its 76% of its ultimate compressive strength within one day, and 97% after three days. However, at later ages up to 42 days, there was no clear trend on the compressive strength development suggesting that the maximum strength was developed during the early days.

Similar results were reported by many researchers. For example, Vroom (1981) reported that 80% of the ultimate concrete strength was developed in one day, and virtually 100% of the ultimate strength was realized after four days. McBee et al. (1983) showed that the sulfur concrete developed about 70% of its ultimate strength within a few hours after cooling, 75 to 85% after 24 hrs at 20°C, and the ultimate strength was commonly obtained after 180 days at 20°C.

4.4 *Effect of curing temperature on compressive strength*

Compressive strength was measurements for SPC samples cured at different temperatures of 26, 40, 60, 80 and 100°C and for different time periods of, one, two and seven days. The variations of compressive strength with temperature and curing time for SPC are shown in Figure 5. The results indicate that for the same curing time as temperature increases compressive strength decreases. A systematic reduction in compressive strength with temperature was obtained specially at high temperatures (80 and 100°C). For the same temperature, as curing time increase, compressive strength increases. For temperatures below 60°C, the strength increase with time ranges between 20 to 30% while for higher temperatures, the rate of strength increase with time is lower. Similar results were reported by McBee and Sullivan (1979) who found that the strength gain is slower at elevated temperatures and faster at lower temperatures. The results also indicate that the compressive strength of SPC for temperatures below 60°C is 20 to 30% more than PCC and for higher temperatures (60 to 100°C), the compressive strength remains within the same range of PCC.

4.5 *Effect of microstructure on compressive strength*

Mechanical strength is directly related to the defects in mortar microstructure, which are generally characterized by SEM. The microscopic studies were determined from a section cut at a distance of 0.5 cm from the surface of SPC samples. Figure 6 shows a comparison between the mortar microstructure of SPC cured after seven days with different temperatures. It is obvious that for SPC cured at 40 and 60°C, the voids entrained within the SPC are small and discontinuous. However, while for SPC cured at 80 and 100°C, the voids are large in sizes without any noticeable cracking. The voids in SPC serve as sites for stress relief and for improving the durability of the material

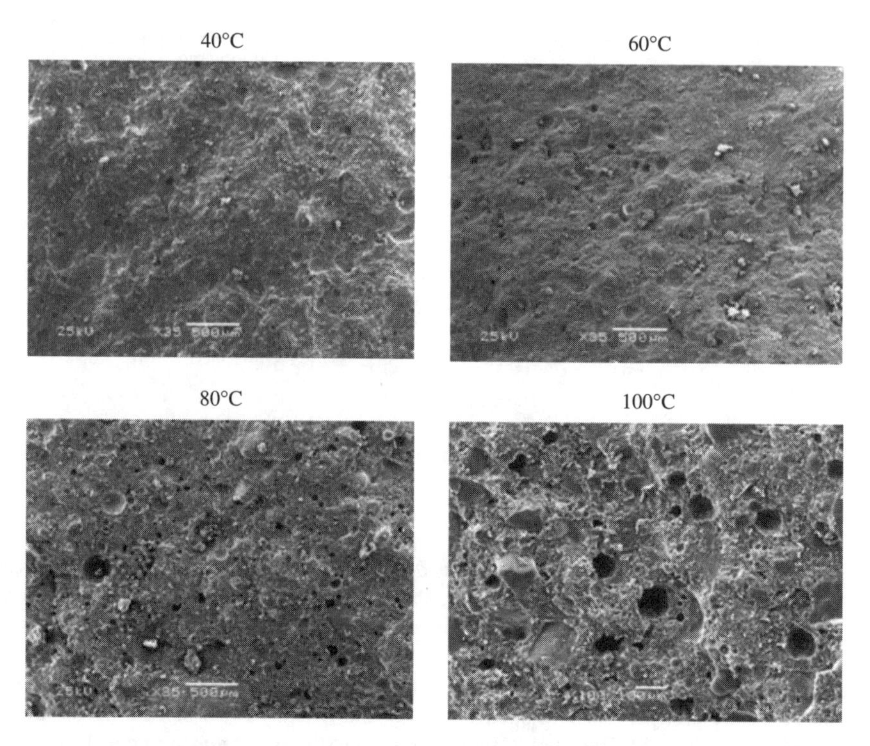

Figure 6. SEM images for sulfur polymer concrete at different temperature.

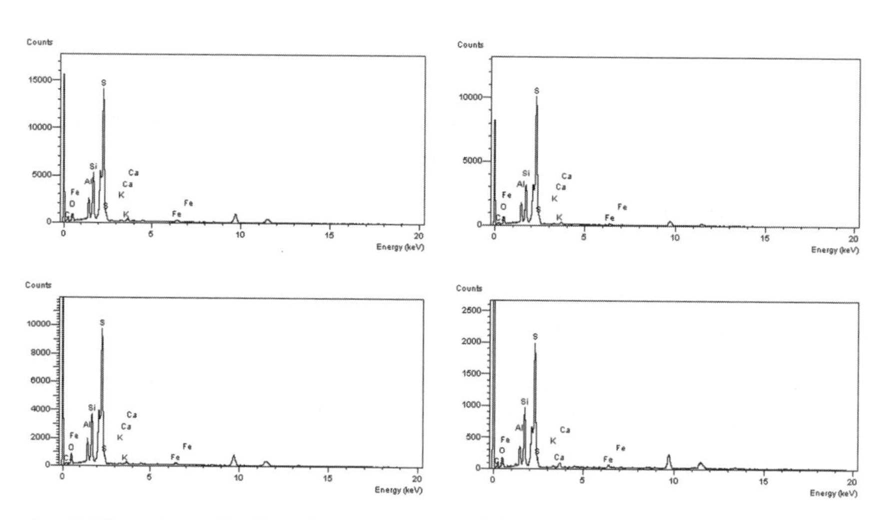

Figure 7. EDX spectrum of sulfur polymer concrete at different temperature, showing the intensity of the residue main chemical elements.

(McBee et al., 1983). Also, the presence of voids in SPC reduces the quantity of sulfur cement required to coat the mineral aggregate thereby minimizing the cement – related shrinkage.

The EDX spectrum of SPC at different temperatures (Figure 7) shows the intensity of the main chemical elements present in SPC. It can be seen that the sulfur peak intensity decreased with

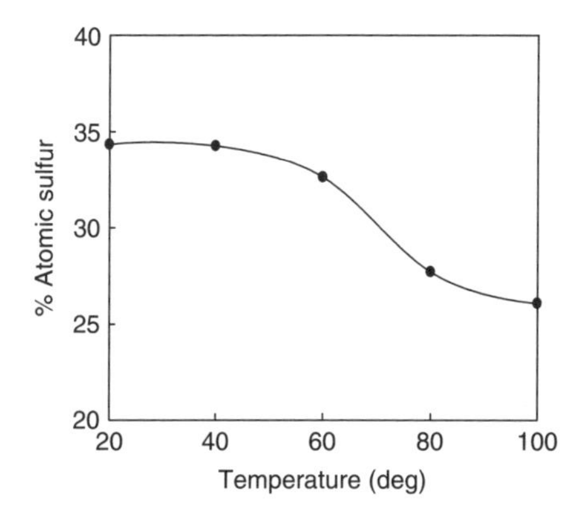

Figure 8. Changing the sulfur percentage of in SPC, varied with temperature, data obtained from EDX results.

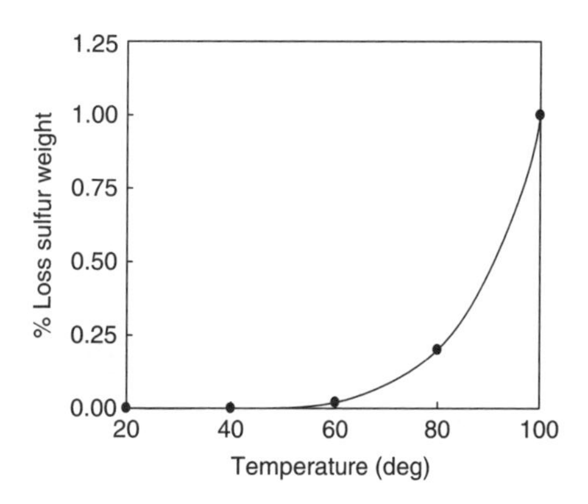

Figure 9. Percentage loss of sulfur weight varied with temperature, data obtained from gravimetric analysis.

increasing temperatures, which in turn confirms the SEM observations. This lower intensity means lower degree of mineral crystalline form, which in turn led to lower compression strength. To illustrate these set of results, the data were presented in terms of % atomic sulfur as a function of temperature as shown in Figure 8. The results indicated that the percentage of atomic sulfur in SPC tended to decrease as the curing temperatures increased, which are compatible with the presence of voids in SPC at elevated temperatures as previously discussed. To further evaluate whether the sulfur was lost due to gas formation or combination with other existing heavy metals to form metal sulfide forms, loss of sulfur weight as a function of temperature was studied as shown in Figure 9. The gravimetric changes in elemental sulfur with temperatures show no loss in sulfur weight at 40 and 60°C, while loss of 0.25% at 80°C and 1% at 100°C after seven days indicating that the possible increase in voids at 80 and 100°C could be attributed to sulfur gas formation.

Furthermore, to evaluate the size of the voids from the results of SEM, we have utilized image analysis software. The results are shown in Figure 10 in terms of the cumulative percentage of voids for SPC at different temperatures. It can be seen that for the same cumulative percent, the diameter Feret ratio increased with increased temperatures indicating that the size of the voids increased as temperature increased. It is worth noting that when diameter Feret ration reaches one, the pore sizes are spherical.

4.6 *Effect of formed minerals on compressive strength*

To evaluate the potential formation of minerals during the solidification/stabilization process of sulfur, modified sulfur, fly ash and desert sand, SPC samples were tested using x-ray diffraction analysis. The x-ray diffraction pattern of SPC at different temperatures is shown in Figures 11a & b. The mineral composition of SPC samples shown in Figure 11a, which subjected to the different temperatures (40°C, 60°C, 80°C, 100°C) and curing time of seven days, made up of major constituents; sulphur ranging between 28.4–31.5 and characterized by decreasing percentage of sulfur with increasing temperature, quartz (SiO_2) ranging between 44.0–50.5 and characterized by increasing percent with increasing temperature, which is approximately the opposite trend of sulfur. The minor constituents involved the following components; aluminium oxide hydrate ($5Al_2O_3!H_2O$) ranging between 1.7–7.9, calcium silicate hydrate ($Ca_{1.5}SiO_{3.5}!xH_2O$) ranging between 1.8–5.0, plagioclase ($CaAlSi_3O_8$) ranging between 2.5–4.7, calcite ($CaCO_3$) ranging between 2.0–3.8, hematite (Fe_2O_3) ranging between 2.2–2.5, dolomite ($CaMg(CO_3)_2$) ranging between 1.5–3.9, and calcium aluminium oxide hydrate ($Ca_3Al_2O_6!H_2O$) ranging between 0–1.2. It is worth noting that the amount of water molecules associated with the formed hydrated compounds is unknown since the process does not utilize water at all. This leads one to pose the following question. Where is the

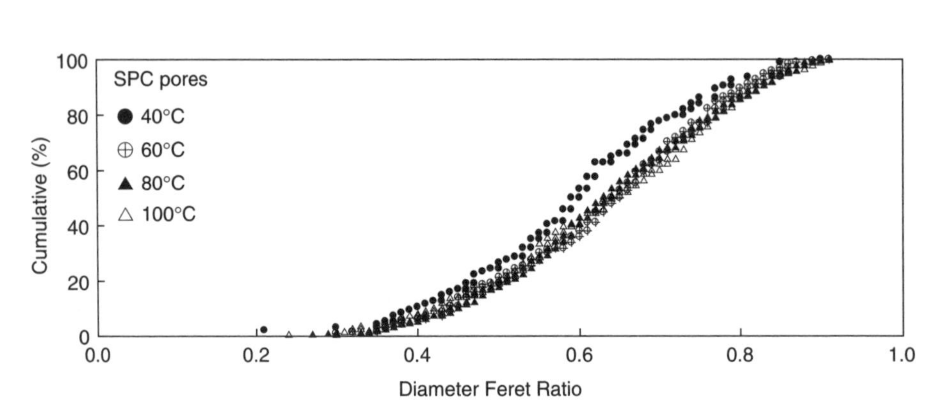

Figure 10. The cumulative percentage of voids, for SPC cured seven days at different temperature, using image analysis program.

water coming from? Is it from the humidity of the atmosphere or from the bonded water in the recycled aggregates? These issues need further investigation.

The lateral variation in the mineral composition of the sulfur samples with temperature after seven days is shown in Figure 11b. The Figure illustrates the distribution of the mineral composition percent of SPC samples with temperature variations (40°C, 60°C, 80°C, 100°C) after seven days. It shows that the percentage of sulfur decreased with increasing of temperature in all samples. An opposite trend of percentage is displayed by those of quartz. Plagioclase takes the trend of sulfur while the other components do not show noticeable variations in concentration throughout different temperatures.

In summary, the analysed SPC samples constitute variable percentages of sulfur, fly ash and sand dunes. The fly ash is by-product of cement and the chemical composition of the fly ash as detected by ICP analysis in decreasing percentage are silica (60%), Al_2O_3 (32.4%), Fe_2O_3 (4.34%), and traces of Mg and Zn. On the other hand, the chemical composition of the sand dunes (ICP analysis) in decreasing percentage made up of major SiO_2 (68%), minor of CaO (10%), MgO (5%) and Al_2O_3 (2.2%), and trace of K_2O (0.8%). It is likely that the source of detected minerals by x-ray diffraction analysis is the following: (1) the silica derived from two sources; quartz (SiO_2), plagioclase (calcium aluminium silicate) and calcium silicate hydrate of fly ash and sand dunes; (2) aluminium is detected in the three phases; aluminium oxide hydrate, calcium aluminium oxide hydrate and plagioclase (calcium aluminium silicate). All derived from fly ash and sand dunes; (3) iron appeared in the form of hematite (Fe_2O_3) and derived from two sources as silica and alumina; (4) calcium appeared in the phase of calcite ($CaCo_3$), dolomite ($CaMg(CO_3)_2$), calcium silicate, calcium aluminium oxide hydrate and plagioclase (Calcium aluminium silicate) and mainly derived from sand dunes; and (5) magnesium found in minor proportion and concentrated mainly in dolomite ($CaMg(CO_3)_2$).

5 CONCLUSION

In this investigation Sulfur Polymer Concrete (SPC) was manufactured and its compressive strength was evaluated as a function of temperature and curing time. In this process sulfur acts as a physical binder which first crystallizes as monoclinic sulfur at 114°C and on further cooling to below 96°C sulfur undergoes a transformation to the stable orthorhombic sulfur at ambient temperatures. The first step was to form modified sulfur cement that remains in the monoclinic sulfur form and does not undergo a phase transformation to the orthorhombic sulfur form upon solidification. SPC samples were prepared and maintained at constant temperatures of 26, 40, 60, 80, and 100°C for different periods of time. The manufactured SPC developed 76% of its ultimate compressive strength after one day, and 97% after three days. No significant strength gain with time was observed. The

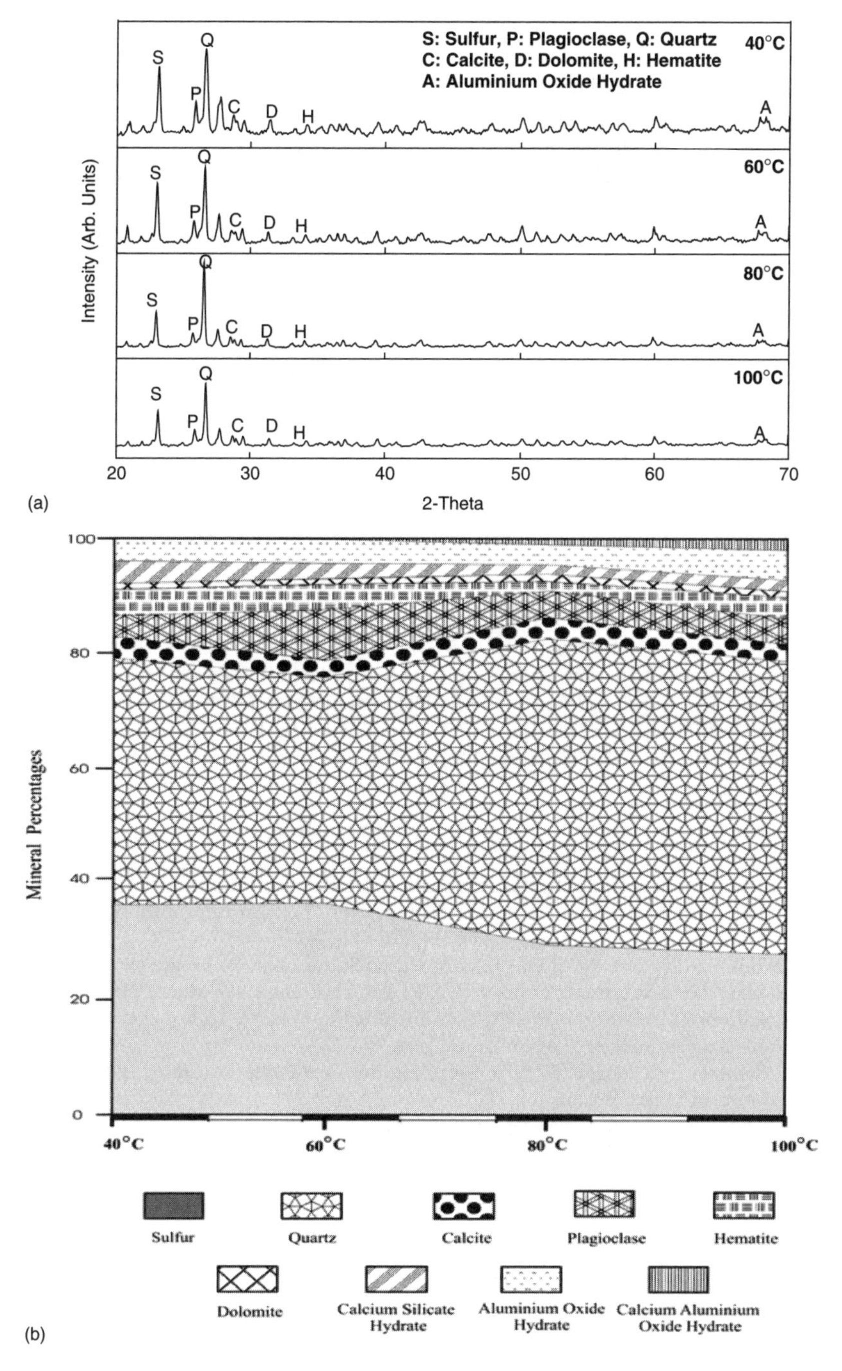

Figure 11. (a) X-ray diffraction of SPC at different temperature, (b) the lateral variation in the temperature of mineral composition of SPC sample cured seven days at different temperature.

compressive strength of SPC was dependent on the temperature at which the material is cured. For temperatures less than 60°C, SPC strength was higher than the Portland Cement Concrete (PCC) by about 20 to 30%. Whilst, for temperatures higher than 60°C and less than 100°C, the compressive strength remained within the range of PCC.

The microstructure analysis of the manufactured SPC revealed that as temperature increased pore sizes increased, which in turn would provide a mechanism for stress release due to thermal stresses, thermal expansion decrease, and shrinkage decrease upon cooling. The mineralogical analysis indicated that the increase of compressive strength of SPC was mainly due to the formation of stable minerals such as quartz, calcium aluminium silicate, calcium silicate hydrate, aluminium oxide hydrate, calcium aluminium oxide hydrate, hematite, calcite, and dolomite. From the mineralogical viewpoint, it is worth noting that SPC does not allow the formation of ettringite, which constitutes a major problem in PCC due to its ability to absorb a large amount water, expand, and produce cracks in PCC.

ACKNOWLEDGEMENTS

The authors would like to acknowledge the funding provided by the Research Affairs Sector, UAE University, for conducting this study. Also, acknowledgements are in order to the Central laboratories Unit (CLU) and Chemical and Petroleum Engineering Laboratories, UAE University, for the use of their facilities; Al Ain Cement Factory for their help in material characterization; and Al Ain Municipality for providing some of the raw materials.

REFERENCES

ACI Committee 548. 1993. Guide for mixing and placing sulfur concrete in construction (ACI 548.2R-93), *American Concrete Institute*, Farmington Hills, Mich.

Beaudoin, J.J. & Feldmant, R.F. 1984. Durability of Porous systems impregnated with dicyclopentadiene-modified sulfur. *The International Journal of Cement Composites and Lightweight Concrete* 6(1): 13–17.

Czarneckt, B. & Gillott, J.E. 1990. Effect of Different Admixtures on the Durability of Sulfur Concrete Made with Different Aggregates. *Engineering Geology*, 28:105–118.

Gemelli, E., Cruz, A.A.F. & Camargo N.H.A. 2004. A study of the Application of Residue from Burned Biomass in Mortars. *Materials Research* 7(4): 545–556.

Gillott, J.E., Jordaan, I.J. & Loov, R.E. & Shrive, N.G. 1980. Sulfur Concretes, mortars and the like, U.S. Patent No. 4, 188, 230.

Leutner, B. & Diehl, L. 1977. Manufacture of sulfur concrete, U.S. Patent No. 4, 025, 352.

Lin, S.L., Lai, J.S. & Chian, E.S.K. 1995. Modification of sulfur polymer cement (SPC) stabilization and solidification (S/S) process. *Waste Management* 15(5/6): 441–447.

Mc Bee, W.C., Sullivan, T.A. & Jong, B.W. 1981a. Modified Sulfur Concrete for use in Concretes, Flexible Paving, Coatings, and Grouts. BuMines Report No. RI 8545, U.S. Bureau of Mines, Washington, D.C., 24.

Mc Bee, W.C., Sullivan, T.A. & Jong, B.W. 1981b. Modified Sulfur Concrete Technology. Proceedings, SULPHUR-81 International Conference on Sulfur, Calgary, 367–388.

Mc Bee, W.C., Sullivan, T.A., Jong, B.W. 1983. Corrosion-Resistant Sulfur Concretes. BuMines Report No. 8758, U.S. Bureau of Mines, Washington, D.C., 28.

Mc Bee, W.C. & Sullivan, T.A. 1979. Development of Specialized Sulfur Concretes. BuMines Report No. RI 8346, U.S. Bureau of Mines, Washington, D.C., 21.

Mc Bee, W.C. & Sullivan, T.A. 1982. Assigned to U.S. Department of Commerce, Modified Sulfur Cement. U.S. Patent No. 4, 311, 826.

Mohamed, A.M.O. & El Gamal, M.M. 2006. Compositional Control on Sulfur Polymer Concrete Production for Public Works. In: Sustainable GCC Environment and Water Resources, edited by Mohamed, A.M.O., Developments in Arid Regions Research, Vol. 3, A.A. Balkema.

Nimer, E.L. & Campbell R.W. 1983. Sulfur cement-Aggregate – Organosilane Composition and Methods for Preparing, U.S. Patent No. 4, 376, 830.

Pickard, S.S. 1985. Sulfur Concrete for Acid Resistance. Chemical engineering 92(15): 77–78.

Syroezhko, A. M., Begak, O. Yu., Fedorov, V. V. & Gusarova, E. N. 2003. Modification of Paving Asphalts with Sulfur. *Russian Journal of Applied Chemistry*. 76(3): 491–496.

Voronkov, M.G., Vyazankin, N.S. & Deryagina, E.N. 1979. Reaction of sulfur with organic compounds, Novosibirsk: Nauka.

Vroom, A.H. 1981. Sulfur Cements, Process for Making Same and Sulfur Concretes Made Thereform. U.S. Patent No. 4, 293, 463.

Woo, G. L. 1983. Phosphoric Acid Treated Sulfur Cement-Aggregate Compositions, U.S. Patent No. 4, 376, 831.

Reclaiming the Desert: Towards a Sustainable Environment in Arid Lands – Mohamed (ed.)
© 2006 Taylor & Francis Group, London, ISBN 0 415 41128 9

Development of modified-sulfur concrete (RECOSUL)

Yuichi Nakano
Research & Development Department, Nippon Oil Corporation, Tokyo, Japan

ABSTRACT: The waste management and recycling has been important in these years. As the population increases and the industry develops, to create a society that generates minimum waste, especially industrial by-product, has been essential.

Sulfur, obtained from crude oil and natural gas through their desulfurization processes, is a useful and important product for the fertilizer, detergent, vulcanized agent, and so on. But, as fuel regulations requires less content of sulfur and the demand of sulfur decreases in the world, sulfur has been in the situation of global structural surplus. From this background, the new utilization of sulfur has been expected.

Our technology, that can be the solution for these various problems, is "Modified-Sulfur Concrete (RECOSUL)". This technology can convert wastes and industrial by-products into useful materials applicable in social infrastructures.

RECOSUL is a new material taking advantages of a character of sulfur that melts at 119–159 degrees centigrade. This consists of sulfur (as a binder) and various industrial by-products (as aggregates). The manufacturing method is as follows. At first, the melting modified-sulfur is prepared at 130–150 degrees centigrade, and then mixed with aggregates, such as stone, sand, Coal fly ash, steel slug. The modified-sulfur permeates into the gaps of aggregates, which makes a robust structure. By cooling and solidifying for a few hours, the mixture turns to dense and stable masses.

RECOSUL has a lot of advantages such as acid and salt resistance, high strength, and rapid setting compared with cement concrete. We have just built operating the pilot plant, which can produce 70 tons of pre-cast products per day last year, and also have been carrying out some field tests to confirm the performance of RECOSUL at empirical scale levels. From these field tests, we have confirmed that RECOSUL is widely applicable to the civil engineering and constructions as an alternative of concrete, especially for some specific applications such as marine constructions and sewerage systems.

1 INTRODUCTION

Waste management and recycling have become increasingly important in recent years. Given the increase in population and the development of industries, it is essential to create a society that generates minimum waste and industrial by-products, in particular.

Sulfur, obtained from crude oil and natural gas through the desulfurization process, is a useful and important product for fertilizers, detergents, vulcanized agents, and other such materials. However,

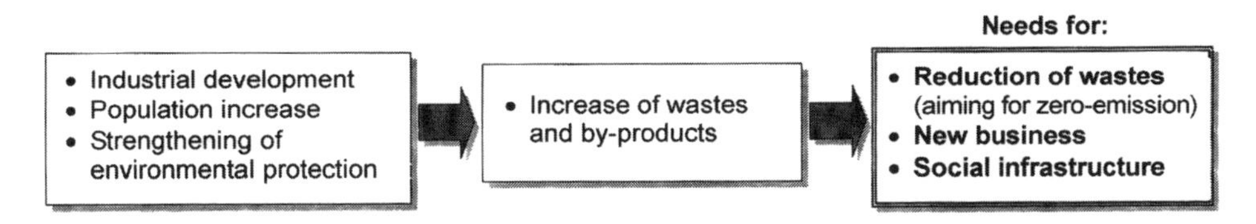

Figure 1. Steps in the recycling of waste products.

fuel regulations are calling for lower sulfur contents and global demand for sulfur is decreasing, thus creating a structural surplus of sulfur throughout the world. Against this background, new utilizations of sulfur are being sought.

Our "Modified-Sulfur Concrete (RECOSUL)" technology offers a potential solution to the above-mentioned problem. It can convert wastes and industrial by-products into useful materials for social infrastructures.

2 TECHNOLOGY

2.1 *Production*

The production method of modified-sulfur concrete is as follows. First, modified sulfur is created by melting solid sulfur flakes at normal temperatures to 120–160°C and adding an additive (polymerization reaction). Then, pre-heated and dehydrated aggregates, such as coal ash, steel slag, or other inorganic substances, are added and mixed at about 130°C. After the mixture is evenly mixed together, it is cast into a mold, where it slowly cools and hardens in a few hours (Fig. 2). The modified-sulfur concrete could be made into any shape by using an appropriate mold at the time it hardens. Fig. 3 shows a comparison of the compositions of ordinary concrete and modified-sulfur concrete.

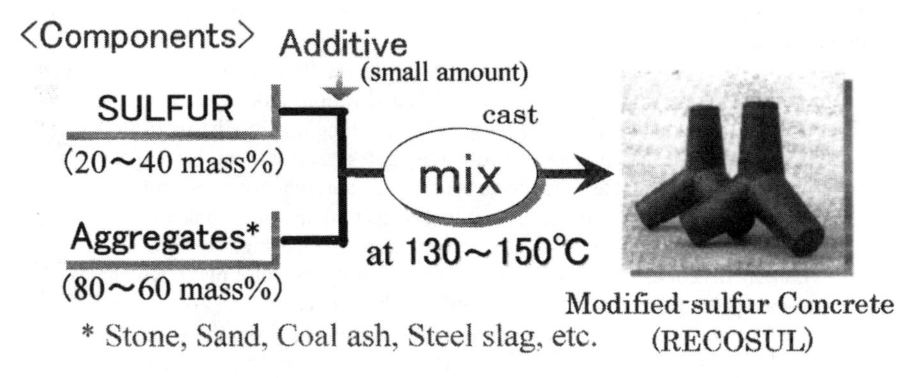

Figure 2. Production method of modified-sulfur concrete.

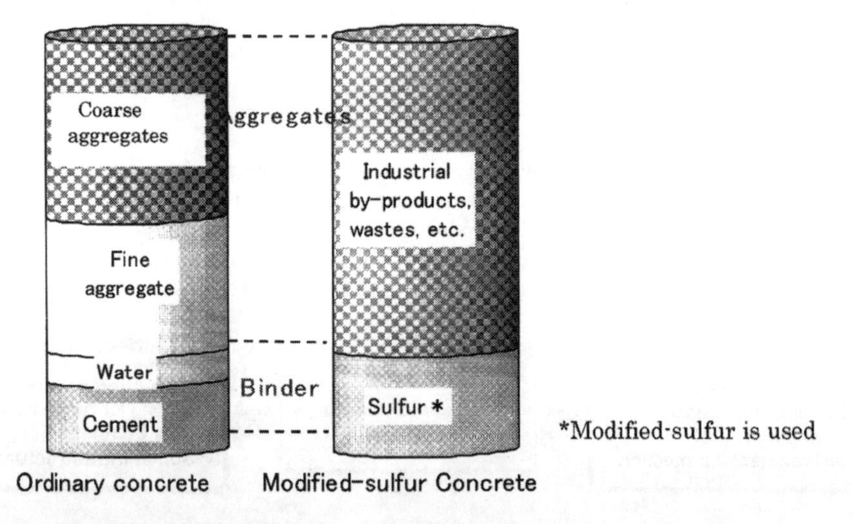

Figure 3. Comparison of the compositions of ordinary concrete and modified-sulfur concrete.

While concrete hardens through a chemical reaction between water and cement, modified-sulfur concrete hardens as the modified sulfur that is used as a binder undergoes a physical change from liquid to solid.

2.2 *Properties*

Modified-sulfur concrete makes full use of the properties of sulfur. At room temperature, sulfur is a hydrophobic solid. On the other hand, at high temperatures (over 120°C), it is a thin liquid, and has the property of sulfurizing and chemically stabilizing heavy metals. With this sulfur as a binder, modified-sulfur concrete takes on an extremely dense structure with modified sulfur penetrating deep in between the aggregates (Fig. 4). By enfolding the aggregates with hydrophobic sulfur, high impermeability is achieved. The properties of modified-sulfur concrete are: (1) high strength (fast curing), (2) high salt tolerance, (3) high acid resistance, and (4) safety.

(1) High strength
Modified-sulfur concrete, with its dense structure, has a compressive strength two to three times greater than ordinary concrete. Its bending strength and tensile strength, two other strength characteristics, also exceed those of ordinary concrete (Table 1). While it takes concrete several days to several weeks for its strength to become evident, it takes only several hours for modified-sulfur concrete to cool and harden and for its full strength to become evident. This fast-curing property contributes to shortening construction period.

(2) Salt tolerance
In accordance with JSTM C 7401 "Concrete chemical resistance test by immersion in liquid solution," a test sample (5 cmφ × 10 cmH) was immersed in a transparent container filled with artificial seawater for 12 months. Fig. 5 shows the result. Like that of ordinary concrete, the compressive strength of modified-sulfur concrete showed no decrease from the start of immersion to an immersion period of 12 months. The fact that it maintained its high strength is an indication that modified-sulfur concrete is salt-tolerant.

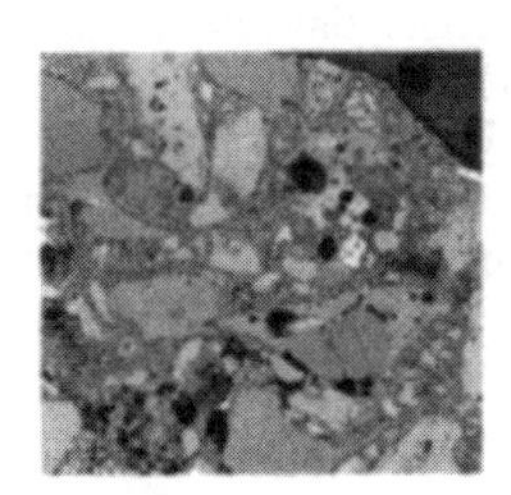

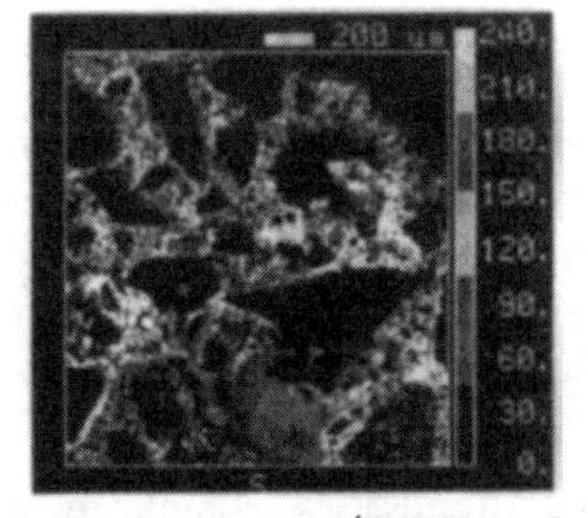

a) Microphotograph b) Sulfur distribution (EPMA analysis)

Figure 4. Structure of modified-sulfur concrete (cross section).

Table 1. Comparison of the basic physical properties of modified-sulfur concrete and ordinary concrete.

	Modified-sulfur concrete	Concrete (specimen age 28 days)	Test method
Density (g/cm^3)	2.5	2.3	
Compressive strength (MN/m^2)	96.7	34.7	JIS A1108
Bending strength (MN/m^2)	10.8	4.2	JIS A1106
Tensile strength (MN/m^2)	5.9	3	JIS A1113

Composition: Modified-sulfur (20%), steel slag (66%), coal ash (14%)

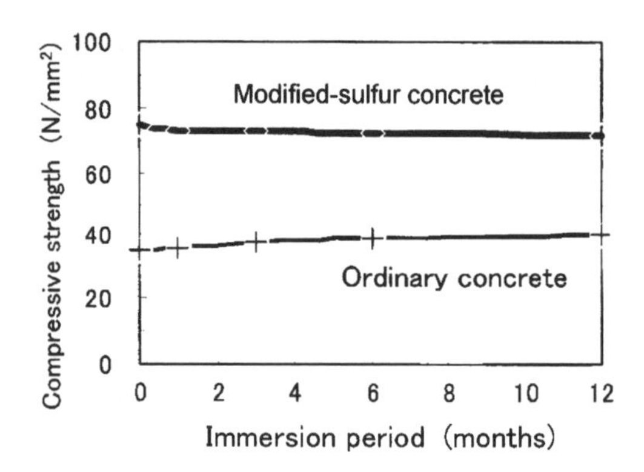

Figure 5. Changes in compressive strength during an artificial seawater immersion test (12 months).

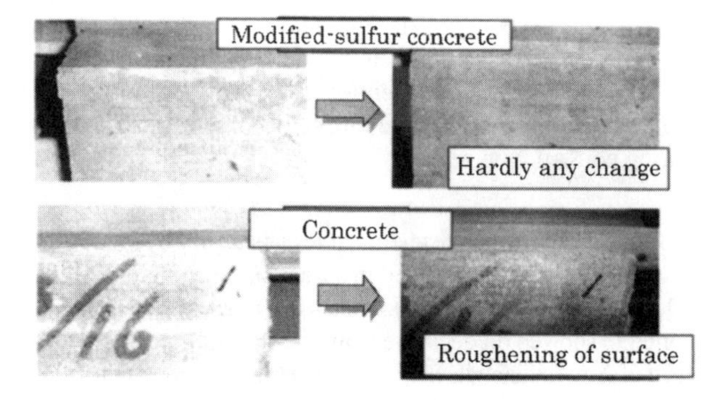

Figure 6. Result of the saltwater spraying test (6 months). (JIS Z 2371 "Saltwater spraying test method").

In order to evaluate surface seawater tolerance at an accelerating rate, a 50 g/L concentrated seawater was continuously sprayed (1–2 ml/hr per 80 cm^2 of horizontal sample surface) onto the surface of a rectangular test sample (15 × 7 × 5 cm) for 6 months in a saltwater spraying device maintained at a uniform temperature of 35°C. The result was observed from outside the device (Fig. 6). Unlike ordinary concrete, modified-sulfur concrete showed no roughness of the surface and hardly any change. The low water absorption and high salt tolerance of modified-sulfur concrete made it difficult for saltwater to penetrate the surface.

(3) Acid resistance
A test sample (5 cmϕ × 10 cmH) was completely immersed in a 10 weight-percent hydrochloric acid solution to examine changes in its strength with time (Fig. 7). Hardly any decrease was observed in the strength of modified-sulfur concrete for 6 months, nor any external change. On the other hand, the compressive strength of ordinary concrete almost reached zero in 3 months, and at 6 months, it no longer retained its original shape. The low water absorption of modified-sulfur concrete made it difficult for hydrochloric acid to penetrate the surface, and the acid-resistant property of the sulfur itself helped modified-sulfur concrete maintain a high acid resistance compared to concrete.

(4) Safety
The most commonly used safety test for recycled materials is the No. 46 dissolution test of the Environment Agency. In this test, samples crushed into particles smaller than 2 mm are shaken for six hours in a flask, and the elements that dissolved into the liquid solution are analyzed. It is one of the most stringent dissolution tests in Japan. The results of the test conducted on five types of sulfur concrete with different aggregate compositions are shown below (Table 2). All samples satisfied the environmental quality standard for soil and showed a high safety factor. This meant that the modified

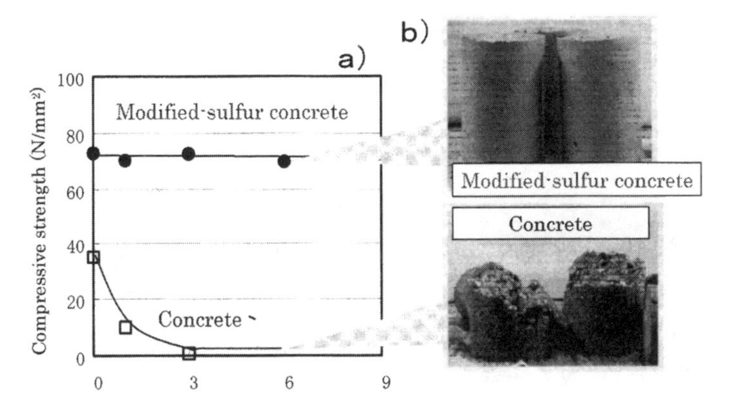

Figure 7. Result of the hydrochloric acid immersion test. a) Changes in compressive strength; b) Photos of test samples after 6 months.

Table 2.

	Mixture ratio (mass %)				Results of Environment Agency No. 46 Dissolution Test (mg/l)					
		Coal ash		Steel slag						
	Sulfur*	Fly ash	Clinker		Total Hg	Cd	Pb	As	Cr(6+)	Se
Fly ash		O			<0.0005	<0.01	<0.01	<0.01	0.12	<0.01
Clinker			O		<0.0005	<0.01	<0.01	<0.01	<0.05	<0.01
Steel slag				O	<0.0005	0.14	<0.01	<0.01	<0.05	0.022
Sample 1	20	14		66	<0.0005	<0.01	<0.01	<0.01	<0.05	<0.01
Sample 2	35	33		32	<0.0005	<0.01	<0.01	<0.01	<0.05	<0.01
Sample 3	40	54		6	<0.0005	<0.01	<0.01	<0.01	<0.05	<0.01
Sample 4	35	33	32		<0.0005	<0.01	<0.01	<0.01	<0.05	<0.01
Sample 5	40			40	<0.0005	<0.01	<0.01	<0.01	<0.05	<0.01
Environmental quality standard for soil					0.0005	0.01	0.01	0.01	0.05	0.01

sulfur penetrated deep into the aggregates and physically blocked off the trace amounts of heavy metals contained therein.

3 POTENTIAL APPLICATIONS AND FIELD TEST RESULTS

The modified-sulfur concrete we are presently researching has entered the demonstration test phase, and is being subject to various field tests. Applications that make full use of the various characteristics of modified-sulfur concrete discussed above are being explored, mainly in areas such as marine structures and sewage systems where temperatures are stable.

For example:

(1) Marine structures: Salt tolerance
(2) Sewage systems: Acid resistance

(1) Marine structures
We are utilizing the salt tolerant property of modified-sulfur concrete and conducting field tests on algal reefs and other marine structures in seven areas throughout Japan. We created a reef in a plant that corresponded to the representative size of algal reefs of $1 \times 1 \times 0.45$ m. Coal ash, steel slag, marine

life shells (scallop and oyster shells), and other inorganic materials could be used as aggregates, as well as aggregates found in the region. Among the field tests, a long-term test has been conducted over a period of approximately four years now, from which we confirmed that living organisms have the same level of affinity toward modified-sulfur concrete as toward ordinary concrete (Fig. 8). The test is

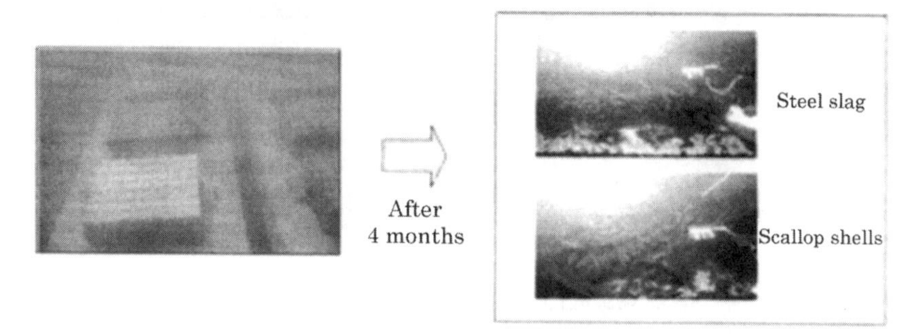

Figure 8. Adhesion of algae.

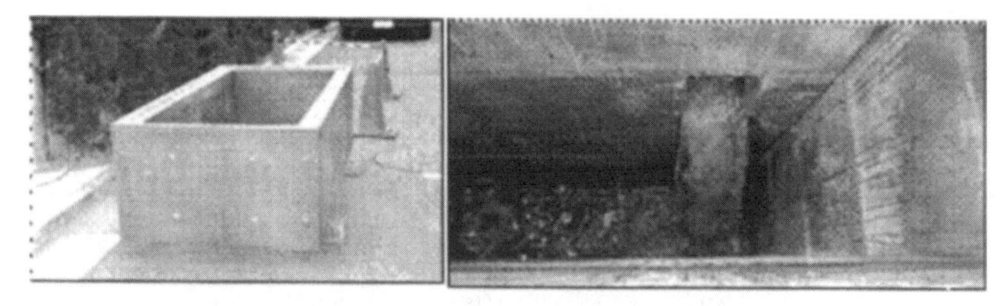

Figure 9. Underground pit.

Figure 10. Demonstration plant (Petroleum Energy Center).

gradually demonstrating the safety and stability of modified-sulfur concrete as marine structures in seawater.

(2) Sewage systems
We conducted a field test on sewage facilities as a joint research project with the construction company Obayashi Corporation and the Bureau of Sewerage of the Tokyo Metropolitan Government. Specifically, we applied modified-sulfur concrete to the frame of the underground pit at the sewage treatment plant. Despite the corrosion environment created by the high concentration of hydrogen sulfide, the return tank showed no corrosion even approximately two years after its construction, and thereby demonstrated high acid resistance. It may also be possible to apply modified-sulfur concrete to sewage pipes and other such sewage facilities.

A demonstration plant capable of manufacturing such pre-cast products was constructed in August 2004 as a grant project of the Petroleum Energy Center. It is equipped with an entire series of facilities, including sulfur receiving tank, aggregate hopper, dehydrator, and mixer, and has a capacity of 70 tons/day. Experiments for mass production technology are daily conducted at this plant.

4 CONCLUSION

The modified-sulfur concrete technology that our company is currently researching is a recycling technology capable of converting industrial wastes and by-products to social infrastructure facilities and achieving a zero emission society. At present, we are conducting field tests and experimenting with various applications that make full use of the characteristics of modified-sulfur concrete. We hope to commercialize it soon so that it could contribute to environmental conservation not only in Japan and the Middle East, but worldwide.

REFERENCES

Hideto FUKUI, Koichi ONO, Kunitomo SUGIURA, Masanari AKIYAMA: Experimental Research on the Dynamic Properties of Solids with Sulfur and Steel Slag, *2000 Annual Japan Society of Civil Engineers National Conference Proceedings, V176, Sept. 2000.*

Hideto FUKUI, Koichi ONO, Kunitomo SUGIURA, Masanari AKIYAMA: Experimental Research on the Basic Physical Properties of Solids with Sulfur and Steel Slag, *2001 Annual Concrete Engineering Conference Submissions, Vol. 23, No. 2, 2001.*

Masanari AKIYAMA, Masashi TAITANI: Basic Research on the Application of SS-Concrete Using Steel Slag and Coal Ash to Marine Structures, *2002 Annual Japan Society of Civil Engineers National Conference Proceedings, V-331.*

Masamichi HIRAKAWA, Arata FUJIWARA: Survey on the Anti-Corrosion Property of Modified-Sulfur Concrete, *Annual Report on the Technical Surveys of the Bureau of Sewerage, Tokyo Metropolitan Government, 2004.*

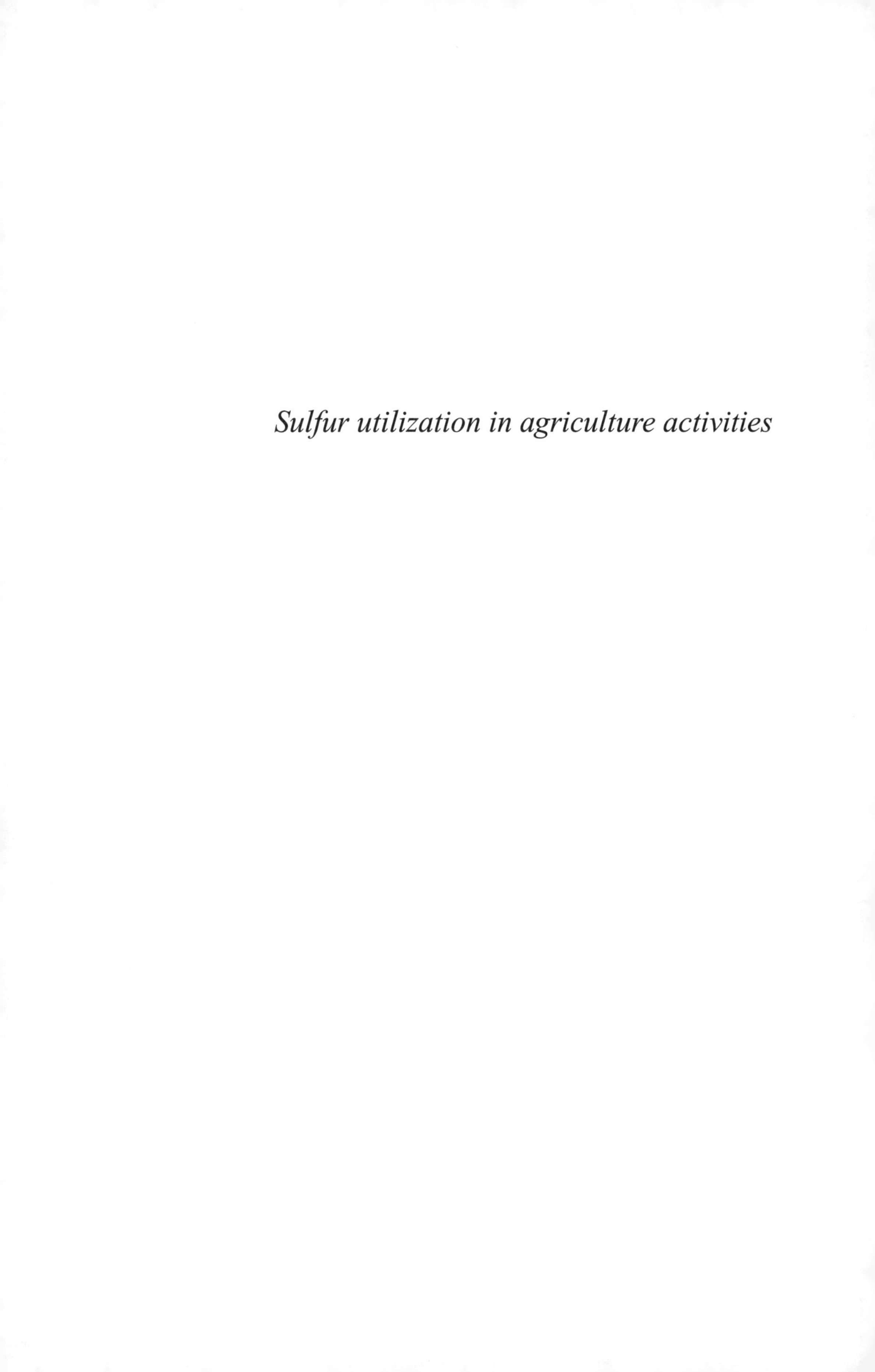

Sulfur utilization in agriculture activities

Reclaiming the Desert: Towards a Sustainable Environment in Arid Lands – Mohamed (ed.)
© 2006 Taylor & Francis Group, London, ISBN 0 415 41128 9

Development and implementation of sulfur amendment for the enhancement of sustainable greenery in Kuwait

R. Al-Daher, M.T. Balba, A. Yateem, H. Al-Mansour & T. Al-Surrayai
Biotechnology Department, Kuwait Institute for Scientific Research, Safat, Kuwait

H. Miyamoto, K. Morimitsu & S. Kamata
Idemitsu Kosan Co., Ltd., Chiba, Japan

ABSTRACT: Soil alkalinity causes an imbalance of multiple nutrients because of reduced availability under alkaline soil pH conditions. Soil applications of developed sulfur amendments were evaluated and used for initial reclamation and long term maintenance of soil quality. A joint research project was conducted in Kuwait Institute for Scientific Research (KISR) in cooperation with Japan Cooperation Center, Petroleum (JCCP) during the period December 1999 and March 2003. The main objective of this project is to develop soil amendment product utilizing byproduct sulfur from hydrodesulfurization processes in Kuwait and an inoculum of sulfur-oxidizing microbes. During this study sulfur-oxidizing bacteria were isolated, production process for soil amendment pellets was developed and the effectiveness of the pellets as soil enhancers was demonstrated. The second phase of this joint project started from April 2003. The main objective of this phase is the large scale production, optimization and implementation of an effective soil amendment product for enhancing desert soil fertility.

1 INTRODUCTION

Serious soil problems exist in many arid and semi-arid regions of the world which limit or prevent crop growth. These are associated with (1) excessive accumulation of soluble salts in the soil and/or (2) excessive replacement of calcium and magnesium on the soil colloids by sodium. Saline is the term applied to the first of these two conditions and alkali is the term applied to the latter. Since both situations may exist in the same soil, however, three categories ordinarily are used – saline, saline-alkali, and non-saline-alkali (Slaton et al., 1997).

Elemental sulfur added to soil undergoes a slow biological oxidation to sulfuric acid. This affects soil first by neutralizing soil bases and lowering pH directly and second by dissolving native soil $CaCO_3$ to form gypsum ($CaSO_4 \cdot 2H_2O$).

The local soil of Kuwait is alkaline with pH normally above 8.00, where it is too high for many important plant species. With the utilization of a newly developed S-based product as soil amendment, it is hypothesized that soil is adequately acidified for improving soil property and thus increasing crop productivity.

In December 1999 jointly research project (phase I) was initiated in cooperation with JCCP and was completed in March 2003. The project aims at the production and development of soil amendment for improving Kuwait's alkaline desert soil and evaluate it in both greenhouse and field conditions. The efforts in formulating the amendment was focused on the use of a locally available organic rich waste- stream, which is municipal wastewater sludge, as a carbon and nutrient source for microbial growth and accurates; a by-product from desulfurization processes, to provide a source of elemental sulfur; and selected sulfur oxidizing bacterial isolates.

During the first phase, the sulfur-oxidizing bacteria have been isolated, the production process for the soil amendment pellets has been developed and the effectiveness of the pellets as soil enhancers have been demonstrated through improved plant growth in greenhouse and field tests.

Because of the promising results, both parties KISR and JCCP had agreed to proceed to the second phase (phase II) which is aimed to a large scale production and optimization of an effective soil amendment product utilizing byproduct sulfur from hydrosulfurization process in Kuwait and an inoculum of sulfur oxidizing microbes for enhancing alkaline desert soil fertility in Kuwait and elsewhere in hot dry arid regions.

A National Greenery Plan (NGP) was prepared by KISR in 1996 and approved by the Public Authority for Agricultural Affair and Fish Resources (PAAF). The plan involves a comprehensive landscaping and greening program of more than 23 hectares over a period of twenty years. Based on the successful results of Phase I of the S-soil amendment project, it is highly recommended that sulfur Amendment is applied in conjunction with the NGP. It is our belief that the implementation of this strategy as an integrated part of the NGP will enhance the implementability and, success of the Plan. The use of the Sulfur amendment will ensure high survival rates of the cultivated plants and reduce cost of re-plantation. In order to produce adequate Sulfur Soil Amendment product for implementation in the NGP, a large-scale production is required.

2 MATERIALS AND METHODS

Phase I of this project comprised three primary activities: formulation of a soil S-amendment, establishment of the role of sulfur-oxidizing microbes in desert soils and product performance testing under both greenhouse and field conditions.

2.1 *Development of soil amendment and evaluation*

Different methods for manufacturing the soil amendment using different supplements and materials with different ratios were investigated. Materials required, such as sulfur and dried sewage sludge, were supplied by the Kuwait National Petroleum Company (KNPC) from the Ahmadi Refinery and the Ministry of Public Works (Al-Rekka sewage treatment plant), respectively. Other supplements, such as bentonite and soil/clay, were supplied from Japan and Kuwait, respectively.

Approximately 1290 kg of sulfur and 430 kg of the dried sewage sludge were crushed at KISR pilot plant no.5 using the vertical cutting mill. At the same time 653 kg of soil, which was collected from the Sulaibiyah experimental site, was sieved.

Desert clay soils were used as an alternative or replacement for the Kuwait soil (from the Sulaibiyah site). They were collected from the Al-Mutla desert area from three different locations and depths with a total volume of 784 kg. The clay was crushed and sieved using the same equipment.

Different trials were made for the preparation of the soil sulfur amendment product by mixing the following: sulfur powder, dried sewage sludge and/or desert clay soil and bentonite (Plate 1).

Sulfur soil amendments from different selected batches were evaluated using different kinds of test plants in the greenhouse at the Sulaibiyah experimental site.

2.1.1 *Microbial component production*

Fifty one soil samples were screened for sulfur-oxidizing bacteria, collected from sulfur contaminated area in Kuwait. The pHs for these samples were determined and were used for the inoculation of Starkey media. The samples were screened for sulfur-oxidizing bacteria at 37°C and 50°C.

Forty five pure bacteria were isolated, 26 isolates at 37°C and 19 isolates at 50°C. All the isolates were gram negative coccobacilli. Ten of the isolates that showed the lowest incubation time, between four to seven incubating days, were selected for further studies. The growth curves and pH curves were studied for these isolates. The maximum growth and the maximum pH drop were determined for each isolate. The maximum growth at 37°C was found with isolate KISR2 (1.8×10^8 CFU), after four days incubation, with a maximum pH drop equal to 3.7. However, the maximum drop in

Plate 1. Mixing different supplement using mixing machine.

Table 1. Optimal condition for biomass production for KNPCF1 and KNPCN13 in starkey medium.

Optimize parameters	KNPCF1	KNPCN13
pH	3	9
Temperature (°C)	45	45
Sulfur concentration (%)	2	2
Yeast extract concentration (%)	0.15	0.15

pH at 37°C was found with KNPCF1 (pH = 2.1), with maximum growth equal to 1.4×10^8 CFU after nine days of incubation. KNPCF1 and KNPCN13 (max growth = 5.1×10^7 CFU, max pH drop = 4.7) were selected for growth condition optimization at 37°C and 50°C, respectively (Table 1).

The optimal condition for KNPCF1 and KNPC13 was optimized using different concentration of elemental sulfur (0–2.5%) and yeast extract (0–0.25%), and different pH (3–9) and temperature (30–60°C); using Starkey media. Both strains had the same optimal temperature (45°C), sulfur concentration (2%) and yeast extract concentration (0.15%). However, they have different optimal growth pH; optimal pH for KNPCF1 is 3 whereas for KNPCN13 is 9.

2.1.2 *Field evaluation of the developed soil amendment*

The sulfur amendment consisted of prepared granules in the following composition: 30% S°, 50% dry sludge, 10% native soil (sand), and 10% Bentonite in addition to the bacteria culture. The S-amendment was thoroughly incorporated into native soil:perlite (1:1, v:v) at different rates dependent on the experimental design of the individual experiment pots. Number of different plant species have been utilized as test crops in this study. These crops include alfalfa, *Prosopis joliflora*, Hibiscus, corn, sunflower and tomato (Plate 2). Several soil and plant tissue samples collected periodically to analyze for pH, EC, SO_4 and other important nutrients. This paper summarizes the results of phase I. During phase II, the large scale production of the soil amendment will be optimized and the product efficacy will be established through the large scale production of the amendment and implementation in National Greenery Plan (NGP). During this phase, appropriate methods for product storage and handling will also be established.

Plate 2. Growth of prosopis juliflora on amended soil at a field test.

3 RESULTS AND DISCUSSION

Increasing application rates of sulfur has significantly decreased pH and increased EC of growth medium under greenhouse conditions. The highest application rate of 200 g S/kg has caused a sharp reduction in the pH of growth medium from 7.5 down to very acidic 3.9 within only 82 days. The pH of growth medium in the next lower sulfur rate (100 g S/kg) had been reduced to only 6.25 within the same period. The EC of growth medium has significantly increased, while dry matter yield of alfalfa has decreased significantly with in the 200 g S/kg S-application rate. Application rates of 1,100 kg/ha of S^o improved rice grain yield at one of three locations in Arkansas, USA, but application of 4,400 kg/ha of S^o resulted in stand loss and reduced yields caused by soil salinity (Slayton et al., 1997).

Application of S in a pelleted form (S^p) has produced significant difference on pH of growth medium than when applied in its elemental form (S^o) in 45 days period. Application of 100 or 200 g S_p/kg growth medium reduced the initial pH from 7.83 to 5.54 or to 4.45, respectively. On the other hand application of the same amounts of S^o reduced initial pH to 7.12 or to 6.73, respectively.

The field evaluation for the S-amendment product has slightly reduced soil pH due probably to the high buffering capacity of the experimental site caused by the relatively high content of $CaCO_3$. Fig. 1 shows the decrease in soil pH in the sunflower plots as the application rate of S increased from 0 to 0.8 kg S/m^2 with concurrent increase of the SO_4 concentration of the soil solution. The SO_4 concentration in sunflower tissues has also increased similarly. The same results were also obtained for corn and alfalfa. The divalent anion SO_4^{2-} is taken up by root system at physiological pH range and long-distance transport of sulfate occurs mainly in the xylem. Reduction of sulfate within the plant is necessary to incorporate sulfur into amino acids, proteins and coenzymes.

Increasing application rates of S has a positive effect on some growth parameters of *Prosopis joliflora*. The height of *P. joliflora* was greater in 0.4 kg S/plant than any of the control or 0.2 kg S/plant treatments. Similar result was obtained for the number of branches of the *P. joliflora*.

Soil amendments such as gypsum and elemental sulfur (S^o) have been used for years (Abrol and Bhumbla, 1979). In some calcareous soils, S-application increased the chemically available P from native soil apatite or added rock phosphate (Garcia and Carloni, 1977). Elemental sulfur which is biologically oxidized (*Thiobacillus thiooxidans*) to H_2SO_4 under aerobic conditions is often applied to reduce soil pH and dissolve insoluble nutrients (i.e. micronutrients and PO_4). The soil is first affected by neutralizing soil bases and lowering pH directly and second by dissolving native soil $CaCO_3$ to form gypsum ($CaSO_4 \cdot 2H_2O$). However, many researchers (Lindemann et al., 1991; Modaihsh et al.,

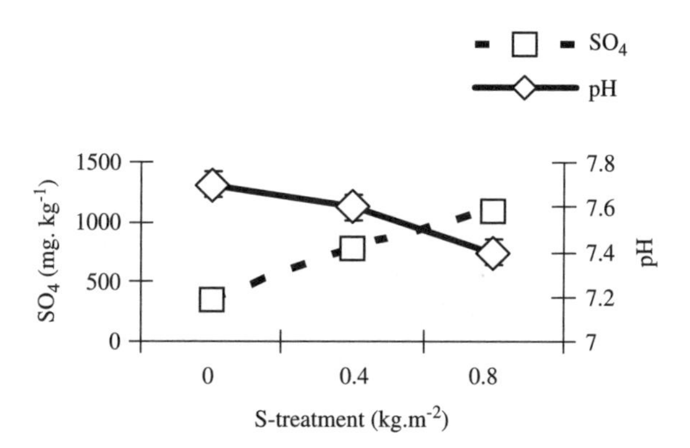

Figure 1. Effect of sulfur amendment on pH and SO_4 concentration of soil at harvest (Sunflower).

1989; Abo-Rady et al., 1988; and Hilal and Abd-Elfattah, 1987) have suggested that response to S° amendment is unpredictable and it is dependent on the buffering capacity of $CaCO_3$. Cifuentes and Lindemann (1993) reported that addition of organic matter to soils amended with S°, significantly decreased soil pH and increased soil SO_4 levels compared with S° alone or organic matter alone. Sulfur oxidation enhancement by sewage sludge has been reported to oxidize >50% of the applied S° within 6 weeks (Cowell and Schoenau, 1995). Falih (1996) reported that amendment of soils with S° led to a large decline in pH and $CaCO_3$, and an evident increase in Mn and Fe, with a slight increase in total soluble salts in most Saudi Arabian soils tested. Besharati and Rastin (1999) found that the application of *Thiobacillus* inoculants along with S° had significant effects on growth parameters of *Zea mays* and on pH of Iranian calcareous soils in pot experiments under greenhouse conditions.

4 CONCLUSION

The results of phase I showed that; the use of the sulfur product has produced somewhat positive responses in terms of changing pH of growth medium in greenhouse and field experimentation as well as the growth responses by many test crops used in this investigation. Further optimization tests for the sulfur product are needed and more crop and/or ornamental species should be included in any future work in phase II.

REFERENCES

Abo-Rady, M.D., O. Duheah, M. Khalil, and A.M. Turjoman. 1988. Effect of elemental sulfur on some properties of calcareous soils and growth of date palm seedlings, Arid Soil Res. Rehabil., 2:121–130.

Abrol, I.P. and D.R. Bhumbla. 1979. Soil Sci. 127:79–85.

Cowell, L.E. and J.J. Schoenau. 1995. Stimulation of elemental sulfur oxidation by sewage sludge, Can. J. Soil Sci., 75:2, 247–249.

Falih, A.M. 1996. Sulfur oxidation in Saudi Arabian agricultural soils. Qatar Univ. Sci. J., 16:2, 297–302.

Garcia, M.E. and L. Carloni. 1977. The effect of sulphur on the solubility and forms of phosphorus in soil. Agrochemica 21:163–169.

Hilal, M.H. and A. Abd-Elfattah. 1987. Effect of $CaCO_3$ and clay content of alkaline soils on their response to added sulfur, Sulfur Agric., 11:15–19.

Lindemann, W.C., J.J. Aburto, W.M. Haffner, and A.A. Bono. 1991. Effect of sulfur source on sulfur oxidation. Soil Sci. Soc. Am. J., 55:85–90.

Modaihsh, A.S., W.A. Al-Mustafa, and A.I. Metwally. 1989. Effect of elemental sulfur on chemical changes and nutrient availability in calcareous soils. Plant Soil, 116:95–101.

Slaton, N.A., R.J. Norman, S. Ntamatungiro, and C.E. Wilson Jr. 1997. Amendment of alkaline soils with elemental sulfur. Arkansas Agricultural Experiment Station, 456:130–136.

Reclaiming the Desert: Towards a Sustainable Environment in Arid Lands – Mohamed (ed.)
© 2006 Taylor & Francis Group, London, ISBN 0 415 41128 9

Lab and field evaluation of sulfur amendment for the improvement of desert soil

H. Miyamoto, S. Kamata and K. Morimitsu
Idemitsu Kosan Co., Ltd., Sodegaura, Chiba, Japan

R. Al-Daher, M.T. Balba, A. Yateem, H. Al-Mansour & T. Al-Surrayai
Kuwait Institute for Scientific Research, Safat, Kuwait

ABSTRACT: A surplus production of sulfur in oil refineries has become a global issue against the background of increase of public concern for environment. In order to address to these problems, an international joint project between Kuwait and Japan on the development of soil amendment using sulfur is being conducted, and the preliminary results are reported in this paper. It was found that the developed sulfur amendment improved soil pH and enhanced plant growth, and that the technology can contribute positively to agriculture and landscape in arid regions. In conclusion, it is strongly suggested that a practical and source-recyclable technology can be established using sulfur which is unfavorably produced in oil refineries.

1 INTRODUCTION

1.1 *Preface*

Sustainable development has become public concern in recent years. In order to achieve sustainable development, cooperation at a global level is important to control the environmental burden caused by human activities to a degree that can be tolerated by the planet. The fossil energy industry is expected, together with other industries, to assess the environmental impact of its resources and take the appropriate measures, as well as to provide a stable supply of energy resources.

Crude oil generally contains sulfur, so emits sulfur oxide, a cause of air pollution and acid rain, when burned intact. Sulfur is therefore removed through a desulfurization process in the oil refinery. The majority of produced sulfur is efficiently reused in industrial and agricultural products. However, in recent years, a volume of global sulfur production has exceeded the consumption, and the excess has been processed as a waste product. Also, it is predictable that sulfur production will increase further because the needs to reduce the sulfur content in the oil product are growing. From the standpoint of the stable use of fossil energy, the development of technologies that offer new methods to effectively reuse excess sulfur is important.

Global warming is another public concern that has recently come to the fore. Therefore demands for technologies designed not only to reduce the carbon emission volume but also to capture and immobilize it by the photosynthesis of plants and the creation of carbon sinks in arid regions would not only contribute to the control of global warming but also to reducing desertification. Furthermore, burning fossil energy is a significant cause of carbon dioxide emissions; therefore the development of the immobilization technology could also help the stable use of petroleum.

The research presented in this paper aims to ensure the effective reuse of sulfur by developing a suitable technology in which excess sulfur is utilized, while also helping the stable supply and use of fossil energy by contributing to the control of global warming.

1.2 *Greening – related issues in the middle east region*

Alkaline soil widely spreads in arid regions. This alkaline nature of soil is a reason for the scarce vegetation in these regions, although a lack of water is also a factor. Soil pH plays a major part in the solubilization of microelements in soil, and this influences the growth and development of vegetation. Although the optimum pH range is generally considered to be 5.5–6.5, it is, in the authors' experience, not rare that soil pH exceeds 8.0 in the Middle East region.

A lack of organic matters causes another problem. Organic matters in the soil play an important role in the fertility of soil, and the slow and continuous release of nutrition to plants. Under normal circumstances with healthy vegetation, microorganisms and animals as well as plants that inhabit the same environment contribute to a sound organic cycle. Without this cycle in place, however, mineralization generally progresses rapidly in arid regions.

1.3 *Project concept*

The utilization of sulfur is a well-known method for reducing the pH level in alkaline soil. In interaction with sulfur oxidizing bacteria, sulfur is oxidized into sulfuric acid, which neutralizes alkaline soil. Generally, sulfur oxidizing bacteria are found in natural soil. However, because micro flora is scarce in the desert soils, technology is proposed utilizing the high performance species isolated and cultured in advance and then re-introduced into the soil together with the sulfur.

Moreover, sulfur is an essential nutrient for plants and is generally absorbed as sulfate ions by plants. Therefore, technologies to transform elemental sulfur into sulfate ions would not only contribute to improvement of soil pH but also to benefits for the soil by supplying essential sulfur nutrient.

In order to resolve the issue of a lack of organic matter, varying composts or peat moss are generally used. These methods, however, are not so cost effective. For this reason, authors focused on the excess sludge generated from the sewage treatment process as the organic matter. The efficacy of sludge as an organic fertilizer has long been recognized in Japan, and its use has proven successful in nurturing fertile farmland. In addition, it is beneficial to consider the reuse of the excess sludge since sewage treatment technologies are common throughout the world and the sludge is easily obtained.

This research studied the possible application of sulfur in plant cultivation through a combination of the other technologies described above. The following points were taken into consideration during this research process.

1) Use of materials and techniques that can be procured locally.
2) Combination of practical technologies to create a single soil amendment in view of convenient application.

1.4 *Project organization*

This research project has been being jointly conducted by Kuwait Institute for Science Research (KISR) and Idemitsu Kosan Co., Ltd. commissioned by Japan Cooperation Center, Petroleum (JCCP), with a subsidy from the Ministry of Economy, Trade and Industry (METI) Japan.

Some of the research results are also presented in the same symposium by KISR. Therefore this paper hereafter focuses on the shelf life of the bacteria and application test of the soil amendment.

2 SHELF LIFE OF SULFUR OXIDIZING BACTERIA

2.1 *Background*

The sulfur amendment that has been developed in the research contains living bacteria. Therefore, it is important to assess the shelf life of the bacteria, in order to develop a viable technology. Shelf life tests were carried out and the results are summarized in the following section.

2.2 *Experimental design*

Preliminary experiments were conducted to select the appropriate weight of sulfur amendment for conducting this study. The results demonstrated the suitability of 1.0 g for each test. Accordingly, it was decided that 1.0 g samples will be used as the starting sulfur amendment weight in the shelf life assessment experiments.

The raw materials for the sulfur amendment originally contain various kinds of microbes. Because it is concerned that these microbes may interfere with counting of sulfur oxidizing bacteria to be conducted after bacteria culture, all the amendment ingredients were sterilized by autoclaving at 121 degree C or UV irradiation. Also, the pellet manufacturing machines were cleaned with anti-microbial soap and 10% of hydrogen peroxide in order to minimize the microbial contamination.

The experiments were designed to assess the shelf life of freshly prepared sulfur amendment after being stored at four different temperatures: 4, 35, 45 degree C and room temperature. One g weight of the amendment was collected from each stored sample and then added to the culture medium. Control treatment without adding sulfur amendment was similarly prepared. The experiments were conducted in triplicates and the flasks were incubated at 37 degree C and for 12 days with rotational shaking. During the incubation period, samples were routinely taken at zero time and after 5, 9 and 12 days for monitoring. The samples were tested for bacterial cell count, pH and sulfate concentration during the course of the experiment. The sampling dates and protocol are summarized in Table 1.

2.3 *Results and discussion*

As mentioned above, in this study, cell count, pH and sulfate concentration in the culture medium were measured after sulfur oxidizing bacteria in the amendment was once cultivated for a certain period in the medium. The most important function that is expected to the amendment is reduction of soil pH, therefore the authors hereafter focused on pH changes. The results of pH measurements of the medium are shown in Table 2. The initial pH of the medium was 8.0, and the pHs in all storage conditions were reduced after the cultivation. However, the degree of the reduction activity tends to have decreased gradually as the storage period became longer. In addition, the bacteria stored at room temperature kept higher activity of pH reduction than the other storage conditions.

Table 1. Sampling schedule.

Start sampling date (Zero day)	End sampling date (12 days)	Amendment storage time (Month)
2 May 05	14th May 2005	0
30 May 05	11th June 2005	1
27 June 05	9th July 2005	2
25 July 05	6th August 2005	3
22 August 05	3rd September 2005	4
19 September 05	1st October 2005	5

Table 2. Changes in the pH of soil amendment during the course of storage.

Storage time	4°C	RT	35°C	45°C
Zero	5.63	5.63	5.63	5.63
1 Month	6.50	5.50	5.87	6.27
2 Months	6.70	5.97	6.27	6.50
3 Months	6.90	5.70	6.70	6.60
4 Months	6.90	6.10	6.47	6.17
5 Months	6.83	6.20	6.73	6.80

pH measured after 12-days cultivation were shown. Starting pH was 8.0.

As a result, the bacteria can keep a certain level of the activity even at 45 degree C, which is considered as a critical temperature for bacteria, and the level is also similar to that of 4 degree C and 35 degree C conditions. This result shows that the bacteria have tolerance to high temperatures. On the other hand, the fact that storage at room temperature (air conditioned) is better in terms of the shelf life than that at 4°C, it is considered as an advantage to distribution after going to market.

Further investigations including application tests to soil and plants are expected to clarify an acceptable range of the activity loss. This test shown here will be continued with longer interval of sampling.

3 APPLICATION TEST AT AN EXPERIMENTAL FIELD

3.1 *Background*

Many application tests had been conducted since this joint project started in 1999, and already reported the results in the joint symposia organized by GCC and JCCP several times. In the early stage of the project the main plants tested were agricultural crops, and the obtained results showed the amendment application promoted growth and harvest of many kinds of plants.

In the present study, further application tests to ornamental plants were carried out at KISR experimental field station.

3.2 *Experimental design*

Four categories of plants and three species from each category, namely 12 plant species, were selected as test plants as shown in Table 3.

The plant seedlings were purchased from local markets. They were transplanted in the field with a random block design with three replicates. At the time of transplanting, 800 g/m^2 of the sulfur amendment was applied to the soil and then additional treatment of same amount of the amendment was made 17 months after transplanting. Drip or sprinkler system is used for the irrigation. The characteristics of used water were brackish with pH level of 7.9 and EC level of 4.0 mS/cm. Manual weeding was applied to maintain a bare area around each block as a buffer zone. Plant height, branch number, stem caliper, plant width and chlorophyll density were measured for trees and shrubs, and plant height was measured for ground cover and grass periodically. In addition, control and treated soils were routinely collected and the chemical profiles including soil pH were analyzed.

3.3 *Result and discussion*

This study is still on going, so preliminarily obtained data to date were analyzed and the effects of the amendment were discussed.

A treatment of the amendment promoted the growth of *Parkinsonia aculeate* and *Ficus infectoria*. The increase of the stem caliper was about 20% and 15% respectively. Also, the content of chlorophyll increased. Some of elemental sulfur taken in by a plant is used for biosynthesis of amino acid, resulting in protein production. It is known that a lack of sulfur causes symptoms of dwarf plants or yellowish leaves. In this study, it is suggested that the amount of sulfur intake by plants increased by the application of the amendment and therefore the color of leaves became darker, as a result that more sulfate ion was supplied in soil from the sulfur amendment.

Table 3. Tree species used at the experimental station.

Category	Species 1	Species 2	Species 3
Tree group	*Conocarpus lancifolius*	*Parkinsonia aculeate*	*Ficus infectoria*
Shrub group	*Bougainvillea* spp.	*Clerodendndrum inerm*	*Vitex agnis*
Ground cover group	*Widelia trilobata*	*Ipomea palmate*	*Carisa grandiflora*
Grass group	*Paspalium vagenatium*	*Pennistum*	*Zosia*

In the case of the shrub group, a stem caliper of *Bougainvillea* spp was increased about 25% by an application of the sulfur amendment. The caliper of *Vitex agnis* was also promoted gradually and the rate of the promotion reached up to 20%, although it was not observed for the first one year. On contrary, significant effects were not obtained on *Clerodendndrum inerm*.

About 10% increase of top height of *Carisa grandiflora*, the ground cover group, was found by the amendment application. In the case of *Widelia trilobata*, a slight growth promotion was observed in the first 10 months, although the growth of untreated ones caught up with the treated ones after that.

Relatively stronger effects were found on all of the grass group. The promotion rates were about 25% in *Paspalium vagenatium*, 10% in *Pennistum* and 50% in *Zosia* respectively.

At the same time of growth measurement, soils were routinely sampled and the pH was measured. As a result, it was found that the pH of treated soil gradually reduced after treatment and reached to the level of neutral, while that of untreated soil didn't change from the beginning level very much. Then the reduced pH gradually increased and reached the level of untreated after one year. The pH was reduced again after another application of sulfur amendment was made.

As discussed above, soil pH was improved by an application of the sulfur amendment and the improvement resulted in plant growth promotion. Although the effect on soil pH disappeared after about one year, additional treatment caused the pH reduction again. However, the effect of the sulfur amendment was observed only in certain species of plants. The reason remained unclear, therefore further investigations, such as tests for number of plant species, are expected in the future.

4 FIELD DEMONSTRATION

Two demonstration tests of the developed technology are being conducted in Kuwait.

The first site is located at Mina Al-Ahmadi Refinery of Kuwait National Petroleum Company (KNPC) with a total area of 4,500 m^2. This land was kindly assigned to the project by KNPC. The field design and the construction were contracted to a landscape company in Kuwait. The area was divided into different blocks where sulfur amendment was treated to some of them and the others untreated. The plantation and amendment application was finished by April 2005. Overall status of the field has been going well, although plant stress was observed because of the harsh environment in summer.

The second site is located at Shaab area with a total area of 10,000 m^2. This area runs along Gulf Road, one of the arterial roads in Kuwait, and is under control by KISR with a tall fence. The field design was drawn by Aridland Agriculture and Greenery Department of KISR, and treated and untreated blocks were prepared. The plantation was completed by June 2005. Therefore, some plants were affected by harsh temperature before the root system was established, and severely damaged. Substitute plantation is currently being conducted, followed by resetting of the blocks.

No results are available yet because they have just been started and the plants had some damage. Monitoring of plants and soils will be continued until the end of this fiscal year and the effects of sulfur amendment will be quantitatively analyzed. Overall views of pictures are shown in Figure 1.

Figure 1. Overall views of demonstration tests. Left: Ahmadi refinery. Right: Shaab.

5 CONCLUSIONS

In this present paper, results of shelf life of sulfur oxidizing bacteria in the amendment and the application test were assessed and discussed as parts of technology development of a sulfur reusable method jointly conducted between KISR and Idemitsu commissioned by JCCP. The initial results suggested that long-term storage with certain care is possible even after this technology steps forward to market and distribution started. The latter result suggested that the developed amendment, which enhanced plant growth, can contribute to increase of agricultural productivity and acceleration of landscaping in arid regions such as Kuwait.

In conclusion, the results strongly suggested that a practical and source-recyclable technology, that solving the problems of global warming, desertification and food supply, can be established using sulfur which is unfavorably produced in oil refinery. In order to examine the feasibility of this approach, more data is required on the application tests and establishing a mass production process with continuous monitoring of the currently ongoing experiments.

REFERENCES

Results of the research conducted to date have already been published with different titles in the following symposia.

"*The First Oman-Japan Joint Symposium on Water Resources and Greening in Desert* (2000)"

"*International Symposium on Elemental Sulfur for Agronomic Application and Desert Greening* (2001)"

"*The Joint State of Kuwait – Japan Symposium in 2003 on Application of New Technologies for Improvement of Desert Environment* (2003)"

"*The Joint Kingdom of Bahrain – Japan Symposium on Challenge on New Horizon toward Managing the Global Environment and Water Resources* (2004)"

Ecology

Reclaiming the Desert: Towards a Sustainable Environment in Arid Lands – Mohamed (ed.)
© 2006 Taylor & Francis Group, London, ISBN 0 415 41128 9

Identify and prioritize objectives for marine conservation in the Kingdom of Bahrain

Adel K. Al-Zayani
Bahrain Centre for Studies and Research, Bahrain

ABSTRACT: The primary aim of this paper is the identification and prioritization of national objectives for marine conservation in Bahrain. The paper tries to give a locally oriented questionnaire (Arabic language) which is essential for any research work especially in the current linguistic and cultural context. In this survey the stakeholders who shared the marine environment in Bahrain identify and prioritize the marine conservation objectives for Bahrain. The response rate of the survey is 83%, a high rate for Bahrain. The survey reveals that most of the objectives presented in the questionnaire are important to the stakeholders. The survey also measures the view of the stakeholders to the two already declared marine reserves viz. Ras Sanad and the Hawar islands in Bahrain. The majority agrees with both these sites with more preference to Hawar islands. The stakeholders also propose 47 sites and site attributes to be protected in Bahrain's territorial waters.

1 INTRODUCTION

Bahrain shares the world's and the Arabian Gulf's marine environment and the latter has its own characteristics, national priorities, and threats. Though most of this coastline is a productive area, with the lowest shore levels and shallow sublittoral regions of high productivity, with food webs based on macroalgae, seagrasses and microphyto benthos, all of which support various stages in the life-cycles of commercially-important fish species (Vousden, 1988 &1995).

The marine territorial waters of Bahrain contain approximately 7510 Km^2 of rich habitats and natural marine resources. The coastline of Bahrain consists of about 590 km of varied and rich environment, including coral reefs, seagrass beds, mangroves and salt marshes, rocky shores, beds of algae, sandy shores, mudflats and mixed shores (IUCN/ROPME/UNEP, 1985; Vousden, 1988 & 1995; Sims and Zinal, 2000).

The important human impact on the marine environment in Bahrain can be divided into two: pollution and physical alteration and destruction of habitats. This human impact can also be divided into land-based sources and sea-based sources. The second human impact on the marine environment in Bahrain is physical alteration and destruction of habitats. Dredging and reclamation are the major and the most environmentally sensitive activities around the coasts of Bahrain. The latest information confirms that the total area of the country increased by just over 8.4% during 1968–2001 due to land reclamation. In addition, dredging from the sea bottom will affect marine life, due to the destruction of whole habitats and increased turbidity in the water column (Madany *et al.*, 1988).

There are several approaches used worldwide to conserve marine biodiversity. The three most widely used approaches are: (1) integrated coastal zone management (Clark, 1992; Mumby *et al.*, 1995); (2) establishment and management of MPAs (Salm and Clark, 1984; Kelleher and Kenchington, 1992; Gubbay, 1995); (3) ecological restoration (Frid and Clark, 1999; Hawkins *et al.*, 1999; Young, 2000).

One of the most powerful approaches that works positively in marine conservation is the establishment of marine protected areas (MPAs). MPAs can be a critical component in conserving marine resources (Allison *et al.*, 1998; NRC, 2001), and maintenance of functional coastal and marine ecosystems (Hockey and Branch, 1997). MPAs contribute significantly to the conservation of marine ecosystems and their species and genetic diversity. They protect marine populations from threatening activities and ensure sustainable living resources, such as fisheries (McNeely, 1994; Agardy, 1999; Roberts, *et al.*, 2001).

Bahrain has declared four MPAs in its territorial waters: the Ras Sanad Mangrove Natural Reserve, the Hawar Islands Natural Reserve Mashtan Island Natural Reserve and Dohat Arad Reserve. These four existing marine reserves are not yet promoted and well-managed, except in a very few limited ways, which do not correspond to regional and international standards. There is no clear conservation plan so far. (Al-Zayani, 2003). Indeed there may be other areas that should be protected in Bahraini waters, a question which will be addressed in the paper. Which are the areas and attributes of Bahrain marine environment that need to be protected? And under which priority sequence?

Before answering these questions and before selecting any conservation approach, national marine conservation goals and objectives need to be identified and prioritized by the local people themselves. Involvement of local people is essential in the marine conservation plan and establishment and management of MPAs (Kelleher, 1999). It seems that local communities' opinions are often not even taken into consideration in the early stages of the conservation process. This paper will seek to involve as broad a range of stakeholders as possible, and make a detailed survey of their objectives, interest and views.

In this study a questionnaire was selected as the most appropriate survey instrument for collecting and measuring data for this study and also for identifying the most useful elements relating to national marine conservation objectives in Bahrain. The main aim of the survey was to identify and prioritise appropriate national marine conservation objectives for Bahrain. There were also two relevant subsidiary aims. The first was to investigate current knowledge and awareness of marine conservation issues among the population in Bahrain. Secondly, other areas that should be protected according to the knowledge and interest expressed by the people of Bahrain (stakeholders) were to be identified.

2 METHODS

The representative respondents selected for the questionnaire survey were people of Bahrain who had a major interest in the marine environment (stakeholders). Organisations which have some interest in the subject and have access to the country's marine environment or resources were also involved.

In order to design the questionnaire a number of steps were taken (Oppenheim, 1992). Firstly, previous work in the area of interest was reviewed. Secondly, a thorough catalogue of questions used in previous research studies related to marine conservation objectives was made. The main focus was on the aims of the study to identify the questions that were to be asked.

Several key questions originated during the process for incorporation into the final questionnaire. These questions were as follows:

1) What is the ranking of the national marine conservation objectives in Bahrain?
2) What is the perceived value of each marine conservation objective?
3) How much knowledge and awareness do the people of Bahrain have about the different aspects in this field?
4) What are the particular areas suggested by the people in Bahrain that need to be protected?

The questionnaire was designed in English, but delivered in the Arabic language since the majority of the targeted population are Arabic speaking, and it would have been very difficult for most of them to respond positively in another language.

The questionnaire used in this study was divided into four parts, each one was designed to cover one or more of the questions mentioned earlier. These four parts had the following themes:

1) Part one: The objectives of marine conservation in Bahrain.
2) Part two: Declared Marine Reserve objectives in Bahrain.

3) Part three: Demographic Data.
4) Part four: Comments on the subject of the questionnaire.

The first and the main part of the questionnaire was designed to investigate how each of the marine conservation objective is evaluated by the respondents. These objectives were selected after reviewing previous research carried out on marine conservation objectives.

The questionnaire lists a total of 13 provisional objectives from a list of global objectives, which were selected from the previous literature and reports (e.g. Kelleher and Kenchington, 1992; Jones, 1994; Kelleher *et al.*, 1995; Salm and Price, 1995) in a format that takes into account the attributes of Bahrain.

The second part of the questionnaire was designed to evaluate the knowledge and awareness of the respondents about marine conservation, and to indicate the objectives behind the declaration of the two existing marine reserves in the country. In addition, this part also explored the other areas that respondents think need to be protected in Bahraini territorial waters.

The third part was designed to determine the characteristics of respondents and tried to produce a profile of the same. This was mainly to find out whether there is any relation between the background of the respondents and the view they expressed.

The final part of the questionnaire was designed to collect additional comments from the respondents on the subject of the questionnaire. At the end of the questionnaire there was provision for the respondents to express their comments in an unsolicited format.

A draft questionnaire in the Arabic language was submitted to two referees who had experience in the use and technical design of questionnaires. Their recommendations were considered at this stage, and changes were made to ensure the development of an effective research instrument.

Furthermore a pilot study was also used to check the consistency and validity of statements, as well as to ensure that they were clearly understood by the respondents in addition to providing information about the time required to complete the questionnaire (Sudman and Bradburn, 1982). This step involved six people from different sample groups in the enquiry area (two fishermen, one researcher, one environmentalist, and two members of the public who used the marine areas for recreation purposes). After this step a few modifications were made to the questionnaire.

The next step, the questionnaire pre-test stage was conducted with an objective to identify any final revision required before adopting the final version of the questionnaire. The pre-test was conducted in Bahrain on nine people from different sample groups. Out of the nine copies of the questionnaire distributed, eight of them were returned and from these responses the necessary final modification was done.

After finalizing the questionnaire, it was distributed to the entire target sample. The total targeted population was 200, of which, 83% (n = 166) returned a completed questionnaire. According to their particular roles in Bahrain, the sample was instead stratified into groups (Cohen, 1996), because there were good reasons to believe that certain sectors of the population would have differing interest in the subject. The sample population was categorised as follows:

1) Government Officials (including the official fisheries authority)
2) Private commerce
3) Non-Governmental Organisations (NGOs)
4) Industry
5) Regional and international organisations
6) Members of the public (including fishermen)

3 RESULTS

The Excel package was selected for the data analysis, in conjunction with other packages such as Minitab and Statgraphics Plus for more advanced analysis. The main analysis focused on the results of the survey of national marine conservation objectives in Bahrain.

3.1 *National marine conservation objectives in the Kingdom of Bahrain*

Figure 1 shows the response scores to the various questions related to the marine conservation objectives. The highest priority was given to fisheries, closely followed by endangered, rare and sensitive habitats and then endangered species. Lowest priority was given to observation/monitoring, regional and international conservation efforts, and historical/cultural. There were consistently high median scores (8–10), therefore some weight had to be given to the lower quartile indicating where some individuals gave very low priorities to a particular objective.

These were for tourism, historical/cultural, international conservation efforts, and observation/monitoring. Statistically there were differences between the responses to the questions (M-W statistic, P-Value) suggesting there were real differences in priorities in the sample group as a whole.

Since the median scores are similar, measure of spread is required. Coefficients of variation (CV) measure the spread in comparative terms by dividing the standard deviation by the sample mean (Chatfield, 1983). The spread of data is measured as a proportion of the mean based on the expected value and the standard deviation of a population response without affecting their ratios. However, the calculated scales would not be true if the scale began at zero or negative values, which is not the case (Chatfield, 1983). Consequently, the CV results can be used to rank agreement.

It is clear from the figure 2 that the respondents closely agreed on the most important objectives (high score: low CV), these were fisheries, habitats and species. At the same time, observation/monitoring, regional and international conservation efforts, historical/cultural and tourism were scored significantly lower by some respondents. This use of CV to identify areas of disagreement is potentially of great interest to planners and managers, and confirm the data displayed in figure 1.

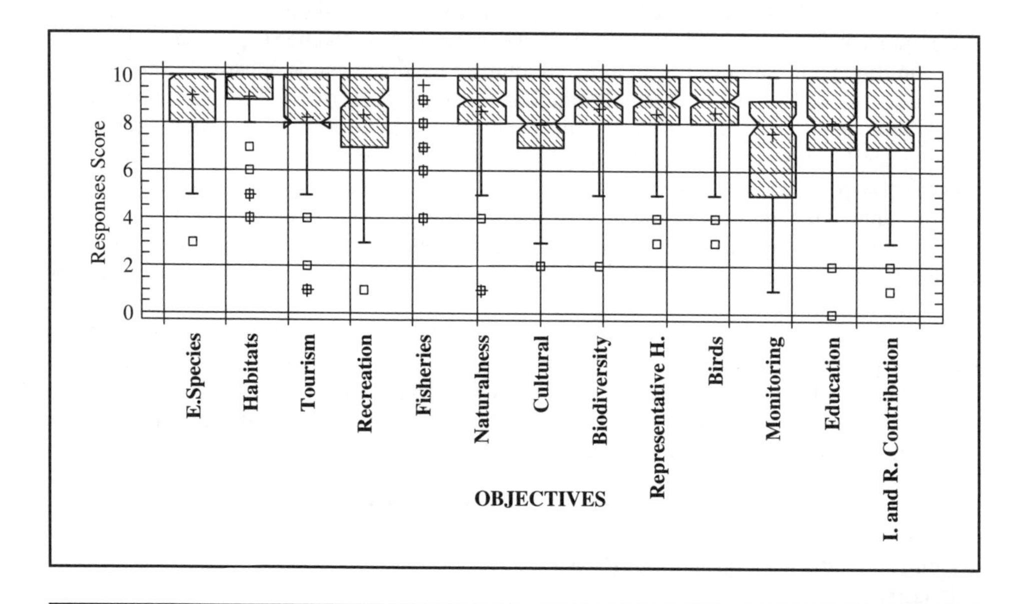

The Box boundaries represent the lower quartile to the upper quartile, covering the centre half of each sample. The centre lines within each box show the location of the sample medians. The + marks the mean. The whiskers extend from the box to the minimum and maximum values in each sample, except for any outside or far outside points, which are plotted separately. Outside points are points which lie more than 1.5 times the inter-quartile range above or below the box (outliers) and are shown as small squares, while far outside points are points which lie more than 3.0 times the inter-quartile range above or below the box and are shown as small squares with plus signs through them.

Figure 1. Box-and-Whisker Plot for the marine conservation objectives in Bahrain.

From the analysis, the value ranking of agreement on national marine conservation objectives in Bahrain, as perceived by the local population, is shown in table 1. A high rank implies a high level of agreement, and a low rank indicates a low level of agreement. In general the highest response score also matches higher level of agreement.

3.2 *Breakdown of responses by different sectors*

Six targeted groups were selected, with altogether thirty-two subgroups. Table 2 shows the average responses of these groups concerning national marine conservation objectives in Bahrain. Though the ranking of different sectors should be done with utmost caution, they help in distinguishing the similar median scores (table 3).

Table 2 shows that the majority of groups attribute the highest scores to fisheries, except for the regional and the international organisations (R. I. Org.), who give endangered species and habitats the highest priority, still giving fisheries a high score. The lowest scores in the groups were variable, but due to the high number of respondents in the official and public groups, observation/monitoring was the lowest objective in the pooled results. Historical/cultural, regional/international contribution and training also were given low priority.

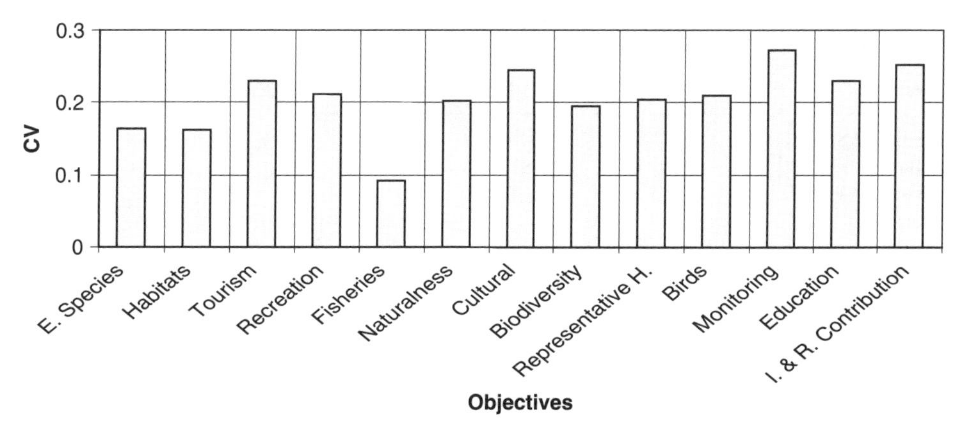

Figure 2. The coefficient of variation CV test for National marine conservation in Bahrain.

Table 1. The ranking level of stakeholders agreement on National Marine Conservation Objectives in Bahrain.

Objectives	Rank
To protect important fisheries habitats	1
To protect sensitive, rare and endangered habitats	2
To conserve rare and endangered species	3
To conserve high biodiversity areas	4
To conserve areas of naturalness	5
To conserve Bahrain's representative habitats	6
To protect areas with local and migratory bird species	7
To protect and promote recreational areas	8
To set up areas for the promotion of training and education	9
To protect and promote tourist attraction areas	10
To protect historical, cultural and traditional areas	11
To contribute to regional and international marine conservation efforts	12
To set up areas for the promotion of observation and monitoring	13

Table 2. The average response scores from respondent groups concerning national marine conservation objectives.

Objectives	Official	Private	NGOs	Industry	R. I. Org.	Public	Pooled
Endangered Sp.	9.2	8	9.8	9.4	*10	8.7	9.1
Habitats	9.2	8.1	*10	9.4	*10	8	9.1
Tourism	7.8	9.5	9.1	#7.5	#8.5	8.6	8.2
Recreation	7.9	*9.5	9.2	*9.5	#8.5	8.7	8.3
Fisheries	*9.5	*9.5	*10	*9.5	9.8	*9.7	*9.6
Naturalness	8.3	8.7	*10	8.2	9.4	8.4	8.5
Historical & cultural	7.5	7.7	9	7.9	8.9	8.2	7.9
Biodiversity	8.5	8.1	9.8	9	9.7	8.4	8.6
Representative Habitats	8.4	7.4	9.8	8.1	9.4	8.2	8.4
Birds	8.1	8.2	9.8	8.2	9.8	8.4	8.4
Observation/Monitoring	#7.3	6.2	9	8	9.2	#7.4	#7.5
Training & Education	7.5	7.4	9.4	8.3	9.1	8.2	8.0
R. & I. Contribution	7.8	#5.5	#8.7	8.4	9	8	7.9

* The highest average score in each group # The lowest average score in each group.

Table 3. The median response scores from respondent groups concerning national marine conservation objectives (ranks for pooled scores given in brackets).

Objectives	Official	Private	NGOs	Industry	R.I. Org.	Public	Pooled
Endangered Sp.	10	8	10	10	10	10	10
Habitats	10	8	10	10	10	9	10
Tourism	8	10	10	8	9	9	8
Recreation	8	10	10	8	8	9	9
Fisheries	10	10	10	9.5	10	10	10
Naturalness	9	9	10	8.5	10	9	9
Historical & cultural	8	8	9	8	8	9	8
Biodiversity	9	9	10	9.5	10	9.5	9
Representative Habitats	9	8	10	8	10	8	9
Birds	9	9	10	8.5	10	9	9
Observation/Monitoring	8	7	10	8	10	8	8
Training/Education	8	8	10	8	10	8	8
Contribution	8	5	10	8	10	9	8

Table 3 shows the median responses of the six targeted groups concerning national marine conservation objectives in Bahrain. According to the questionnaire criteria all the median responses in the table are important. The overall responses regarding the thirteen objectives, which are reflected by the poled results can be categorised into three priority groups.

1) Group one (objectives of very high value): Endangered sp., Habitats and Fisheries.
2) Group two (objectives of high value): Recreation, Naturalness, Biodiversity, Representative Habitats and Birds.
3) Group three (objectives of slightly lower value): Tourism, Historical and Cultural, Observing/Monitoring, Training/Education and R & I Contribution.

The three priority groups above match the results recorded in table 1.

3.3 *Views on the declared marine reserves in Bahrain*

The majority of respondents agreed with the country's already declared two marine sites (Ras Sanad and the Hawar Islands) with a slight preference for the Hawar Islands compared to Ras Sanad (Figures 3&4).

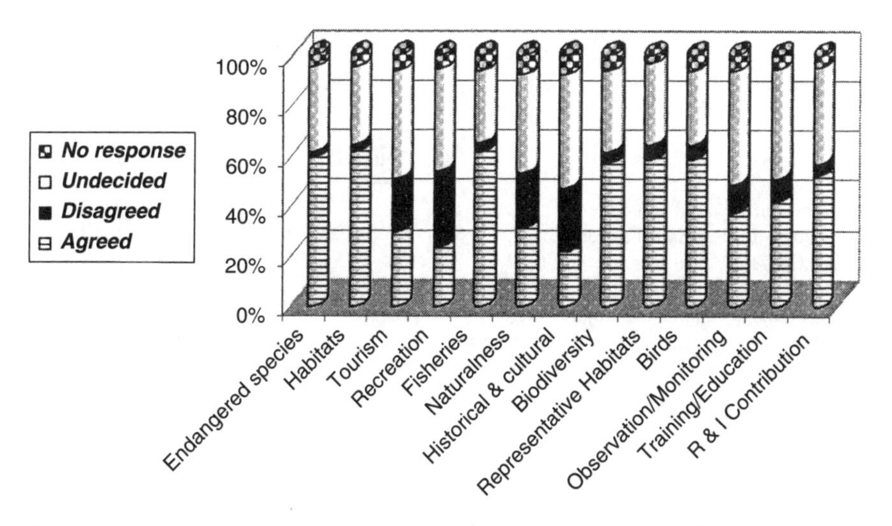

Figure 3. Questionnaire responses match of objectives for the Ras Sanad: percentage agreeing, disagreeing, undecided or not responding.

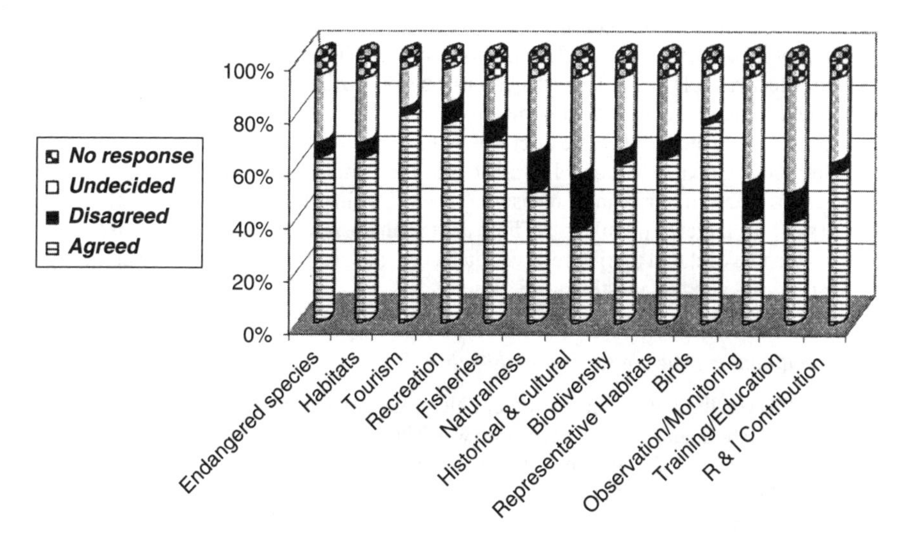

Figure 4. Questionnaire responses match of objectives for the Hawar Islands: percentage agreeing, disagreeing, undecided or not responding.

This is mainly due to the unfamiliarity with the Ras Sanad reserve where Hawar is better known to Bahrainis. Lack of access to Ras Sanad clearly prompted the low agreement with the recreational objective. A high level of agreement with match to objectives concerning endangered species, habitats, fisheries, biodiversity, representative habitats and birds was given for both reserves.

These figures can also be used as a basis for understanding the knowledge and awareness of the respondents with respect to the status of the marine conservation in Bahrain and their objectives. It reflects the fact that the respondents are not very aware of the role of regional and international agreements on marine conservation.

Table 4. Response rates to qualitative questions.

Question summary	Responses	
	Num. of responses (Total 166)	%
Other marine conservation objectives	39	23
Other objectives for Ras Sanad Reserve	19	11
Other objectives for Hawar Reserve	20	12
Other marine conservation areas proposed	87	52
Comments on the questionnaire subject	74	44

4 QUALITATIVE RESPONSES

Qualitative responses were gathered and analysed, and the results were used to clarify, support and enhance the more quantitative scores.

As can be seen from table 4 there were five questions that elicited unsolicited responses in the questionnaire. There were different response rates for each question, the highest response (52%) coming from the written question related to proposing other areas for protection as marine conservation areas in Bahraini territorial waters; the second highest response (44%) was from comments on the questionnaire subject, which elicits comments on the questionnaire subject. Other written responses were quite low: On other marine conservation objectives (23%), on other objectives for Ras Sanad (11%) and on other objectives for the Hawar Islands (12%). These are questions related to marine conservation objectives in Bahrain. This result is probably due to the limited knowledge of the respondents in the field of marine conservation objectives.

The responses came more to the question which the respondents could convey knowledge of and could answer. Thus, the knowledge and awareness of the sample groups relating to marine conservation objectives in the country can be measured. There are other factors which can help to estimate the awareness of the targeted population. These are the analysis of the groups' response results, the responses on the two declared marine reserves questions, and the relationship between the organisation they came from and the questionnaire subject.

4.1 *Other national marine conservation objectives*

The question "Do you think that there are other objectives, which are not included in the table?" was included in the questionnaire. The response rate was 23%, as shown in table 4. Some of these objectives may logically be added, others not. Table 5 shows the additional objectives suggested by the respondents and the number of times these objectives were proposed.

In fact, the thirteen objectives mentioned in the table above can be sorted into three groups as follows:

1) New objectives which can be considered to be practical.
2) Objectives, which are already included in one way or another in the listed objectives.
3) Objectives which can be ignored as being impractical.

The new practical objectives which should be considered are numbers 1, 3, 11 and 13 (Table 5), because these objectives focus on particular areas which are not included in the main objectives mentioned in the questionnaire. Objective one, which occurred ten times, will be incorporated into the objective related to training and education. Objectives 3 and 11 also would be considered in the implementation stage. Unfortunately, objective 13 cannot be applied, because there was no data available for implementing such objectives, apart from some information concerning mangrove plants (*Avicennia marina)* as being important for medical purposes, as it contains valuable medical organic compounds (Zahran *et al.,* 1983; Abul-Razik, 1994).

Table 5. Additional objectives suggested by the respondents.

No	Objective	Number of responses
1	To promote environmental awareness	10
2	To protect the marine environment from pollution	7
3	To protect aquaculture areas	3
4	To develop the legal instruments for marine conservation	3
5	To conserve natural resources	2
6	To protect the coastal environment	2
7	To protect high economic value areas	2
8	To conserve genetic resources	2
9	To protect fishing areas	2
10	To promote sustainable development	1
11	To protect areas where there are desalination plants	1
12	To promote institutional involvement	1
13	To protect areas where substances with medicinal value can be found	1

There are several objectives which are already included in the main list. Objectives numbers 4, 5, 6, 7, 8, 9 and 12 in table 5 appear in another format in the original list provided.

Objectives no 2 and 10 are considered as the umbrella or the overall aim for the establishment of marine conservation. These two objectives should be implemented in the designation and management process after selecting the most valuable areas for conservation.

4.2 *Other proposed marine conservation areas by respondents in Bahrain*

The highest qualitative responses were to question to suggest other marine conservation areas in Bahrains territorial waters that should be protected (52%). The reason for this high response is because the respondents have some knowledge about the marine areas and their importance, though they may not have the understanding of the objectives for marine conservation. Some 47 sites were proposed as protected areas (table 6). Thirty-one of these sites are specific areas and the rest are area attributes only.

5 DISCUSSION AND CONCLUSION

Clearly defined MPA goals and objectives are an essential step before any other steps for defining and selecting MPAs (Agardy 1994 and 1997; Salm and Price 1995; Kelleher, 1999; NRC, 2001). Local people and stakeholder opinions are highly recommended for consideration in this step (Kelleher, 1999), to benefit from their contribution and involvement in the implementation process, using their experience and knowledge of the areas which should be conserved.

The use of a questionnaire to survey the population in Bahrain on a national scale is a new approach used for the first time to determine marine conservation objectives or MPAs. Most of the previous work carried out in this field involved selecting MPA objectives through consultation with either environmental or/and wildlife conservation agencies or fisheries (Jones, 1994; WWF, 1998; NRC, 2001) without exploration of other sectors or the general public.

It is very important to establish criteria that cover all the relevant sectors' uses and interests, to ensure overall marine environment protection. Therefore, the identification of a comprehensive set of national marine conservation objectives is a key factor.

In this study the local peoples' opinions were surveyed to identify the priority conservation goals and area attributes requiring protection in Bahraini territorial waters. The questionnaire survey aimed to rank marine conservation objectives in the area of study. The above data analysis relating to national marine conservation objectives, which were gathered by the questionnaire, reveals that most of the objectives presented in the questionnaire were important to the respondents. According to their responses, these objectives were classified, categorised, weighted and ranked.

Table 6. The areas and attributes of MPAs proposed by the respondents.

Area proposed	Responses	Attributes proposed	Responses
Tubli Bay	21	Coral reef areas	20
Fasht Al Adhm	16	Fisheries nursery areas	12
Fasht Al Jarim	7	Pearl oyster habitats	8
Ras Al Barr	7	The Islands	6
Fasht Al Dibal	6	Al Herrat areas	6
Suhalah North of Budaiya	5	Mudflat habitats	4
Al Muhammadiyah Island	3	Rocky habitats	4
Abu Thamah	3	Seagrass habitats	3
The West coast of Bahrain	3	Mangrove habitats	3
Khawr Fasht	3	Dugong areas	3
Jiddah	3	Sea bird habitats	2
Ras Hayan	3	Algae habitats	2
Umm An Nasan	3	Tourism and recreational beaches	2
Al Shakh Island	3	Shallow water	1
East Coast of Bahrain	2	Turtle areas	1
Muharraq surrounding areas	2	All the bays	1
Shtayeh	2		
Al Zallaq	1		
From Jaw to Mashtan	1		
Askar & Jaw	1		
Al Jazayer Beach	1		
Abu Sabaa Beach	1		
Qummais	1		
Abu Al Dajal	1		
Abu Suor	1		
Ras Al Qrain	1		
Hawar Islands	1		
Ras Sanad	1		
Qetat Jaradah	1		
South Coast of Bahrain	1		
Dry dock beach	1		

Three columns of table 7 were organised to identify marine conservation objectives in Bahrain. The first column shows the value of each objective and they are categorised according to the level of importance, as reflected by the opinions of the stakeholders deemed. All the objectives were deemed important, but the levels of importance are different. The second column reflects the weight of each value. The third and the last column in table 6 reflects the overall ranking of the objectives according to priority, which is concluded from the survey analysis.

The fisheries protection objective is an essential objective according to worldwide experience (Agardy, 1994, 1997 and 1999; Hockey and Branch, 1997; NRC, 2001). The results of this survey also support this view. It is clear from table 7 that protecting important fisheries habitats comes first. This objective is at the top of the ranking list and has very high value, because the local community in Bahrain depends on fish and shrimp in their diet; they consume about 93% of the total fish landings (Abdulqader, 1994). Moreover, there is perhaps an awareness that their fisheries resources are under threat, which could affect their socio-economic and food security. Protection from all threatening activities is the only way to ensure that these resources will be sustained.

Conservation of sensitive, rare and endangered habitats, and of rare and endangered species objectives obtained a very high value level, while objectives to conserve high biodiversity areas, areas of naturalness, representative habitats and protect areas with local and migratory bird species obtained a high value level.

The objective related to protecting and promoting recreational areas also obtained a high value level, and this could be related to the respondents concern about their recreation as a priority.

Table 7. Overall assessment of the responses.

Objectives	Value	Weight	Rank
To protect important fisheries habitats	VHV	3	1
To protect sensitive, rare and endangered habitats	VHV	3	2
To conserve rare and endangered species	VHV	3	3
To conserve high biodiversity areas	HV	2	4
To conserve areas of naturalness	HV	2	5
To conserve Bahrain's representative habitats	HV	2	6
To protect areas with local and migratory bird species	HV	2	7
To protect and promote recreational areas	HV	2	8
To set up areas for the promotion of training and education	SHV	1	9
To protect and promote tourist attraction areas	SHV	1	10
To protect historical, cultural and traditional areas	SHV	1	11
To contribute to regional and international marine conservation efforts	SHV	1	12
To set up areas for the promotion of observation and monitoring	SHV	1	13

VHV = Very High Value, HV = High Value, SHV = Slightly High Value.

The rest of the objectives are considered as one group (insignificant difference between them). These objectives are categorised as slightly high value. The results of the analysis show that the respondents are concerned about these objectives less than VHV and HV objectives. This is absolutely logical because SHV objectives may be considered luxury items compared with the previous key objectives.

The very high value objectives, as indicated by respondents, were ranked 1, 2 and 3 and given a weighting of 3. The high value objectives (ranked 4 to 8) were given a weighting of 2, and only slightly high valued objectives (9 to 13) were given a weighting of 1.

In general, the results obtained in this survey do not differ much from those of previous studies (e.g. Joens, 1994; Agardy, 1994, 1997 and 1999; Salm and Price, 1995; Hockey and Branch, 1997; Kelleher, 1999; Salm *et al.*, 2000; NRC, 2001) or from those presented in international organisation reports (e.g. WWF, 1998; WWF/IUCN, 1998), with regard to marine conservation objectives priorities. In this study, objectives are ranked according to national priorities, giving more emphasis on the implementation stage, in keeping with the status of the region and its requirements (Salm and Price, 1995).

Hence, the overall impression about the target respondents in this survey is that they are highly aware of their marine environment and have a good knowledge in terms of conserving their environment. This is clearly defined from the similarity of ranking results and responses to each question addressed in this survey. One of the results is the 47 areas and attributes proposed for conservation by the respondents.

The selection of national marine conservation objectives and their prioritisation is the first basic step and very important as far as the remaining steps in marine conservation. According to these objectives, classification and prioritisation, planning, designing and implementation of marine conservation in Bahrain are the necessary actions which should be taken for a better and healthier environment.

REFERENCES

Abdulqader, E.A.A. 1994. *Bahrain benthic resources.* FAO Indian Ocean fishery commission, Committee for the development and management of the fishery resources of the Gulfs. Eighth session. IOFC: DMG/94/Inf.11, 19.

Abul-Razik, M.S. 1994. *Mangrove plants (Avicennia marina): General study and experiments in Qatar (in Arabic).* Qatar: University of Qatar.

Agardy, T. 1994. Advances in marine conservation: the role of marine protected areas. *Trends in Ecology & Evolution* 9(7): 267–270.

Agardy, T. 1997. *Marine protected areas and ocean conservation.* USA: Academic Press.

Agardy, T. 1999. Global trends in marine protected areas. *Proceedings of the workshop on Trends and future challenges for U. S. National Ocean and coastal policy.* In Billana Cicin-Sain, Robert W. Knecht & Nancy

Foster (eds.). Washington DC: Centre for the study of marine policy, University of Delaware and National Ocean Service, NOAA.
Al Zayani, A.K. 2003. *The selection of marine protected areas (MPAs): A model for the Kingdom of Bahrain.* UK: University of Southampton.
Allison, G.W., Lubchenco J. & Carr, M.H. 1998. Marine reserves are necessary but not sufficient for marine conservation. *Ecological Applications* 8: S79–S92.
Chatfield, C. 1983. *Statistics for technology – A Course in Applied Statistics*. London: Chapman & Hall.
Clark, J.R. 1992. *Integrated management of coastal zones*. Fisheries technical papers No. 327, Rome: Food and Agriculture Organization of the United Nations (FAO).
Cohen, L. & Holliday, M. 1996. *Practical statistics for students*. London: Paul Chapman Publishing Ltd.
Frid, C.L.J. & Clark, S. 1999. Restoring aquatic systems: an overview. *Aquatic Conservation: Marine and Freshwater Ecosystems* 9: 1–4.
Gubbay S. (ed.) 1995. *Marine Protected Areas: Principles and techniques for management.* UK: Chapman & Hall.
Hawkins, S.J., Allen, J.R. & Bray, S. 1999. Restoration of temperate marine and coastal ecosystems: nudging nature. *Aquatic Conservation: Marine and Freshwater Ecosystems* 9(1): 23–46.
Hockey, P.A.R. & Branch, G.M. 1997. Criteria, objectives and methodology for evaluating marine protected areas in South Africa. *South African Journal of Marine Sciences* 18: 369–383.
IUCN/ROPME/UNEP 1985. *An ecological study of sites on the coast of Bahrain.* UNEP Regional Seas reports and studies No. 72: 14.
Jones P.J.S. 1994. A Review and analysis of the objectives of marine nature reserves. *Ocean & Coastal Management* 24: 149–178.
Kelleher, G. & Kenchington, R. 1992. *Guidelines for Establishing Marine Protected Areas*. A Marine Conservation and Development Report. Gland, Switzerland: IUCN.
Kelleher, G., Bleakley C. & Wells, S. (1995). *A global representative system of marine protected areas*. The Great Barrier Reef Marine Park Authority, The World Bank, The World Conservation Union (IUCN), The World Bank, Washington, D.C., U.S.A.
Kelleher G. 1999. *Guidelines for Marine Protected Areas*. Gland, Switzerland and Cambridge, UK: IUCN.
Madany, I.M., Ali, S.M. & Akhter, M.S. 1988. *A review of dredging and land reclamation activities in Bahrain. Proceeding of the ROPME Workshop on Coastal Area Development.* UNEP Regional Seas Reports and Studies No. 90. UNEP.
McNeely J.A. 1994. Protected areas for the 21st century: working to provide benefits to society. *Biodiversity and conservation* 3: 390–405.
Mumby, P.J., Raines, P.S., Gray, D.A. & Gibson J.P. 1995. Geographic information systems: a tool for integrated coastal zone management in Belize. *Coastal Management* 23: 111–121.
NRC (National Research Council) 2001. *Marine Protected Areas: Tools for sustaining ocean ecosystems.* Washington DC: National Academy Press.
Oppenheim, A.N. 1992. *Questionnaire design, interviewing and attitude measurement.* London: Printer.
Roberts C.M., Bohnsack J.A., Gell F., Hawkins J.P. & Goodridge, R. 2001. Effects of marine reserves on adjacent fisheries. *Science* 294: 1920–1923.
Salm, R. & Clark, J. 1984. *Marine and coastal protected areas: A guide for planners and managers*. Switzerland: International Union for Conservation of Nature and Natural Resources (IUCN).
Salm, R. & Price, A. 1995. Selection of marine protected areas. In Gubbay S. (ed.), *Marine Protected Areas: Principles and techniques for management.* UK: Chapman & Hall.
Salm, R.V. & Clark, J. 2000. *Marine and Coastal Protected Areas: A guide for planners and managers.* Washington DC: IUCN.
Sims, R., Zainal & A.J.M. 2000. *Marine environment geographic information system (MAREGIS*). Bahrain: Bahrain Centre for Studies and Research.
Sudman, S. & Bradburn, M.N. 1982. *Asking questions: A practical guide to questionnaire design.* San Francisco: Jossey-Bass.
Vousden, D.H. 1995. *Bahrain marine habitats and some environmental effects on Seagrass beds*. UK: University of Wales (Bangor).
Vousden, D.H. 1988. *The Bahrain marine habitat survey. A study of the marine habitats in the waters of Bahrain and their relationship to physical, chemical, biological and anthropogenic influences. Volume 1.* Bahrain: Environmental Protection Technical Secretariat.
WWF 1998. *Marine protected areas: WWF's role in their future development.* Gland, Switzerland: WWF.
Young, T.P. 2000. Restoration ecology and conservation biology. *Biological Conservation* 92: 73–83.
Zahran, M.A., Younes H.A. & Hajrah H.H. 1983. On the ecology of mangal vegetation of the Saudi Arabian Red Sea coast. *Journal of the University of Kuwait (Science)* 10(1): 87–99.

Reclaiming the Desert: Towards a Sustainable Environment in Arid Lands – Mohamed (ed.)
© 2006 Taylor & Francis Group, London, ISBN 0 415 41128 9

Overcoming innate dormancy in the indigenous forage grass *cenchrus ciliaris* seeds

A. El-Keblawy & M.Y. Ibrahim
Department of Biology, Faculty of Science, UAE University

ABSTRACT: *Cenchrus ciliaris* (buffel grass), indigenous to Arabian Peninsula, has been identified as one of the most tolerant forage grass to drought during vegetative and reproductive stages in the field and is currently being successfully used as a fodder under experimental conditions in the UAE and in some arid regions all over the world. However, propagation of this species from seeds is a problem because they have innate dormancy. Different treatments that might help in breaking the innate dormancy of *C. ciliaris* seeds were evaluated. The results showed that presence of floral structures enclosing the seeds significantly retarded both the germination percentage and germination rate. Storage for 4 months at room temperature significantly increased germination rate and final germination percentage of *C. ciliaris* seeds and spikelets. Farther storage for 9 months resulted in germination deterioration of the seeds, but not for spikelets. Soaking seeds stored for 9 months at 30°C and 40°C significantly increased their germination (95.6% and 87.8%, respectively), but soaking them at 50°C significantly reduced their germination (16.1%), compared to control (non-soaked seeds, 38.3%). All concentrations of gibberellic acid had the ability to significantly break down the innate dormancy of *C. ciliaris* seeds, but none of the concentrations of kinetin and thiourea succeeded to do that. In fact, the higher concentrations of kinetin reduced seed germination.

1 INTRODUCTION

The Arabian Peninsula experiences some of the most extreme climatic conditions found on the Earth. It is characterized by low erratic rainfall, high evaporation rates, extremely high temperatures during summer, and high soil salinities. A sustainable livestock industry in the Arabian Peninsula requires sustainable systems of both grazing and the production of cheap fodder plants. Currently, the main fodder crops in the area are alfalfa and Rhodes grass, which are not adapted the prevailing conditions and require vast quantities of water (up to 48,000 m^3/ha/yr), which often derived from non-renewable groundwater sources (Boer 1997, Peacock *et al.* 2003). The production of these forages has resulted in abandonment of many farms due to the increase in soil salinity. The utilization of adopted indigenous forage species to replace the exotic ones could be the solution for such problem (Peacock *et al.* 2003). *Cenchrus ciliaris* (Buffel grass), indigenous to Arabian Peninsula, has been identified as one of the most tolerant forage grass to drought during vegetative and reproductive stages in the field and is currently being successfully used as a fodder under experimental conditions in the UAE and in parts of Australia, South Africa and Pakistan under commercial conditions (Yadav *et al.* 1997, Peacock *et al.* 2003). In addition, this species is a highly nutritious grass and is valued for its high yields of palatable forage and intermittent grazing during droughty periods in the tropics in addition to its low cost of establishment, tolerance to drought conditions and crop pests, and its ability to withstand heavy grazing and trampling by livestock (J.A Duke, unpublished data).

Developed seeds would be ready for germination following dispersal or would have a kind of dormancy that may delay their germination until the arrival of the favorable season for seedling survival. The main causes of seed dormancy are (1) a physiological inhibiting mechanism of germination in the embryo (=physiological dormancy), (2) seed coat impermeable to water (=physical dormancy), or (3) underdeveloped embryo (=morphological dormancy) (Baskin and Baskin 1998, 2004).

Seed dormancy can influence patterns of plant distribution, recruitment dynamics and persistence in the plant community (Harper 1977). The advantages of dormancy is to enable seeds to accumulate in the soil seed bank and protect plants from expanding their entire reproductive output at a given time (Koller 1969). On the other hand, maintenance of seed dormancy when conditions are optimal for germination can be disadvantage, as the seeds are exposed to lethal environmental factors such as granivory and extreme temperature for long time (Owens *et al.* 1995).

Seed dormancy is generally due to several factors either internal and/or external. The internal factors include the impermeability of the teguments to water or oxygen, presence of some inhibitors in the seed coat, changes in internal hormonal balance or physiological immaturation of embryos (Khan 1977). Several factors have been reported to overcome seed dormancy including storage under different conditions (Simpson 1990, Ralowicz and Mancino 1992, Corbineau *et al.* 1992, Gutterman 1993, 1996 and Colbach *et al.* 2002), removal of bracts or dispersal organs enclosing the seeds (Trethowan *et al.* 1993, Osman and Ghassali 1997, Al-Charchafchi and Clor 1989, Khan and Ungar 2000), presoaking the seeds in warm water (Hashim 1990, Wardale *et al.* 1991, Warrag 1994, Thapliyal and Naithani 1996 and Leskovar *et al.* 1999) and treatment of seeds with different dormancy regulating substances (Khan and Ungar 1997, 2000, 2001, 2002, El-Keblawy *et al.* 2005).

The dispersal units (spikelets) of *C. ciliaris* consists of one or two, rarely more, caryopses (hereafter termed seeds) enclosed by paleas, lemmas and glumes, surrounded by an involucre of bristles. At harvest time, buffel grass seeds have low germination percentage, which gradually increase during dry storage (Winkworth 1971). Seeds of *C. ciliaris* from India stored for one year showed significantly higher germination (32.4%) than freshly harvested seeds (20%) (Yadav *et al.* 2001). In addition, dry heat treatment at 40°C for 10 days significantly improved germination in most of the accessions of *C. ciliaris* (Jethani *et al.* 2002). Furthermore, seeds of 30 ecotypes of *C. ciliaris* collected from USA, Mexico, Pakistan, Australia, Kenya, Tanzania, Botswana and South Africa showed greater germination when shelled or washed in running water for 2.5 h than untreated seeds (Venter and Rethman 1992).

Seed dormancy in buffel grass is greatly depending on the origin. For example, Hacker and Ratclif (1989) reported great variation in dormancy attributes between 15 accessions of buffel grass from regions with different latitude (near-equatorial and near-tropical). Germination was higher from equatorial than from near tropical accessions and from high than from low rainfall provenances. In addition, four weeks after harvest, germination ganged from 0 to 5.5%, depending on the accessions, and increased to 30% after storage at high temperature for16 weeks in accessions of low rainfall regions and 4 weeks in accessions of high rainfall regions (Hacker and Ratclif 1989). The aim of the present study was to assess dormancy level through the estimation of final germination percentage and germination rate as a result of (1) presence and absence of floral parts around the seeds (i.e., dispersal units or spikelets vs. naked seeds), (2) presoaking the seeds in water for different durations and different temperatures, (3) seed storage for 4 and 9 months at room temperature, (4) treatment of the seeds with different dormancy regulating substances (GA3, kinetin and thiourea). The results of the study would develop a strategy for re-seeding degraded arid rangelands in the Gulf regions. Sowing the seeds with appropriate levels of dormancy ensure both satisfactory regeneration and some persistence of seeds in the soil seed bank (Sharif-Zadeh and Murdoch 2001).

2 MATERIALS AND METHODS

Cenchrus ciliaris spikelets of the MAK 3 accession were collected from the experimental field of International Centre for Biosaline Agriculture (ICBA), Dubai, on June 2004 and January 2005 and also from different natural habitats including Al Ain, Diba, Fujairah, and Al Ain-Dubai road on April, 2005. Spikelets were threshed to separate caryopses (hereafter termed seeds) by using a hand-made rubber thresher. In order to assess the effect of storage and the presence of floral parts (i.e., naked seeds vs. spikelets) on germination of *C. ciliaris*, the germination of dispersal units (spikelets) were compared with naked seeds for the MAK 3 accession collected from ICBA in July 2004 and Jan 2005 (4 and 9 months of storage). The seeds were stored at room temperature in brown paper bags. Germination of this experiment occurred in September 2005.

The effect of seed pre-soaking in water at different temperatures for different durations was performed on seeds collected from ICBA in May and stored in room temperatures for 9 months. Seeds were soaked in water baths adjusted at 30°C, 40°C, and 50°C for 15, 30 and 60 minutes.

The effects of different concentrations of different dormancy regulating chemicals were assessed on a native accession from Al Ain-Dubai road collected in April 2005. Seeds were germinated in three concentrations of each of gibberellic acid (0.3 mM, 1.5 mM, and 3 mM), kinetin (0.05 mM, 0.25 mM, and 0.5 mM), and thiourea (5 mM, 10 mM, and 15 mM) and in distilled water (control). Germination of this experiment was carried out in May 2005, one month after seed collection.

The germination was conducted in 9-cm Petri-dishes containing one disk of Whatman No. 1 filter paper, with 5 ml of distilled water or test solutions. Three replicate dishes were used for each treatment, each with 20 seeds. Dishes were then organized randomly in an incubator adjusted at 15/25°C with the higher temperature coincide with continuous light from cool two 40 W daylight fluorescent tubes and two 100 W incandescent lamps. Seeds were considered to be germinated with the emergence of the radicals. Germinated seedlings were counted and removed every alternate day for 14 days following seed sowing.

The rate of germination was estimated using a modified Timson index of germination velocity $= \Sigma G/t$, where G is the percentage of seed germination at 2d intervals and t is the total germination period (Khan and Ungar, 1984). The maximum value possible here using this index was $700/14 = 50$. The greater the value, the more rapid is the germination.

Three-way analyses of variance (ANOVAs) were carried out to test effects of main factors (seed storage, sowing temperature and sowing duration) and their interactions on the final germination percentage and germination rate. Two-way ANOVAs were performed to evaluate the effects of fruit structure (spikelets vs. naked seeds) and seed age (or storage) on final germination percentage and one-way ANOVA was carried out to assess the difference between the different dormancy regulating substances on both germination percentage and rate. The same test was used to asses the difference between the different concentrations of each dormancy regulating chemicals. Tukey test (Honestly significant differences, HSD) was used to estimate least significant range between means. The germination percentages were arcsine transformed to meet the assumptions of ANOVA. The transformation improved normality of distribution of data. All statistical methods were performed using SYSTAT, version 11.0.

3 RESULTS

3.1 *Effects of seed structure and storage*

Two-way ANOVA showed significant effects for fruit structure (glumed spikelets vs. caryopses or seeds, $F = 118.0, P < 0.001$), storage ($F = 12.3, P < 0.01$), and their interaction ($F = 28.1, P < 0.01$) on final germination percentages of *C. ciliaris*. Overall final germination and germination rate were significantly greater for seeds than for spikelets. Final germination of fresh harvested naked seeds increased from 18.3% to 71.7% after 4 months of storage and then decreased to 38.3% after 9 months of storage. On the other hand, germination of fresh harvested spikelets increased from 3.3% and 5%, after harvest and 4 months storage, respectively to 13.3% after 9 months of storage.

Two-way ANOVA also showed a significant effect of fruit structure on germination rate ($F = 13.5$, $P < 0.01$), but insignificant effects for storage ($F = 0.83, P > 0.05$), and their interaction ($F = 0.69$, $P > 0.05$). Overall germination rate of naked seeds was significantly greater than that of spikelets.

3.2 *Effects of seed soaking*

Two-way ANOVA showed significant effects for temperature of soaking ($F = 256.4, P < 0.001$) and the interaction between soaking temperature and soaking duration ($P = 4.27, P < 0.05$) on final germination of *C. ciliaris* seeds stored for 9 months. Seeds soaked at 30°C attained significantly greater germination (95.6%) than those soaked at 40°C and both attained significantly greater germination than those soaked at 50°C (16.1%). Compared to control (non-soaked seeds), soaking the seeds at 30°C and 40°C significantly increased their germination, but soaking them at 50°C significantly reduced their germination. For all soaking periods, germination of seeds soaked at 30°C and 40°C was significantly greater than for those soaked at 50°C, so the difference was more pronounced after 60 minutes (Fig. 1). Germination of seeds soaked at 30 and 40°C was significantly greater than that of control at all soaking periods. However, germination of seeds soaked at 50°C was lower than that of control at all soaking periods, so the differences were significant after 30 and 40 minutes. For example, germination of seeds soaked for 15, 30 and 60 minutes increased over that of control by 139%, 152% and 156%, respectively at 30°C, and by 113%, 130% and 143%, respectively at 40°C, but decreased than the control by 26%, 78% and 70% at 50°C (Fig. 1).

The analysis of variances showed significant effect for temperature of soaking ($F = 13.13$, $P < 0.001$) on germination rate of *C. ciliaris* seeds stored for 9 months. Seeds soaked at 30 and 40°C attained significantly greater germination rate (39.3 and 40.2, respectively) than those soaked at 50°C (31.1).

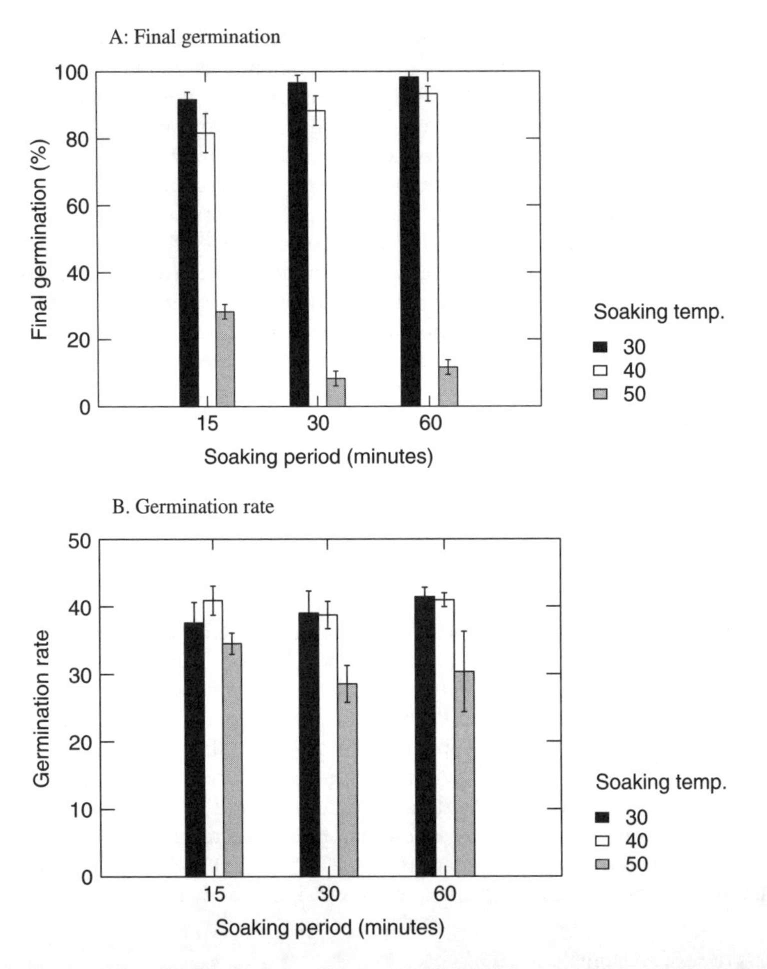

Figure 1 Effects of soaking temperature and duration on (a) final germination percentage and (b) germination rate of *Cenchrus ciliaris* seeds stored for 9 months. Final germination and germination rate of non-treated seeds (control) = 38.3% ± 3.3 and 44.1 ± 1.0, respectively.

Table 1. Effects of different dormancy regulating compounds on final germination percentage and germination rate (means ± SE) for fresh seeds of *Cenchrus ciliaris* collected from natural habitats on April 2005 and germinated on May 2005.

Treatment	Concentration mM	Mean Final germination (%)	Germination rate
Control		48.3 ± 4.4	30.4 ± 1.1
GA3	0.3	93.3 ± 1.7	35.0 ± 2.8
	1.5	91.7 ± 4.4	32.0 ± 0.9
	3	93.3 ± 1.7	34.4 ± 1.4
	Overall	92.8 ± 1.5	33.8 ± 1.7
Kinetin	0.05	51.7 ± 1.7	35.2 ± 2.6
	0.25	25.0 ± 5.8	30.2 ± 2.6
	0.5	26.7 ± 4.4	32.6 ± 3.5
	Overall	34.5 ± 3.97	32.7 ± 2.9
Thiourea	10	33.3 ± 3.3	28.3 ± 1.5
	15	43.3 ± 1.7	28.0 ± 0.9
	5	38.3 ± 6.0	30.8 ± 1.2
	Overall	38.3 ± 3.67	29.0 ± 1.2

3.3 *Effects of dormancy regulating chemicals*

3.3.1 *Effects on final germination*

One-way ANOVA showed a significant difference in final germination percentage of *Cenchrus ciliaris* treated with different dormancy regulating chemicals ($F = 81.1$, $P < 0.001$). Tukey test showed that seeds treated with gibberellic acid attained significantly greater germination than seeds of the control and those treated with kinetin and thiourea. There was no significant difference between control and seeds treated with both kinetin and thiourea (Table 1).

One way ANOVAs also showed significant differences in final germination between different concentrations of gibberellic acid ($F = 14.3$, $P < 0.01$) and kinetin ($F = 10.83$, $P < 0.01$), but not for thiourea ($F = 2.38$, $P > 0.05$). There was no significant difference between the three concentrations of gibberellic acid (0.3, 1.5 and 3 mM) and all of them attained significantly greater germination than control. However, there was no significant difference between control and 0.05 mM kinetin and both attained significantly greater germination than higher concentrations of kinetin (0.25 and 0.5 mM) (Table 1).

3.3.2 *Effects on germination rate*

One-way ANOVA showed a significant difference in germination rate of *C. ciliaris* between different dormancy regulating chemicals ($F = 3.08$, $P < 0.05$). Seeds treated with gibberellic acid attained significantly faster germination (33.8) than seeds treated with thiourea (29.0). No significant differences were observed between the control and the different dormancy regulating chemicals. The ANOVA also showed insignificant difference between the different concentrations of the different dormancy regulating chemicals (Table 1).

4 DISCUSSION

The results of the present study showed that the removal of all floral parts from around the seeds significantly increased final germination and germination rate of *Cenchrus ciliaris*. This could be attributed to some substances present in the floral parts that would act as germination inhibitors. Parihar and Patil (1984) analysed spikelet leachate of *C. ciliaris* after soaking in distilled water for 24 hour and found that it is composed of cyanidin-3-arabinoside, acylated with caffeic acid. These phenolic and commarin substances are often reported as inhibitors for seed germination in some species (Mayer and Poljakoff-Mayber 1975).

Innate dormancy produced by the presence of endogenous inhibitory compounds has been reported in bracts or other structures enclosing the seeds in some species (wheat and rye, Trethowan *et al. 1993; Atriplex halimus* and *Salsola vermiculata*, Osman and Ghassali 1997; *A. canescens*, Al-Charchafchi and Clor 1989; *A. griffithii*, Khan and Ungar 2000).

The present study also showed that presoaking seeds stored for 9 months in water at 30 and 40°C resulted in a significant increase in final germination than un-soaked seeds. However, soaking the seeds at 50°C resulted in the deterioration of their viability, especially for seeds stored for 9 months. Similarly, in five *Acacia* species, the increase of water temperature promoted seed germination up to a critical temperature beyond which a decline in final germination took place (Clemens *et al.* 1977). The enhancement of final germination following seed soaking support the presence of some inhibitory substances in the seed coat in addition to their presence in the spikelet structures. Similarly, Venter and Rethman (1992) showed that seeds of 30 ecotypes of *Cenchrus ciliaris* showed greater germination when shelled or washed in running water for 2.5 h than untreated seeds. Innate dormancy produced by the presence of endogenous inhibitory compounds has been detected in the seed testa of many species including *Cardus nutans* (Wardale *et al.* 1991), *Prosopis juliflora* (Warrag 1994), *Nyctanthes arbor-tristis* (Thapliyal and Naithani 1996), *Spinacia oleracea* (Leskovar *et al.* 1999).

The role of inhibitory substances in the germination of seeds of some plant species seem to have ecological implications as it provides information on the environmental conditions in which they inhabit (Rojas-Arechiga and Vazquez-Yanes 2000). Such autotoxic effects may control germination in concert with external conditions, e.g., inhibit premature germination when the seed is on the mother plant, extend germination over longer period of time, or allow germination to occur only after sufficient amount of rainfall to secure seedling establishment (Friedman 1995). In addition, Seeds may use environmental signals to inform the timing of their germination and thereby use dormancy as a mechanism of habitat choice (Preston and Baldwin 1999).

Dormancy is normally overcome by after-ripening, a process occurs when dormant seeds are exposed during storage to a set of environmental conditions (Bewley and Black 1982). The effects of storage on seed after-ripening have been reported from many species of grasses including *Avena fatua* (Simpson 1990), *Hilariu behngeri* (Ralowicz and Mancino 1992), *Bromus rubens* (Corbineau *et al.* 1992), *Hordeum spontaneum* (Gutterman 1993), *Schismus arabicus* (Gutterman 1996) and *Alopecurus myosuroides* (Colbach *et al.* 2002). The results of the present study showed that storage for 4 months at room temperature significantly increased germination rate and final germination percentage of *C. ciliaris* seeds. Farther storage for 9 months resulted in germination deterioration. McIvor and Howden (2000) arrived to a similar result for *C. ciliaris* introduce to Australia. The germination increased from 5% for fresh seeds to 55% after 16 weeks and then decreased again to 6% after 32 weeks of seed storage on soil surface (McIvor and Howden 2000). The mechanism whereby after-ripening promotes the capacity to germinate would include an alteration in permeability of seed coat membrane (Hartmann *et al.* 1990), a decrease in inhibitors content and increase in the rate of non-enzymatic oxidation reactions, which must occur before the seed can germinate (Karssen 1970).

GA3 can overcome physiological dormancy of seeds and stimulate the germination of seeds with dormant embryo (Hartmann *et al.* 1990). The promotion effects of GA3 on germination has been documented in several species (e.g., *Zygophyllum simplex*, Khan and Ungar 1997; *Echinacea purpurea*, Kochankova *et al.* 1998. Fernández et al (1997) studying the quantification of different GAs and their metabolism in dormant and non-dormant beechnuts (*Fagus sylvatica* L.) and found that both quantitative differences between dormant and non-dormant seeds in the analysed Gas and the capacity of non-dormant seeds to carry out metabolic conversions when labeled GA20 was injected into the seeds revealing a dynamic role of GAs in dormancy release (Fernández *et al*, 1997). In addition, it has been shown that GA3 introduced at the onset of imbibition markedly stimulated germination of henbit (*Lamium amplexicaule* L.) seeds (Taylorson & Hendricks, 1976).

In *C. ciliaris*, the effect of GA3 on final germination varied in different studies. For example, hydration of seeds of two *C. ciliaris* varieties with 100 ppm of GA3 for 18 h resulted in only 25.3 and 27.3% germination for the two varies (Bhatt, 2001). Similarly, Yadav *et al.* (2001) indicated that the application of 300 ppm GA3 significantly increased seed germination to only 28.9%, compared to 22.1% for the control. However, treatment the seeds of *C. ciliaris* from Indian deserts with 5 ppm GA3 resulted in 98.3% germination (Shahi and Sen 1991). In the present study, application of both 0.3 and 3 mM GA3 resulted in significantly greater germination (93.3%), compared to control (48.3%, Table 1).

The application of different concentrations of both kinetin and thiourea did not improve germination of *C. ciliaris* than control. Actually, 0.5 mM kinetin decreased significantly the final germination than the control (Table 1). Kinetin was very effective in alleviating innate dormancy in *Zygophyllum simplex* (Khan and Ungar, 1997), but not in *Sporobolus arabicus* (Khan and Ungar 2001) and *Prosopis juliflora* (El-Keblawy *et al.* 2005). Also, the effect of thiourea in alleviating innate dormancy was effective in *Z. simplex* (Khan and Ungar, 1997) partially effective in *Atriplex griffithii* (Khan and Ungar 2000) and *S. arabicus* (Khan and Ungar 2001), but not effective in *P. juliflora* (El-Keblawy *et al.* 2005).

Many factors influence the breaking of dormancy and germination of seeds. To isolate the effect of single factors may be difficult in the field because of the lack of control in field environments and because several factors may together determine the overall response of the seeds (Chaharsoghi and Jacobs 1998). The experiments described in the present study determined from a series of controlled treatments which of a number of factors would probably influence germination in the field. According to these experiment, *Cenchrus ciliaris* seeds could be used in restoration or rehabilitation of arid rangelands or growing it as fodder species but after storing the seeds for 9 months prior to their use, (2) use the naked seeds, rather than seeds enclosed within floral parts, (3) soak the seeds in water at 30°C or 40°C for 15 to 30 minutes, (4) treat the seeds with gibberellic acid.

REFERENCES

Al-Charchafchi, F.M. & Clor, M.A. (1989). Inhibition of germination and seedling development of *Atriplex canescens. Annals of Arid Zone* 28: 113–116

Baskin J.M. & Baskin, C.C. (2004). A classification system for seed dormancy. *Seed Science Research* 14: 1–16

Baskin, C.C. & Baskin, J.M. (1998). *Seeds: Ecology, Biogeography and Evolution of dormancy and germination.* San Diego: Academic Press

Bewley, J.D. & Black, M. (1982). *Physiology and Biochemistry of Seeds in Relation to Germination-Viability, Dormancy and Environmental Control.* Berlin: Springer-Verlag.

Bhatt, R.K. (2001). Seed treatment for enhancement of germination and crop establishment in *Cenchrus ciliaris* L. *Range Management and Agroforestry* 22: 122–124

Boer, B. (1997). An introduction to the climate in the UAE. *Journal of Arid Environment* 35: 3–16

Chaharsoghi, A.T. & Jacobs, B. (1998). Manipulating dormancy of capeweed (*Arctotheca calendula* L.) seed. *Seed Science Research* 8:139–146

Cheam, A.H. (1986). Patterns of change in seed dormancy and persistence of *Bromus diandrus* Roth. (great brome) in the field. *Australian Journal of Agricultural Research* 37: 471–481

Clemens, J., Jones, P.G. & Gilbert, N.H. (1977). Effects of seed treatments on germination in *Acacia. Australian Journal of Botany* 25: 269–276

Colbach, N., Dür C., Bruno, C. & Richard G. (2002). Effect of environmental condition on *Alopecurus myosuroides* germination. II. Effect of moisture conditions and storage length. *Weed Research.* 42: 222–230

Corbineau, F., Belaid, D. & Come, D. (1992). Dormancy of *Bromus rubens* L. seeds in relation to temperature, light and oxygen effects. *Weed Research.* 32: 303–310

Duke, J.A. Handbook of Energy Crops (unpublished) @ http://www.hort.purdue.edu/newcrop/duke_energy/Cenchrus_ciliaris.html

El Keblawy, A. & Al Rawai, A. (2004). Effects of salinity, temperature, and light on germination of invasive *Prosopis juliflora* (Sw.) D.C. *Journal of Arid Environment* 61: 555–565

El Keblawy, A., Al Ansari, F. & Al Rawai, A. (2005). Effects of dormancy regulating chemicals on innate and salinity induced dormancy in the invasive *Prosopis juliflora* (Sw.) D.C. Schrub. *Plant Growth Regulation* 46: 161–16

Fernández, H., Doumas, H., P. & Bonnet-Masimbert, M. (1997). Quantification of GA1, GA3, GA4, GA7, GA8, GA9, GA19 and GA20 metabolism in dormant and non-dormant beechnuts. *Plant Growth Regulation* 22: 29–35

Friedman, J. (1995). Allelopathy, autotoxicity, and germination. In J. Kigel and G. Galili (eds). *Seed Development and Germination*. New York: Marcel Dekker, Inc. pp 629–643

Gutterman, Y. (1993). Seed germination in desert plants. Adaptations of desert organisms. New York: Springer.

Gutterman, Y. (1996). Temperatures during storage, light and wetting affecting caryopses germinability of *Schismus arabicus,* a common desert annual grass. *Journal of Arid Environments* 33: 73–85

Hacker, J.B. & Ratclif, D. (1989). Seed dormancy and factors affecting dormancy breakdown in buffel grass accessions from contrasting provenances. *Journal of Applied Ecology* 26: 201–212

Harper, J.L. (1977). *Population Biology of Plants*. New York: Academic Press

Hartmann, H.T., Kester, D.E. & Davies, F.T. (1990). *Plant Propagation Principles and Practices*. New Jersey: Prentice-Hall International.

Hashim, I.M. (1990). Abundance, seed pod nutritional characteristics, and seed germination of leguminous trees in South Kordofan, Sudan. *Journal of Range Management* 4: 333–335

Jethani, I., Vari, A.K. & Mitrabinda (2002). Seed germination studies in three pasture grass species grown in North India. *Range Management and Agroforestry* 23: 105–109

Karssen, C.M. (1970). The light promoted germination of the seeds of *Chenopodium album* L. IV. Effect of the photoperiod during growth and development of the plants on the dormancy of the produced seeds. *Acta Bot. Neerl.* 19: 81–94

Khan, A.A. (1977). Seed dormancy: changing concepts and theories. In: Khan, A.A. (ed). The physiology and biochemistry of seed dormancy and germination. Amsterdam: Elsevier, north Holland biomedical Press. pp 09–50

Khan, M.A. & Ungar, I.A. (1984). The effect of salinity and temperature on the germination of polymorphic seeds and growth of *Atriplex triangularis* Willd. *American Journal of Botany* 71: 481–489

Khan, M.A. & Ungar, I.A. (1997). Alleviation of seed dormancy in the desert Forb *Zygophyllum simplex* L. from Pakistan. *Annals of Botany* 80: 395–400

Khan, M.A. & Ungar, I.A. (2000). Alleviation of innate and salinity-induced dormancy in *Atriplex griffithii* Moq. var. *stocksii* Boiss. *Seed Science & Technology* 28: 29–37

Khan, M.A. & Ungar, I.A. (2001). Effect of germination promoting compounds on the release of primary and salt-enforced seed dormancy in the halophyte *sprobolus arabicus* Boiss. *Seed Science & Technology* 29: 299–306

Khan, M.A. & Ungar, I.A. (2002). Influence of dormancy regulating compounds and salinity on the germination of *Zygophyllum simplex* L. seeds. *Seed Science & Technology* 30: 507–514

Klips, R.A. & Penalosa, J. (2003). The timing of seed fall, innate dormancy, and ambient temperature in *Lythrum salicaria. Aquatic botany* 75: 1–7

Kochankova, V.G., Grzesik, M., Chojnowski, M. & Nowak, J. (1998). Effects of temperature, growth regulators and other chemicals on *Echinacea purpurea* (L.) Moench seed germination and seedling survival. *Seed Science & Technology* 26: 547–554

Koller, D. (1969). The physiology of dormancy and survival of plants in desert environments. In: *Dormancy and Survival Symposium.* Cambridge society of Experimental Biology 23: 449–469

Leskovar, D.I., Esensee, V, & Belefant-Miller, H. (1999). Pericarp, leachate, and carbohydrate involvement on thermo inhibition of germinating spinach seeds. *Journal of American Society of Horticultural Sciences* 124:301–306

McIvor, J.G. & Howden, S. M. (2000). Dormancy and germination characteristics of herbaceous species in the seasonally dry tropics of northern Australia. *Austral Ecology* 25: 213–222

Osman, A.E. & Ghassali, F. (1997). Effects of storage conditions and presence of fruiting bracts on the germination of *Atriplex halimus* and *Salsola vermiculata. Experimental Agriculture* 33: 149–155

Owens, M.K., Wallace, R.B. & Archer, S. (1995). Seed dormancy and persistence of *Acacia berlandieri* and *Leucaena pulverulneta* in a semi-arid environment. *Journal of Arid Environments* 29: 15–23

Peacock, J.M., Ferguson, M.E., Alhadrami, G.A., Mc Cann, I.R., Al Hajoj, A., Saleh, A., and Karnik, R. (2003). Conservation through utilization: a case study of the indigenous forage grasses of the Arabian Peninsula. *Journal of Arid Environments* 54: 15–28

Preston, C.A. & Baldwin, I.T. (1999). Positive and negative signals regulate germination in the post-fire annual, *Nicotiana attenuata. Ecology* 80: 481–494

Parihar, S.S. & Patil, B.D. (1984). Seed germination studies with *Cenchrus ciliaris* L. II. Isolation and characterization of germination inhibitors from the spikelets. *Current Science* 53: 387–388

Qaderi, M.M., Presti, A. & Cavers, P.B. (2004). Dry storage effects on germinability of Scotch thistle (*Onopordum acanthium*) cypselas. *Acta oecologica* 27: 67–74

Ralowicz, A.E. & Mancino, C.F. (1992). After ripening in curly mesquite seeds. *Journal of range management* 45: 85–87

Rojas-Arechiga, M. & Vazquez-Yanes, C. (2000). Cactus seed germination: a review. *Journal of Arid Environments* 44: 85–104

Shahi, A.K. & Sen, D.N. (1991). Effect of GA on seed germination of aromatic and some fodder grasses of Indian desert. *Research and Development Reporter* 8: 80–81

Sharif-Zadeh, F. & Murdoch, A.J. (2000). The effects of different maturation conditions on seed dormancy and germination of *Cenchrus ciliaris*. *Seed-Science-Research*. 10: 447–457

Silock, R.G. & Smith, F.T. (1990). Viable seed retention under field conditions by western Queensland pasture species [grasses; legumes]. *Tropical Grasslands* (Australia). 24: 65–74

Simpson, G.M. (1990). Seed dormancy in grasses. Cambridge: Cambridge University Press.

Taylorson, R.B. & Hendricks, S.B. (1976). Interactions of phytochrome and exogenous gibberellic acid on germination of *Lamium amplexicaule* L. seeds. *Planta* (Historical Archive) 132: 65–70

Thapliyal, R.C. & Naithani, K.C. 1996. Inhibition of germination in Nyctanthes arbortristis (Oleaceae) by pericarp. *Seed Science & Technology* 24: 67–73.

Trethowan, R.M, Pfeiffer, W.H., Pena R.J. & Abdalla, O.S. (1993). Preharvest sprouting tolerance in three triticale biotypes. *Australian Journal of Agricultural Research* 44: 1789–1798

Venter, P.S., Rethman, N.F.G. (1992). Germination of fresh seed of thirty *Cenchrus ciliaris* ecotypes as influenced by seed treatment. *Journal of the Grassland Society of Southern Africa* 9: 181–182

Wardale, D.A., Ahmed, M. & Nicholson, K.S. (1991). Allelopathic influence of nodding thistle (*Cardus nutans* L.) seeds on germination and radical growth of pasture plants. *New Zealand Journal of Agricultural Research* 34:185–192

Warrag, M.O.A. (1994). Autotoxicity of mesquite (*Prosopis juliflora*) pericarps on seed germination and seedling growth. *Journal of Arid Environments* 27: 79–84

Winkworth, R.E. (1971). Longevity of buffel grass seed sown in an arid Australian range. *Journal of Range Management* 24: 141–145

Yadav, M.S., Rajora, M.P., Sharma, S.K. (2001). Effect of storage and seed treatments on germination in buffel grass. *Range Management and Agroforestry* 22: 1–5

Reclaiming the Desert: Towards a Sustainable Environment in Arid Lands – Mohamed (ed.)
© 2006 Taylor & Francis Group, London, ISBN 0 415 41128 9

Effects of camel versus oryx and gazelle grazing on the plant ecology of the Dubai desert conservation reserve

D.J. Gallacher & J.P. Hill
Zayed University, Dubai, United Arab Emirates

ABSTRACT: Grazing of the Dubai inland desert has changed substantially over the last century, and particularly over the last three decades. Populations of oryx, ostriches and gazelles have been replaced by an increased camel herd, which is at least 2.5 times historical levels. Camel grazing patterns differ to smaller herbivores, affecting plant species composition. Camels are given supplementary feed, so their population is not limited by seasonal availability of vegetation. Desert plants face longer periods of heavy grazing from a larger camel population, and shorter periods for recovery. Plant chemical defenses may also be less effective from the different grazing regime.

Although widely considered to be overgrazed, there is little information in the UAE on appropriate stocking levels for purposes of ecological sustainability or for maximizing pastoral production. The effect of grazing on vegetation was studied within the recently formed Dubai Desert Conservation Reserve (DDCR). Camel farms on the DDCR release camels during the day, allowing them to graze natural vegetation within the Reserve. It also contains an inner enclosure of five years in which camels were replaced by oryx and gazelles, separated by a 20 km fence. Fence line studies were made of (1) small (<1 m high) perennial plants, (2) seedling emergence during the winter of 2004/5, and (3) size and distribution of large shrubs (>1 m high). In addition, telephone surveys were conducted on DDCR farmers, and spatial distribution of trees was recorded.

Heavy grazing in the DDCR has reduced the cover of small perennial plant species, reducing their capacity for annual forage production. The extent of overgrazing on gravel substrata was severe, but it was also significant on sand substrata. There was some evidence of localized dune stabilization in the camel exclosure, due to increased vegetation. Germination density of perennial species was greater in the camel exclosure, probably caused by higher seed production of the larger plants. Germination density of annual plant species was not affected by grazing, but was much greater in closer proximity to established small shrubs. No germination was observed to be associated with feces of camels, oryx, gazelle or dhub lizards (*Uromastyx aegyptiaca*). Plants reached reproductive maturity at a height of 10 cm or less. Large shrubs were differentially impacted by grazing systems. *Calligonum comosum* was devastated by camel grazing. *Leptadenia pyrotechnica* and *Lycium schawii* were substantially reduced in size, though their long term impact is not yet known. Two species benefited from heavy grazing; the large shrub *Calotropis procera* and the sedge *Cyperus conglomeratus*. Among trees, the regeneration of *Prosopis cineraria* appears to have been severely reduced by herbivory at the small shrub stage, but no evidence of effects on *Acacia tortilis* was recorded.

Observed vegetation differences were primarily due to a greater level of grazing in the DDCR than the exclosure, but the ecological impact of camel grazing differs to that of oryx and gazelles. Rapid recovery within the camel exclosure indicates that plant species are well adapted to periods of heavy grazing, and ecological degradation in the DDCR is reversible. Nevertheless, complete recovery of plant species composition may take decades after a reduction of stocking rate. Recovery would benefit native wildlife, and also farmers by reducing their reliance of supplementary feed. Several options for reducing the impact of camel grazing are considered.

1 INTRODUCTION

The United Arab Emirates is a rapidly developing country with an equally rapid change in land use, largely through urban development of open desert. Plants growing on undeveloped land face an increasing density of livestock herbivores. Recreational off-road vehicles are a further threat in some locations. Animal populations have been affected by the building of highways and fences, which may have subdivided the populations of particular species.

A conservation zone of 225 km^2 was established in Dubai emirate in 2003, encompassing sand dune and intermittent gravel substratum ecological zones. The Dubai Desert Conservation Reserve (DDCR) has the role of preserving these ecological zones for future generations while providing a resource for recreational use, in particular for international tourists. However, the DDCR still contains farms that produce livestock, horticultural products and forage crops. Understanding the relationship between camel management and desert ecology is essential to optimize desert forage production, and to preserve native species. In this paper we will argue that policies related to the management of camels should be revised. Better management of forage through reduced exposure to camel grazing would increase overall forage production, increase plant species richness, and enable native animal populations to recover.

Camels have been significant to the local desert ecology at least since the Bedouin camel-herding lifestyle emerged, but possibly also before then as wild animals. Up until half a century ago the inland desert was grazed by oryx (*Oryx leucoryx*), now extinct in the wild; gazelles (*Gazella* spp.), whose numbers have plummeted, and the Arabian ostrich (*Struthio camelus syriacus*), which became extinct by late 1940 (Gross and Jongbloed, 1996). Hunting of gazelles was banned in 1983, though enforcing the ban has been difficult (Hellyer, 1996) and numbers have not recovered. Grazing by smaller species such as rodents, lizards and the Cape Hare (*Lepus capensis*) continues to fluctuate with the seasons, though overall numbers are probably reduced. Plant species became adapted to grazing by all these herbivores through physical and biochemical defenses, and physiological adaptations.

Recent changes in herbivory are threefold. Firstly, there has been a shift away from a mix of herbivore species toward a single species; the camel. Larger herbivores expend more energy to move, and so are less selective of their food source (Murray, 1991). Camels tend to graze an area heavily and then not return for some time, whereas smaller herbivores graze fewer species over a wider area and may remove less herbage from each plant. Secondly, stocking rates are no longer limited by desert plant production. Camels can survive and even reproduce during drought years from the supplementary feed provided by farmers. Their population has increased accordingly, causing an increased proportion of time in which plants are subjected to heavy grazing. Thirdly, livestock are less restrained by plant chemical defenses against herbivory since they can mix desert plants and supplementary feed in their stomachs.

A plant ecosystem can benefit from herbivory in several ways. Many plant species rely on herbivores for seed dispersal. Removal of above ground biomass reduces the surface area for transpiration, which might lead to a higher survival of grazed plants in dry seasons. Moderate grazing usually increases biodiversity through trampling and the removal of dominant plant species (Fernandez-Gimenez and Allen-Diaz, 1999), but this may not apply to habitats with a low biomass production such as the UAE (Oba et al., 2001). Heavy stocking rates reduce annual plant biomass production by limiting plant size and frequency. This can reduce feed available to camels, as well as to the small wild herbivores. Lower populations of rodents, lizards and cape hare sustain fewer carnivores such as cats, the Arabian red fox (*Vulpes vulpes arabica*), and birds of prey.

1.1 *Study site and objectives*

The DDCR fence line completely surrounds an inner enclosure (Al Maha) of 27 km^2. Camels were removed on completion of the fence in July 1999, and several species of oryx and gazelle were introduced, some indigenous but others exotic to the area. Al Maha stocking rates in April 2005 were approximately 0.092 oryx and 0.075 gazelles per hectare. The DDCR still contains approximately 960 camels (0.043 camels / ha) as well as other livestock that are restricted to the farms.

Since April 2005 it also contains some *Oryx leucoryx* and gazelles that were released from the Al Maha enclosure. The site has provided a unique opportunity for the authors to study the recovery of a large expanse of sand desert under a different grazing regime. Previous studies in the Gulf region have reported a rapid regeneration of vegetation from livestock exclosures (Khan, 1980a, 1981, Oatham et al., 1995, Zaman, 1997, Barth, 1999). However, exclosure of large herbivores is neither practical nor desirable in the DDCR, or in open UAE rangeland. Instead, it is necessary to study the relationship between herbivore and plant ecology, to optimize the production of both. The Al Maha fence is 24.12 km in length, of which 150 m crosses gravel substratum and the rest lies on sand that varies from very stable to actively moving. The natural botany of the area is similar to the Prosopis-Calligonum vegetation type described by Ghazanfar (2004) in Oman.

The authors have conducted several studies toward determining how the two management systems affect plant ecology. Three studies were conducted on each side of the Al Maha fence, including:

- Ecology of small (<1 m high) perennial plants, of which most were dwarf shrubs.
- Seedling emergence and survival after the 2004 winter rains.
- Size and spatial distribution of large shrubs (*Calligonum comosum*, *Leptadenia pyrotechnica*, *Lycium schawii*, and *Calotropis procera*). This work is ongoing.
- Two further studies pertain to the DDCR as a whole.
- Telephone survey of farmers. Interviews were conducted, translated and transcribed by Nasra Al Juma, a Zayed University graduate.
- Spatial distribution of *Prosopis cineraria* (ghaf) and *Acacia tortilis*.

Rainfall was unusually low from the period the Al Maha fence was constructed until the 2004 winter, but rains in 2004 produced a mass germination event. The closest rainfall data comes from the Dubai International Airport, which indicates an average annual rainfall of 93.8 mm that falls mostly between December and April (World Meteorological Organization, 2005), but which fluctuates widely both temporally and spatially.

1.2 *Camels in the UAE*

Agricultural settlement on the Arabian peninsula began around 2500 to 2000 BC with the domestication of the date palm (*Phoenix dactylifera*), which enabled an oasis based lifestyle (Potts, 2001). Dromedary camels (*Camelus dromedarius*) were probably first domesticated in southern Arabia, with their use spreading northward to reach Syria around 1100 BC (Bulliet, 1990). Fossil evidence of domesticated camels has been attested for the Iron Age (1200 to 300 BC) in the UAE (Stephan, 1995), but may have occurred as early as the 4th millennium BC (Peters, 1997). Use of the camel for transport, along with *falaj* irrigation technology, enabled a rapid increase in the number of settlements in the region (Potts, 2001). However, the greatest impact on plant ecology of the inland desert would have occurred with the broad scale emergence of nomadic camel herding. This lifestyle probably emerged later than the spread of settlements, but could still be thousands of years old. Camels can be used to provide milk, meat, transport of people or goods, mechanical power, entertainment, products from their leather or hair, and can be a store of wealth (Bulliet, 1990).

During winter if the feed was good, camels could obtain most of their water requirements by eating plants. Their herders could also survive temporarily without fresh water by drinking camel milk (Heard-Bey, 2001). People would move camp throughout the season each time feed supply for camels dwindled. Wells were dug in frequently visited places, enabling people to extend their nomadic range. In some places permanent settlements were established, such as Liwa (Abu Dhabi emirate) which has been continuously occupied since at least the 16th century (Heard-Bey, 2001). By these mechanisms the entire inland sand desert of the UAE became exposed to camel grazing, though it was not evenly distributed. Land surrounding permanent settlements were more heavily grazed, while land surrounding semi-permanent settlements were periodically grazed and then given time to recover. Other areas were labeled as restricted (Arabic: *harim*), and only accessed during drought periods when no other feed was available (Aspinall, 2001). Tribes used these areas to survive irregular weather patterns that could produce a drought for several years at a time.

Following unification of the Emirates in 1971, the government instituted policies to increase agricultural production and to establish permanent settlements and income for nomadic people. The density of wells was dramatically increased, and semi-permanent farms or occasionally used areas became permanent year-round farms. Camel numbers initially waned after unification, falling from approximately 97 000 to 39 500 in 1976, but they then steadily increased to 250 000 today (FAOSTAT, 2004). Other livestock industries have increased even faster since unification, notably goats (125 000 to 1 450 000) and dairy cattle (5 000 to 115 000). However, only camels are permitted to graze open desert throughout the Emirates. Irrigated forage production has increased commensurately, but the livestock sector still relies on imported feed. At 2.99 camels / km^2, density in the UAE is second only to Qatar and far higher than that of Saudi Arabia (3.36 and 0.12 camels / km^2 respectively) (FAOSTAT, 2004). Within Dubai emirate, camel farms displaced by urban expansion were moved deeper into the emirate, thereby increasing camel density on the remaining range.

Every camel in the UAE is owned by somebody. Farms are scattered throughout the open range, typically with an employee in residence and an owner who visits weekly. Land tenure is not always clear. Most camels are allowed to graze the desert on an 'open access' basis that is common throughout West Asia (Ferguson et al., 1998). DDCR farms usually feed camels at the farm in the afternoon (3 to 4 pm) and perhaps also in the morning (7 to 8 am), so most camels return to the farms at these times. Some camels stay away for days at a time if desert forage is sufficient, and some stay permanently at the farm if it isn't.

Camel production in the UAE today is primarily for racing. There are 14 000 actively racing camels competing on 15 race tracks throughout the country (Anonymous, 2005), and the camel population is distributed around racetracks rather than food or water sources (Yagoub and Hobbs, 2003). Females comprise 74% of the current herd (FAOSTAT, 2004).

2 PLANT RESPONSES TO GRAZING

Plant categories used in this study are based primarily on size. Although taxonomically arbitrary, this classification is useful for grouping species according to their rooting depth and their exposure to grazing. Ephemeral species use temporary, shallow moisture and minimize grazing exposure by producing large numbers of plants in a short time. Germination occurs only in Spring when moisture is more likely to last long enough for seedling growth. Trees can use permanent groundwater and grow above the reach of herbivores. Shrubs and perennial grasses are thought to tap water of medium depth, though knowledge of root depth is based on observations of above ground growth rather than direct measurement. Small shrubs are permanently exposed to grazing by all herbivores, while large shrubs can outgrow all herbivores except camels.

The sand surface rapidly dries after rain, but then acts as a protective layer for preserving deeper moisture (Al Wadie, 2002). Changes in water table depth are known from farm wells, but usage and relative permanence of plant-available water above this level is not known. Roots are generally deeper in arid areas where surface water is unreliable, and these conditions provide an advantage for trees and woody shrubs over herbaceous species (Abd el-Ghani, 2000).

2.1 *Large shrubs*

Most vegetal biomass within the DDCR occurs as *Leptadenia pyrotechnica* shrubs, one of four species that grows higher than 1 m but in which most or all photosynthetic tissue is within reach of camels (approximately 310 cm, based on browse line measurements of 55 trees). Response to the different grazing regimes is species specific. *Calotropis procera* benefits from heavy camel grazing, but *Calligonum comosum* is severely reduced in number while *Lycium schawii* and *Leptadenia pyrotechnica* are reduced in size.

Leptadenia pyrotechnica comprised 90.1% of all large shrubs growing near the Al Maha fence. Palatability is moderate (Ould Soulé, 1998) and it contains latex, but camels, oryx and gazelles all graze it. Livestock graze it to a low hedge in high traffic areas. Farmers in the DDCR considered it valuable as fodder and for killing intestinal worms of livestock. They commented that plant numbers have decreased in recent decades, but six years without camels in Al Maha did not produce significantly different numbers. This indicates that population shifts occur on a longer time scale. However, camel exclosure resulted in a greater median height (2.8 and 3.4 m in the DDCR and Al Maha respectively) and canopy cover (10.0 and 15.2 m^2). Shrubs with a large canopy (e.g.; >50 m^2) contained noticeably more gazelle resting sites underneath them. Large canopies were common in Al Maha (6% of the population) but the largest observed in the DDCR was just 48 m^2. Large canopy plants appear to provide a unique microhabitat, though it is not known if the habitat favors survival of other species. The authors disturbed one hare (*Lepus capensis*) and one fox (*Vulpes vulpes arabica*) while measuring 897 *L. pyrotechnica* shrubs, so there were insufficient observations of small mammals for comment.

The species is well adapted to heavy grazing, as evidenced by survival in the DDCR. Maximum plant herbage production probably requires some grazing, though how much is not known. Roots occur at whatever depth contains moisture, and have been recorded to 11.5 m (Batanoun and Wahab, 1973), enabling it to grow through the summer.

Calotropis procera roots reach 1.7–3.0 m in sandy desert soils, with few to no roots near the surface (Sharma, 1968). It is a succulent species that absorbs a lot of water after rainfall (Aziz and Khan, 2003). Plant numbers usually increase with camel density through the removal of competition (Khan, 1980b, Al Wadie, 2002). DDCR farmers unanimously reported that its numbers had increased over the last decade, particularly on the sand substratum, and it has become the dominant large shrub species in some areas of the DDCR. Latex consumption causes nervousness, frequent urination, frothing at the mouth, dyspnoea and diarrhea in goats (El Badwi et al., 1998). Ungulates graze on it sparingly, but it can be combined with other feed sources to prevent harmful effects (Nehra et al., 1987, Abbas et al., 1992). Over time, gazelles in Al Maha removed all leaves and flowers within reach. Plants in Al Maha therefore became tall (>2.5 m) and thin (<10 m^2 canopy), while plants in the DDCR exhibited a wider range of both variables (0.5–4.0 m height, 0–30 m^2 canopy). As gazelle numbers increase in the DDCR, recruitment of new plants can be expected to decline.

Calligonum comosum is one of the most important plants in UAE bedu folklore (Khan, 1979). It is common in Al Maha but only rarely seen in the DDCR, due to it being a favored food source of camels. It typically grows on sandy substrate where underground water has low salinity (Brown, 1978, Khan, 1979), though roots also absorb rainwater quickly (Asher, 1996). It is eaten readily by camels and less readily by sheep and goats (Ould Soulé, 1998), but plants within Al Maha indicate that oryx and gazelles browse it only very lightly. Camels eat all non-woody tissue, making plants appear lifeless and unrecognizable until producing limited foliage and flowers in February and March (Khan, 1980b, Western, 1988). DDCR farmers agreed that it had decreased to very small numbers, but its rapid recovery in Al Maha indicates that plants still exist as stumps or rootstock.

The authors have observed just three plants in the DDCR, though none were seen during a structured study of fence line vegetation. Two plants were single-trunk trees with a camel height browse line, while the other was a small sprout from rootstock. Growth form in Al Maha was a woody shrub of up to 4.5 m height, often containing multiple trunks. A strong relationship between height and canopy indicated that plant growth was well proportioned.

Lycium schawii occurs across both Al Maha and the DDCR, but its growth form is very different between enclosures. Camel grazing removes the apical dominance of leading stems, which then produce thick, woody side branches. This produces a thickly hedged shrub in which leaves and flowers are protected by a thorny, woody exterior. Plants in Al Maha showed almost no grazing from oryx and gazelles, hence the leading stems grew long and spindly. Remnants of the former hedging structure could be seen on most plants, dating from before camels were removed from the enclosure. The species is moderately palatable and difficult to propagate from seed (Heywood, 2004).

Ten of the 64 observed *L. schawii* plants were found growing within the canopy of *L. pyrotechnica* shrubs, indicating a link among the two species. Eight of the ten plants were located within two clusters in Al Maha, both in areas where *L. pyrotechnica* was densely populated. We hypothesize that seeds were deposited by birds that eat the fruit of *L. schawii* but nest in *L. pyrotechnica*. Other authors have noted *L. schawii* shrubs growing under the protection of *Acacia tortilis* (Western, 1983, Jongbloed, 1996), but we haven't observed this in the DDCR.

2.2 *Trees*

The tree species of greatest importance to the natural ecology of the UAE are *Prosopis cineraria* and *Acacia tortilis*, both of which commonly occur in clusters within this habitat. There are no clusters within Al Maha, so a comparison of management systems is not possible. Other tree species exist in anthropogenically modified locations. The two *Tamarix aphylla* trees are unlikely to have occurred naturally in this environment (Jongbloed et al., 2003). Several individuals of the invasive exotic *P. juliflora* exist and the species is very common in the towns surrounding the DDCR, raising the risk of invasion. There is a possibility that *P. juliflora* could crossbreed with *P. cineraria* (Pasiecznik *et al.*, 2004), though this has not been observed. *Acacia tortilis* grows mostly on gravel substratum, while *P. cineraria* grows on sand substratum bordering gravel. DDCR farmers reported that both species have decreased in number, but the extent of decline is not known. Gravel plains are preferred locations of farms and towns, which would contribute to the decline of trees.

The DDCR contains 771 *Prosopis cineraria* trees and shrubs in four distinct clusters, and approximately 30 that are associated with current or abandoned farms. The species can reproduce vegetatively from root suckers (Brown, 1988) but it is not known if clustering exists entirely because of asexual reproduction, or because only certain habitats are suitable for growth. Flower and seed production in the DDCR appear low, but seedling production is difficult to quantify. Most trees within a cluster appear to be a similar age, since trunk diameters fit a logarithmically normal distribution. Shrubs of 2–3 m height occur in a section of just one cluster that was previously fenced to exclude camels. Sprouts of up to 50 cm length were observed in three of the clusters, but none were observed to survive over summer. These observations indicate that regeneration of the species is severely limited under current herbivory levels. Death occurs at a size in which plants are vulnerable to all herbivores, though it is likely that camel density is the main cause.

Acacia tortilis is limited to the eastern side of the DDCR. Of the 184 observed trees, 181 were in 11 clusters ranging in size from 2 to 50 plants. Eight clusters contained just one dominant tree surrounded by many smaller trees, each of which could have emerged from the roots of the dominant tree. Farmers reported the existence of two subspecies in the UAE; *salam* (possibly *A. tortilis* ssp. tortilis) and *samer* (possibly *A. tortilis* ssp. raddiana),. *Salam* is smaller with more stems, and more common in the DDCR. Livestock readily eat leaves of *samer,* but not *salam*, which might explain the difference in growth form. Fruit of both subspecies is eaten by livestock. Livestock will graze on *salam* leaves when other sources are depleted (Hobbs, 1989). Rapid growth tends to occur after rainfall periods when herbivores have many other species on which to graze (Springuel et al., 1995).

Large mammalian herbivores are the main means of seed dispersion (Rohner and Ward, 1999). Animals improve seed survival by carrying them away from insect seed predators that surround the parent tree, and germination is facilitated by passage through the gut. Passage through a camel gut takes longer than through an oryx or gazelle, and thus germination may be improved more by camels than other livestock. However, a high camel density might impact seedling survival through grazing and trampling. Large trees exhibit hydraulic lift in wetter seasons, though this may not benefit understory vegetation (Ludwig et al., 2003).

2.3 *Small perennial plants*

Results of the authors' work on this plant category has been reported (Gallacher and Hill, in press). Plots were chosen to represent localized maximum plant density to reduce the confounding effect of spatial variation in plant cover. Within Al Maha, plots contained almost three times the median plant cover, higher species richness and biodiversity, and a 36% higher median number of plants. Regeneration within Al Maha was substantial on all substrates (gravel, stable sand, and semi-stable sand) but was greatest on the gravel substrate.

Camels probably graze gravel substrate areas more than smaller herbivores because they provide a higher ratio of energy gain to energy expenditure (see Murray, 1991). The DDCR gravel substrate appeared to be devoid of plants in this category, but closer inspection revealed that plants of several perennial species existed with just a few leaves and a much larger root structure. This ability to survive under heavy grazing has enabled vegetation to recover quickly on the Al Maha gravel substratum. Al Maha vegetation on this habitat is dominated by *Heliotropium kotschyi*, most likely representing a seral transitional state.

Cyperus conglomeratus was the only species of this category to benefit from heavy camel grazing. Most other species exhibited reduced plant size, with some also exhibiting reduced number or range. *C. conglomeratus* has previously been reported as a disturbance species and an indicator of excessive grazing (Ferguson et al., 1998, Barth, 1999). Nevertheless, its use as a plant indicator of disturbance is limited to extreme cases, since dune microenvironments regularly face disturbance under natural conditions.

2.4 *Germination events*

Germination in the DDCR occurs less that once a year, since it requires both the correct season (Spring) and sufficient soil moisture. The authors studied one event only, after the rains of winter 2004. Spatial variation in germination density was very large, as has been reported in other arid environments (Guo et al., 2000, Brown, 2003, Robinson, 2004). Highest densities in an area always occurred at the base of the wayward wind side of steep dunes. Rain water penetrated deeper into the sand in these locations and remained there for longer. Almost no germination occurred on top of active dunes, though valleys in active dune areas still contained seedlings. Sand plots were dominated by *Eremobium aegyptiacum*, *Arnebia hispidissima*, *Cyperus conglomeratus*, *Indigofera colutea* and *Silene villosa* while gravel plots were dominated by *A. hispidissima*, *Monsonia nivea*, *Dichanthium foveolatum* and *Neurada procumbens*.

Camel grazing had no impact on annual species, but reduced the number of seedlings of perennial plants ($P < 0.001$). This supports the theory that annual species outgrow the rate of grazing. The fall in perennial seedlings is likely to reflect lower seed production in heavily grazed areas and restricted seed transportation. Seedling densities of *Haloxylon salicornicum* and *Indigofera* spp. were noticeably highest near established plants.

The authors observed feces of camels, oryx, gazelles and dhub (*Uromastyx aegyptiaca*) lizards, but found no germination in association with them. *Acacia tortilis* seeds are likely to be transported through animal consumption, but at a very low frequency. Most other species appear not to use this mechanism. Middens were noticeable for their lack of germination, rather than their facilitation of it.

3 DISCUSSION

3.1 *Impact of grazing in the DDCR*

Heavy grazing in the DDCR has reduced the cover of small perennial plant species, reducing their capacity for annual forage production. The extent of overgrazing on gravel substrata is severe, but it is also significant on sand substrata. Large shrubs have been differentially impacted. *Calligonum comosum* has been devastated by camel grazing. *Leptadenia pyrotechnica* and *Lycium schawii* have been substantially reduced in size, though their long term impact is not yet known. Two species have benefited from heavy grazing; the large shrub *Calotropis procera* and the sedge *Cyperus conglomeratus*. The desert squash *Citrullus colocynthis* also appears to flourish in areas of high camel traffic and defecation, though the authors have not yet conducted studies on this species. Among trees, the regeneration of *Prosopis cineraria* appears to have been severely reduced by herbivory at the small shrub stage, but no evidence of effects on *Acacia tortilis* has been recorded. Grazing did not have any direct impact on ephemeral species.

The main cause of observed differences among enclosures (DDCR stocked with camels vs Al Maha stocked with oryx and gazelles) is the amount of forage removed by livestock. Lower grazing intensity within Al Maha has enabled almost all perennial species to show signs of recovery. However, camels impact plant species differently to oryx and gazelles. They decimate plants on gravel substrata, and facilitate the increase of *C. procera* and *C. conglomeratus* on sand substrata through removal of competing species. Furthermore, camels affect the shape of some larger species differently due to their greater height. Trees must grow twice as high before vegetation is above the browse line of camels, and *L. pyrotechnica* shrubs are modified. Substituting some of the camels with other herbivores would enable a higher sustainable stocking rate, while lessening the impact on plant species composition.

Structure of *L. pyrotechnica* showed clear differences between enclosures. The increased canopy in Al Maha provided sites for sheltering gazelles and oryx, and provided a greater choice of nesting sites for birds and small mammals. Whether the smaller animals benefited significantly from this is not known, but use of the larger Al Maha shrubs by oryx and gazelles was clear. Camels reduced height of shrubs more than canopy cover, since they can graze from above. Size reduction of *L. schawii* limited the availability of fruit for consumption by wildlife.

3.2 *Models of recovery*

Ecological degradation is frequently blamed on overgrazing, but its effect on irreversible decline of plant communities is often overrated (Dean and Macdonald, 1994). Existing shrub and perennial grass species in Al Maha recovered substantially during five years of drought. Similar observations have been reported in camel exclosures throughout the region (Khan, 1980a, 1981, Oatham et al., 1995, Zaman, 1997, Barth, 1999). This indicates that species are adapted for survival under heavy grazing, as would have occurred during the frequent multiple-year droughts. It is possible that less adapted plant species have become extinct in the area, but the authors have found no supporting evidence. Irreversible decline of rangelands is mainly caused by a change to soil structure or infiltrability (Wilson and Tupper, 1982), to which sand substrata are relatively resistant (Scoones, 1992). Significant ecological decline has occurred, but most is reversible by changing land management. The only proven irreversible degradation is the extinction of the Arabian Ostrich (*Struthio camelus syriacus*), though genetic erosion of other species is likely.

The current species mix in Al Maha probably represents a seral transitional state. Species life span affects the time taken for a natural vegetation state to emerge. In a habitat dominated by perennial shrubs this may be 20–50 years (Allen-Diaz and Bartolome, 1998, Todd and Hoffman, 1999). Even so, the vegetation state to emerge may be unnatural if vegetation dynamics follow a non-equilibrium model. The authors have previously argued that small perennial plants follow an equilibrium model (Gallacher and Hill, in press), and that stable natural vegetation states are likely to emerge with appropriate livestock management.

The authors have not found evidence of population shifts among plant groups (annual/perennial, herbaceous/woody, shrub/tree), but some can be expected. Expansion of large shrubs in Al Maha is likely to partially displace smaller perennial plants through competition for water, and trees could return. Perennial grasses are expected to increase since they are intolerant to heavy grazing (Jeffries and Klopatek, 1987, Beeskow et al., 1995, Todd and Hoffman, 1999), but this has not yet been observed. *Pennisetum divisum* increased in canopy cover, but not plant number, and its recovery was less than the recovery of many dwarf shrub species. Annual species are unlikely to be affected, since they are not impacted by grazing in this habitat.

3.3 *Solutions*

The DDCR is chronically stressed from overgrazing by camels, but has the capacity to recover. Most plant species exhibit reduced size and/or number. Reduced foliage has a direct impact on the populations of small (non-livestock) herbivores, which then affects the populations of carnivores. Direct evidence of wild animal population decline is currently limited to Dhub lizards (*Uromastyx aegyptiaca*), whose population increased rapidly within Al Maha. However, Al Maha unarguably has a higher capacity to support wildlife.

Rangeland resources are also being mismanaged from a pastoralist viewpoint. Reduced plant size of most species is so severe that the carrying capacity of the desert is clearly reduced. Vegetation cover on the gravel substratum increased by 100 fold within Al Maha, enabling a much greater annual herbage production when camels were excluded. Changes on sand substrata were less dramatic, but still significant. Farmers could lower their feed costs by reducing the exposure of desert plants to camel grazing, since larger plants would have a higher annual herbage production.

The impact of camel grazing could be addressed in several ways, each with its own policy and ecological considerations:

- Reduction of the national herd. This would have no effect in the short term, since farmers would compensate by reducing supplementary feed. There would probably be no difference to desert ecology until camel numbers were reduced back to around 100 000, or 40% of current numbers. It could be achieved in the longer term by reviewing agricultural subsidies and market guarantees, though not all farmers will be sensitive to economic constraints.
- Allocation of rangeland to farmers or farmer groups. This could be politically difficult, but would enable people to take greater responsibility for the land they use. This approach would need to be coupled with research and farmer education programs, and would move camel husbandry further away from traditional methods. Hence it would not be appropriate for preserving culture, but should enable a return of natural ecology in the medium term.
- Reduction of camel exposure to open rangeland. Assuming a stable population of camels, rangeland grazing could be reduced by limiting the number of camels allowed to exit a farm each day. Given this option, farmers are likely to rotate the animals they release, since native plants are widely believed to improve the health of livestock. This system could be difficult to enforce, but would enable the most rapid recovery of desert ecology once implemented. The diet of fewer camels with a lot of time on rangeland will differ from the diet of more camels with less time on rangeland, thus affecting plant species composition. The former would probably enable greater recovery of *Calligonum comosum*, for example. However, reduction of total camel exposure to desert rangeland is currently more important than the method of reduction.
- Protection of most threatened habitats. In the DDCR, gravel substrata have been decimated by camel grazing. Protection of these habitats would be expensive but possible, but doing so would put greater pressure on the remaining unprotected areas. This is a temporary option that could be considered in conjunction with other options.

ACKNOWLEDGEMENTS

The authors are indebted to Greg Simkins and Husam el Alqamy, Dubai Desert Conservation Reserve, for their advice, cooperation and support. This work was made possible through the financial and administrative support of Zayed University, for which the authors are grateful.

REFERENCES

Abbas, B., El Tayeb, A.E., Sulleiman, Y.R., 1992. Calotropis procera: feed potential for arid zones. Veterinary Record 131, 132.

Abd el-Ghani, M.M., 2000. Floristics and environmental relations in two extreme desert zones of western Egypt. J. Global Ecology and Biography 9, 499–516.

Al Wadie, H., 2002. Floristic Composition and Vegetation of Wadi Talha, Aseer Mountains, South West Saudi Arabia. OnLine Journal of Biological Sciences 2, 285–288.

Allen-Diaz, B.H., Bartolome, J.W., 1998. Understanding Sagebrush-Grass vegetation dynamics. Ecological Applications 8, 795–804.

Anonymous, 2005. Camel Racing News: Information and Resource Guide to Camel Racing. www.zipzak.com

Archer, S., 1996. Assessing and interpreting grass-woody plant dynamics. in: Hodgson, J., Illius, A.W. (Eds). The ecology and management of grazing systems. CAB International, Wallingford, Oxon, UK. pp. 101–134.

Asher, M., 1996. Phoenix Rising: The United Arab Emirates Past, Present & Future. The Harvill Press, London, UK.
Aspinall, S., 2001. Environmental Development and Protection in the UAE. in: Al-Abed, I., Hellyer, P. (Eds). United Arab Emirates: A New Perspective. Trident Press, Bookcraft, UK. pp. 277–304.
Aziz, I., Khan, M.A., 2003. Proline and water status of some desert shrubs before and after rain. Pakistan Journal of Botany 35, 911–915.
Barth, H.-J., 1999. Desertification in the Eastern Province of Saudi Arabia. Journal of Arid Environments 43, 399–410.
Batanoun, K.H., Wahab, A.M.A., 1973. Eco-physiological studies on desert plants 8. root penetration of Leptadenia-Pyrotechnica (Forsk) Decne in relation to its water balance. Oecologia 11, 151–161.
Beeskow, A., Elissalde, N.O., Rostagno, C.M., 1995. Ecosystem changes associated with grazing intensity on the Punta Ninfas rangelands of Patagonia, Argentina. Journal of Range Management (U.S.A) 48, 517–522.
Brown, G., 2003. Species richness, diversity and biomass production of desert annuals in an ungrazed Rhanterium epapposum community over three growth seasons in Kuwait. Plant Ecology 165, 53–68.
Brown, G., Porembski, S., 1998. Flora and vegetational aspects of miniature dunes in a sand-depleted Haloxylon salicornicum community in the Kuwait desert. Flora 193, 133–140.
Brown, J.N.B., 1978. Natural Vegetation and Reafforestation in Abu Dhabi. Emirates Natural History Group Bulletin 4, 31–32.
Brown, K., 1988. Ecophysiology of Prosopis cineraria in the Wahiba Sands, with reference to its reafforestation potential in Oman. Journal of Oman Studies Special Report No. 3, 257–270.
Bulliet, R.W., 1990. The Camel and the Wheel. Columbia University Press, New York, USA.
Dean, W.R.J., Macdonald, I.A.W., 1994. Historical changes in stocking rates of domestic livestock as a measure of semi-arid and arid rangeland degradation in the Cape Province, South Africa. Journal of Arid Environments 26, 281–298.
El Badwi, S.M.A., Adam, S.E.I., Shigidi, M.T., Hapke, H.J., 1998. Studies on laticiferous plants: Toxic effects in goats of Calotropis procera latex given by different routes of administration. Deutsche Tierartzliche Wochenschrift 105, 425–427.
FAOSTAT, 2004. FAOSTAT - Agriculture. www.fao.org
Ferguson, M., McCann, I., Manners, G., 1998. Less Water, More Grazing. ICARDA Caravan 8, 9–11.
Fernandez-Gimenez, M.E., Allen-Diaz, B.H., 1999. Testing a non-equilibrium model of rangeland vegetation dynamics in Mongolia. Journal of Applied Ecology 36, 871–885.
Gallacher, D.J., Hill, J.P., in press. Effects of camel grazing on the ecology of small perennial plants in the Dubai (UAE) inland desert. Journal of Arid Environments ##, ##-##.
Ghazanfar, S.A., 2004. Biology of the Central Desert of Oman. Turkish Journal of Botany 28, 65–71.
Gross, C., Jongbloed, M., 1996. Traditions and Wildlife. in: Vine, P.J., Al-Abed, I. (Eds). Natural Emirates: Wildlife and environment of the United Arab Emirates. Trident Press, London, UK.
Guo, Q., Brown, J.H., Valone, T.J., 2000. Abundance and distribution of desert annuals : are spatial and temporal patterns related? Journal of Ecology 88, 551–560.
Heard-Bey, F., 2001. The Tribal Society of the UAE and its Traditional Economy. in: Al-Abed, I., Hellyer, P. (Eds). United Arab Emirates: A New Perspective. Trident Press, Bookcraft, UK. pp. 98–116.
Hellyer, P., 1996. The Natural History Movement. in: Vine, P.J., Al-Abed, I. (Eds). Natural Emirates: Wildlife and environment of the United Arab Emirates. Trident Press, London, UK.
Heywood, V., 2004. Egypt: Conservation and Sustainable Use of Medicinal Plants in Arid and Semi-Arid Ecosystems. Project brief. UNDP Global Environment Facility fund.
Hobbs, J.J., 1989. Bedouin Life in the Egyptian Wilderness. University of Texas Press, Austin, Texas, USA.
Jeffries, D.L., Klopatek, J.M., 1987. Effects of Grazing on the Vegetation of the Blackbrush Association. Journal of Range Management (U.S.A) 40, 390–392.
Jongbloed, M., 1996. Plant Life. in: Vine, P.J., Al-Abed, I. (Eds). Natural Emirates: Wildlife and environment of the United Arab Emirates. Trident Press, London, UK.
Jongbloed, M., Feulner, G.R., Boer, B., Western, A.R., 2003. The comprehensive guide to the wild flowers of the United Arab Emirates. Environmental Research and Wildlife Development Agency, Abu Dhabi, UAE.
Khan, M.I.R., 1979. Taming the Abu Dhabi Desert. Emirates Natural History Group Bulletin 8, 19–22.
Khan, M.I.R., 1980a. Al Bujair Nursery in the Western Region. Emirates Natural History Group Bulletin 10, 16–18.
Khan, M.I.R., 1980b. Natural Vegetation of the UAE. Emirates Natural History Group Bulletin 11, 13–20.
Khan, M.I.R., 1981. Al Babha Plantation. Emirates Natural History Group Bulletin 13, 17–19.
Ludwig, F., Dawson, T.E., Kroon, H.d., Berendse, F., Prins, H.H.T., 2003. Hydraulic lift in Acacia tortilis trees on an East African savanna. Oecologia 134, 293–300.

Murray, M.G., 1991. Maximizing energy retention in grazing ruminants. Journal of Animal Ecology 60, 1029–1045.
Nehra, O.P., Oswal, M.C., Faroda, A.S., 1987. Management of fodder trees in Haryana. Indian Farming 37, 31–33.
Oatham, M.P., Nicholls, M.K., Swingland, I.R., 1995. Manipulation of vegetation communities on the Abu Dhabi rangelands. I. The effects of irrigation and release from longterm grazing. Biodiversity and Conservation 4, 696–709.
Oba, G., Vetaas, O.R., Stenseth, N.C., 2001. Relationships between biomass and plant species richness in arid-zone grazing lands. Journal of Applied Ecology 38, 836.
Ould Soulé, A., 1998. Mauritania country profile. FAO Crop and Grassland Service, Rome, Italy.
Pasiecznik, N.M., Harris, P.J.C., Smith, S.J., 2004. Identifying Tropical Prosopis Species: A Field Guide. HDRA Publishing, Coventry, UK.
Peters, J., 1997. The dromedary: ancestry, history of domestication and medical treatment in early historic times. Tierarztl Prax Ausg G Grosstiere Nutztiere 25, 559–565.
Potts, D.T., 2001. Before the Emirates: an Archaeological and Historical Account of Developments in the Region c. 5000 BC to 676 AD. in: Al-Abed, I., Hellyer, P. (Eds). United Arab Emirates: A New Perspective. Trident Press, Bookcraft, UK. pp. 28–69.
Robinson, M.D., 2004. Growth and abundance of desert annuals in an arid woodland in Oman. Plant Ecology 174, 137–145.
Rohner, C., Ward, D., 1999. Large mammalian herbivores and the conservation of arid Acacia stands in the Middle East. Conservation Biology 13, 1162–1171.
Scoones, I., 1992. Land degradation and livestock production in Zimbabwe's communal areas. Land Degradation and Rehabilitation 3, 99–113.
Sharma, B.M., 1968. Root systems of some desert plants in Churu, Rajasthan. Indian Forester 94, 240–246.
Springuel, I., Shaheen, A.S., Murphy, K.J., 1995. Effects of Grazing, Water Supply, and Other Environmental Factors on Natural Regeneration of Acacia raddiana. in: An Egyptian Desert Wadi System Ed. Neil E. West. In Rangelands in a Sustainable Biosphere. Proceedings of the Fifth International Rangeland Congress, Vol.1:529–530. Publ. by Society for Range Management, Colorado, U.S.A.
Stephan, E., 1995. Preliminary report on the faunal remains of the first two seasons of Tell Abraq/Umm al Quwain/United Arab Emirates. in: Buitenhuis, H., Uerpmann, H.-P. (Eds). Archaeozoology of hte Near East II. Backhuys Publishers, Leiden, The Netherlands.
Todd, S.W., Hoffman, M.T., 1999. A fence-line contrast reveals effects of heavy grazing on plant diversity and community composition in Namaqualand, South Africa. Plant Ecology 142, 169–178.
Western, R.A., 1983. Vegetation of the Arabian Gulf Coast of the UAE. Emirates Natural History Group Bulletin 21, 2–11.
Western, R.A., 1988. Adaptations of Plants to a Desert Environment. Emirates Natural History Group Bulletin 36, 17–23.
Wilson, A.D., Tupper, G.J., 1982. Concepts and factors applicable to the measurements of range condition. Journal of Range Management (U.S.A) 35, 684–689.
World Meteorological Organization, 2005. World Weather Information Service - Dubai. www.world-weather.org
Yagoub, M.M., Hobbs, J.J., 2003. Geographic information system applications for camels: the case of Al-Ain, UAE. Arab World Geographer 6, 101–111.
Zaman, S., 1997. Effects of rainfall and grazing on vegetation yield and cover of two arid rangelands in Kuwait. Environmental Conservation 24, 344–350.

Reclaiming the Desert: Towards a Sustainable Environment in Arid Lands – Mohamed (ed.)
© 2006 Taylor & Francis Group, London, ISBN 0 415 41128 9

Effect of grazing on vegetation and soil characters of the rangelands in the United Arab Emirates

K.H. Shaltout
Botany Department, Faculty of Science, Tanta University, Tanta, Egypt

A.A. El-Keblawy & M.T. Mousa
Biology Department, Faculty of Science, UAE University, Al-Ain, UAE

ABSTRACT: The present study was dealing with the effect of grazing on vegetation composition and phytomass in the rangelands of three sites Hili (Abu Dhabi), Biat (Ajman) and Madam (Sharjah) in the United Arab Emirates. Remarkable increases were recorded in total density and cover of species as a result of protection, while some species exhibited negative responses. Species richness was higher inside, while relative concentration of dominance was higher outside. Soil salinity and important soil nutrients e.g. K, Mg and Na are significantly higher in the free grazing area. In general the overall variation in phytomass due to grazing for most species was significant. The total annual consumption ranged between 272 and 748 kg/ha, while the annual production ranged between 202 and 277 kg/ha. The relationship between annual production and consumption by domestic animals showed that annual consumption is higher than production in most locations. Carrying capacity was between 0.18 and 0.25 head/ha.

1 INTRODUCTION

Rangelands are a vital component of global food and wood production. If properly managed, rangelands and the livestock they support can provide useful products. However, the history of overgrazing has resulted in permanent loss of vegetation, erosion, desertification, wildlife extinction and drop in water table (Dean et al. 1995, Kassas 1995, Dun widdie 1997, Holechek et al. 2001, Janssen & Anderies 2004). Neither over-grazing nor complete protection of the rangelands is recommended (Shaltout & El-Ghareeb 1985, El-Keblawy 2004). In fact, proper range management could benefit the range plants in many ways. For example, livestock can help seeds spread and fertilizes soil. Also the browsing of some parts of the plants reduce the intensity of competition between the remaining parts for the limited amounts of moisture and nutrients, especially under the condition of extreme arid environments as those found in UAE (El-Keblawy 2003). Under proper grazing management, rangelands could remain productive despite aridity (Shaltout et al. 1996, Hill & Braaten 2003). Proper grazing management requires careful control of the number of animals on a piece of land and this consequently, requires the determination of the carrying capacity. Determination of the carrying capacity requires information about productivity and nutritive value of the palatable species all over the year (Le Houérou 1970, 1972a, 1972b, 1975, 1977, 1980, El-Kady 1980, 1987, Heneidy 1986, 1996, 2002, Barrs 2002).

2 STUDY AREA

The study area lies in the western part of Abu Dhabi Emirate between 23° and 25° latitude and 52° and 56° longitudes. Semi-mobile dunes are the dominant visual feature, with a relatively high water table resulting in evaporative crusts in many depressions. The western region of Abu Dhabi contains the bulk of high, unstable aeolian dunes of mixed fine-grained sand in the country. There is little or no horizon differentiation and plants have a precarious existence on the steeper slopes. Low rainfall and high temperatures characterize the climate of the UAE. Absolute maximum temperatures rise to 49°C. In January temperatures can be as low as 5°C, though this is rare on the coast because of the moderating influence of the sea. The mean maximum for Abu Dhabi in July is 40.1°C, and the mean minimum in August is 32.2°C. Most precipitation occurs between December and April, the mean annual precipitation is around 75 mm (Ministry of Agriculture of UAE 1996).

3 MATERIALS AND METHODS

3.1 *Vegetation measurements*

3.1.1 *Sampling methods*

181 stands each of 20 × 20 m were selected during spring 2001 at the following locations Hili (Abu-Dhabi), Madam (Sharjah), Biat (Ajman) (Fig. 1) to represent grazing conditions, protection period in Hili, Biat and Madam were 15, 3 and 5 years, respectively. Nomenclature was according to Mandaville (1990) and Migahid (1990), cover of species by line-intercept method (Canfield 1941), density by quadrate method and the size index of species was calculated as the mean of height and diameter.

3.1.2 *Standing crop phytomass*

The standing crop phytomass of the above-ground parts of the common species in the three locations was estimated during spring 2001 using 120 randomly distributed 4 m^2 quadrates (40 quadrates per location). In each quadrate, all the above-ground organs were harvested except the shrubby species *Haloxylon salicornicum*, whose phytomass (TAGB) were estimated according to Mousa (2005) see equation (1), *Acacia tortilis ssp. tortilis* and *Prosopis cineraria* were estimated

$$\text{TAGB} = 1.2032 + 0.7671 \log \text{Diameter} \quad (1)$$

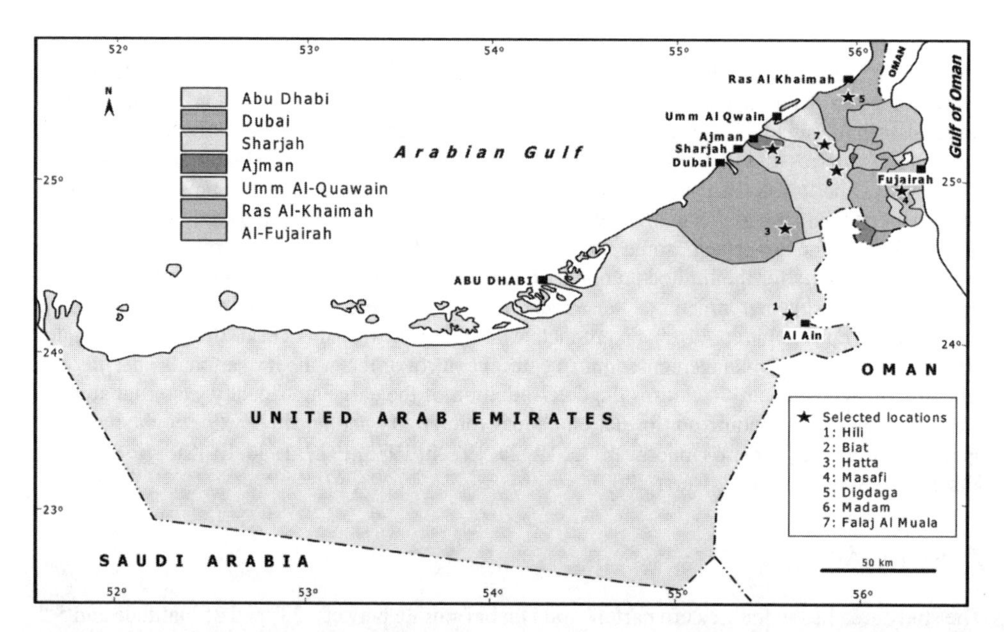

Figure 1. Location map of the United Arab Emirates, simplified from National Atlas of UAE (Embabi et al. 1993).

according to El-Juhany et al. (2002), see equation (2). The annual production of the vegetation in the present study was calculated using a turnover rate of 0.37 for some desert ecosystems (see Shaltout and El-Ghareeb 1985).

$$TAGB = 25.59 + 0.418647 \text{ Diameter} \tag{2}$$

3.1.3 *Animal consumption*

Animal consumption was estimated by bite method. This technique was applied by El-Kady (1980 & 1987) and Heneidy (1986). The stocking rate of animals in the area (3.2, 2.2 and 1.1 head/ha in Hili, Biat and Madam, respectively). Carrying capacity was calculated as the ratio between the estimated annual production and the annual consumption per head.

3.2 *Soil analysis*

Three soil samples were collected from each stand. Soil texture and organic matter were determined by hydrometer and loss-on-ignition. Determination of calcium carbonate was carried out using calcimeter. Soil-water extracts (1–5) were prepared for the determination of soil salinity (EC) and soil reaction (pH) using direct indicating conductivity and pH-meter. Flame photometer was used for the determination of K, Ca and Na. P and N spectrophotometrically and Mg using atomic absorption (Allen et al. 1974).

3.3 *Data analysis*

Species richness for each habitat and vegetation group was calculated as the average number of species per stand. Species turnover was calculated as the ratio between the total number of species recorded in a certain vegetation cluster and its alpha-diversity (Whittaker 1972, Wilson & Shmida 1984). Relative concentration of dominance was measured using Simpson index. The significance of variation between the means of absolute cover of species and soil variables of protected and grazed area was tested by one way ANOVA. These techniques were according to SPSS release 11.1.5 (Nie et al. 2001).

4 RESULTS

4.1 *Effect of grazing on diversity indices*

When compared with the free grazing sites, the vegetation inside the protected sites in Hili has high total species (48), species richness (4.7 species/stand), relative evenness (3.1) and total density (187 ind./100 m^2), but low species turnover (9.2) (Table 1). In Biat most vegetation variables are higher inside than outside except species turnover which is higher outside (8) than inside (7.3). Vegetation variables of the protected sites in Madam are also higher inside than outside (e.g. relative evenness, relative concentration of dominance and total density).

Table 1. Values of some vegetation variables inside and outside the protected sites in three locations.

	Hili			Biat			Madam		
Vegetation variable	Pr	Gr	Pvalue	Pr	Gr	Pvalue	Pr	Gr	Pvalue
Total species*	48	28	<0.001	74	28	<0.05	51	24	<0.001
Species richness**	4.7	2.6	<0.05	8.9	4.1	<0.01	4.1	3.5	<0.05
Species turnover**	9.2	9.7	<0.05	7.3	8.0	<0.05	5.8	11.4	<0.01
Relative evenness**	3.1	2.5	<0.05	3.5	2.6	<0.01	3.5	2.6	<0.05
Relative concentration of dominance**	7.4	17.5	<0.01	12.0	30.0	<0.01	11.2	25.9	<0.001
Total cover (m/100 m)**	42.4	31.1	<0.01	69.8	14.9	<0.001	60.7	21.4	<0.001
Total density (ind./100 m^2)**	187.0	97.5	<0.001	552.9	97.7	<0.001	355.5	165.8	<0.001

Pr: protected, Gr: grazed, *tested by χ^2, **tested by one-way ANOVA

4.2 *Effect of grazing on cover and density*

The total number of species recorded in Hili is 54 species. 53 in the protected sites, one in the grazed and 22 in both. The total absolute density is much higher in the protected (187 ind./100 m^2) than in the free grazing sites (97.5 ind./100 m^2) (Table 2). Some species have significant higher density in the protected site (e.g. *Stipagrostis plumosa* and *Prosopis juliflora*); while some others have the reverse (e.g. *Fagonia indica* and *Zygophyllum mandavillei*). The total absolute cover is higher in the protected (42.4 m/100 m) than in the free grazed sites (31.1 m/100 m). Some species have significant higher cover in the protected site (e.g. *Indigofera articulata* and *Arnebia hispidissima*); while some others have the reverse (e.g. *Cyperus conglomeratus* and *Tribulus terrestris*).

The total number of species recorded in Biat is 77 species, 40 species of them are in the protected site, 3 species in the grazed and 33 in both. The total absolute density is much higher in the protected site (552.9 ind./100 m^2) than the free grazing sites (97.7 ind./100 m^2) (Table 3). Some species have significant higher density inside (e.g. *Eragrostis turgida* and *Stipagrostis plumosa*); while some others have the reverse (e.g. *Chenopodium murale* and *Euphorbia granulata*). The total absolute cover is higher in the protected (69.8 m/100 m) than in the free grazing site (14.9 m/100 m). Some species have significant higher cover inside (e.g. *Acacia tortilis ssp. tortilis* and *Stipagrostis plumosa*); while some others have the reverse (e.g. *Paronychia arabica* and *Medicago polymorpha)*.

The total number of recorded species in Madam is 55 species, 31 of them is in the protected sites, and 4 in the grazed and 20 in both. The total absolute density is much higher in the protected (355.4 ind./100 m^2) than in the free grazing sites (165.8 ind./100 m^2) (Table 4). Some species have significant higher density inside (e.g. *Stipagrostis plumosa*, *Acacia tortilis ssp. tortilis* and

Table 2. Effect of grazing on mean absolute density (ind./100 m^2) and cover (m/100m) in Hili.

	Absolute density				Absolute cover			
Species	Pr	Gr	P-value	RID	Pr	Gr	P-value	RID
Rhazya stritica	3.3	3.6	<0.05	−0.08	0.6	1.6	<0.05	−0.63
Stipagrostis plumosa	2.4	1.5	<0.05	0.60	1.2	0.8	<0.05	0.50
Tamarix arabica	2.0	6.4	<0.01	−0.69	1.5	2.6	<0.05	−0.42
Neurada procumbens	3.2	4.3	<0.05	−0.26	1.2	1.0	>0.05	0.20
Dipterygium glaucum	5.0	3.3	<0.05	0.52	1.1	0.8	<0.05	0.38
Cassia italica	3.0	4.0	<0.05	−0.25	0.6	1.2	<0.05	−0.50
Calotropis procera	2.0	1.0	<0.05	1.00	1.4	0.2	<0.01	6.00
Indigofera articulata	1.5	1.0	<0.05	0.50	0.9	0.1	<0.01	8.00
Zygophyllum mandavillei	3.2	6.4	<0.01	−0.50	0.8	1.5	<0.05	−0.47
Fagonia indica	3.5	7.0	<0.01	−0.50	1.6	2.9	<0.05	−0.45
Prosopis cineraria	1.3	13.5	<0.01	−0.90	1.3	2.8	<0.05	−0.54
Tribulus terrestris	4.4	2.5	<0.05	0.76	1.2	2.2	<0.01	−0.45
Cyperus conglomeratus	4.0	3.0	<0.05	0.33	0.0	1.1	<0.05	−1.00
Acacia tortilis spp. tortilis	3.9	6.8	<0.01	−0.43	1.7	2.9	<0.05	−0.41
Euphorbia granulata	2.3	4.0	<0.05	−0.43	0.1	0.2	<0.05	−0.50
Arnebia hispidissima	2.6	1.3	<0.05	1.00	0.9	0.2	<0.01	3.50
Haloxylon salicornicum	2.8	3.9	<0.05	−0.28	1.8	2.1	>0.05	−0.14
Savignya aegyptiaca	1.4	3.0	<0.05	−0.53	0.4	0.5	<0.05	−0.20
Centropodia forsskaolii	2.6	6.0	<0.01	−0.57	0.5	2.9	<0.01	−0.83
Orobanche cernua	4.0	6.0	<0.05	−0.33	0.1	0.1	>0.05	0.00
Pennisetum divisum	2.0	1.0	<0.05	1.00	0.2	0.2	>0.05	0.00
Prosopis juliflora	7.0	2.0	<0.05	2.50	3.6	3.2	>0.05	0.13
Others	119.6	6.0			19.7	1.1		
Total	187.0	97.5			42.4	31.1		

Pr: protected, Gr: grazed, *tested by χ^2, **tested by one-way ANOVA and RID: relative change

Centropodia forsskaolii); while some others have the reverse (e.g. *Tribulus terrestris* and *Cyperus conglomeratus).* The total absolute cover is much higher in the protected (60.7 m/100 m) than in the free grazed sites (21.4 m/100 m). Some species have significant higher cover inside (e.g. *Stipagrostis plumosa* and *Crotalaria aegyptiaca*); while some others have the reverse (e.g. *Cyperus conglomeratus* and *Heliotropium digynum*).

4.3 *Effect of grazing on size index*

Most common palatable species (Figure 2) have higher sizes in the protected sites compared the grazed sites In Hili *Haloxylon salicornicum* among all has the highest size index in the protected site. The size of all species in the grazed sites is smaller than in the protected sites, except that of *Cyperus conglomeratus*, *Arnebia hispidissima* and *Calotropis procera.*

Table 3. Effect of grazing on mean absolute density (ind./100m^2) and cover (m/100m) in Biat.

	Absolute density				Absolute cover			
Species	Pr	Gr	P-value	RID	Pr	Gr	P-value	RID
Acacia tortilis ssp. tortilis	9.9	6.0	<0.01	0.65	3.0	1.2	<0.01	1.50
Emex spinosa	4.6	3.0	<0.01	0.53	0.2	0.1	<0.05	1.00
Paronychia arabica	12.7	1.0	<0.01	11.70	0.3	0.4	>0.05	−0.25
Citrullus colocynthesis	2.5	2.2	>0.05	0.14	0.6	0.2	<0.05	2.00
Tribulus terrestris	5.2	4.5	>0.05	0.16	0.9	0.6	<0.05	0.50
Setaria verticillata	4.6	4.3	>0.05	0.07	1.1	1.1	>0.05	0.00
Eragrostis turgida	23.4	4.3	<0.001	4.44	1.6	1.2	<0.05	0.33
Arnebia hispidissima	3.8	3.4	>0.05	0.12	0.6	0.5	>0.05	0.20
Crotalaria aegyptiaca	5.2	1.0	<0.05	4.20	0.8	0.5	<0.05	0.60
Cyperus conglomeratus	3.4	1.0	<0.01	2.40	0.7	0.1	<0.05	6.00
Neurada procumbens	13.6	2.0	<0.001	5.80	1.1	0.3	<0.01	2.67
Haloxylon salicornicum	30.0	2.3	<0.01	12.04	5.6	1.7	<0.01	2.29
Plantago boissieri	13.0	3.3	<0.01	2.94	0.7	0.2	<0.05	2.50
Solanum incanum	6.6	2.2	<0.01	2.00	0.4	0.1	<0.05	3.00
Medicago polymorpha	7.7	1.9	<0.01	3.05	0.5	0.6	>0.05	−0.17
Sporobolus spicatus	1.2	3.0	<0.01	−0.60	0.1	0.1	>0.05	0.00
Lotus schimperi	11.2	3.5	<0.01	2.20	0.8	0.5	<0.01	0.60
Centropodia forsskaolii	3.9	2.1	<0.05	0.86	0.2	0.1	<0.05	1.00
Astragalus annularis	5.8	1.0	<0.05	4.80	1.9	0.6	<0.01	2.17
Euphorbia granulata	3.1	6.0	<0.05	−0.48	1.7	0.4	<0.01	3.25
Launaea capitata	3.3	1.0	<0.01	2.30	0.3	r	<0.05	5.00
Stipagrostis plumosa	23.7	3.5	<0.001	5.77	1.6	0.2	<0.05	7.00
Erucaria hispanica	19.0	3.1	<0.01	5.13	1.9	0.9	<0.05	1.11
Tephrosia purpurea	10.9	3.5	<0.01	2.11	1.7	0.8	>0.05	1.13
Chenopodium murale	2.3	8.0	<0.01	−0.71	0.7	0.5	<0.05	0.40
Erodium malacoides	5.3	3.7	<0.05	0.43	0.6	0.4	<0.05	0.50
Heliotropium digynum	6.0	3.0	<0.01	1.00	1.8	0.5	<0.001	2.60
Senecio glaucus	1.0	4.0	<0.05	−0.75	0.4	0.1	<0.05	7.00
Cleome rupilca	6.7	1.6	<0.01	3.19	1.2	0.4	<0.05	2.00
Convolvulus pilosellifolius	1.0	1.0	>0.05	0.00	1.2	r	<0.01	23.00
Portulaca oleracea	4.0	1.0	<0.01	3.00	0.2	r	<0.05	1.00
Centaurea pseudosianica	2.0	1.0	<0.05	1.00	0.2	0.3	<0.05	−0.33
Dipterygium glaucum	1.0	4.0	<0.05	−0.75	0.3	0.2	<0.05	0.50
Silene villosa	19.0	1.3	<0.001	13.62	0.4	0.2	<0.05	1.00
Others	276.3				35.0			
Total	552.9	97.7			69.8	14.9		

Pr: protected, Gr: grazed, *tested by χ^2, **tested by one-way ANOVA and RID: relative change

Table 4. Effect of grazing on mean absolute density (ind./100 m^2) and cover (m/100m) in Madam.

	Absolute density				Absolute cover			
Species	Pr	Gr	P-value	RID	Pr	Gr	P-value	RID
Acacia tortilis ssp. tortilis	15.1	4.7	<0.001	2.21	2.8	1.0	<0.05	1.80
Crotalaria aegyptiaca	9.1	6.8	<0.01	0.34	2.6	0.6	<0.05	3.33
Euphorbia granulata	3.6	2.0	<0.05	0.80	2.5	0.2	<0.01	11.50
Indigofera articulata	2.8	2.3	>0.05	0.22	0.5	0.2	<0.05	1.50
Centropodia forsskaolii	16.5	11.7	<0.05	0.41	0.5	0.3	<0.05	0.67
Heliotropium digynum	1.7	2.3	>0.05	−0.26	0.1	0.4	<0.05	−0.75
Aerva javanica	3.3	6.6	<0.05	−0.50	2.7	1.1	<0.05	1.45
Haloxylon salicornicum	12.1	5.2	<0.05	1.33	1.6	1.2	<0.05	0.33
Panicum turgidum	4.7	2.9	<0.05	0.62	1.3	0.6	<0.05	1.17
Prosopis cineraria	9.5	3.1	<0.01	2.06	3.0	0.2	<0.05	14.00
Fagonia indica	3.0	8.4	<0.01	−0.64	0.6	0.1	<0.05	5.00
Tribulus terrestris	2.3	20.0	<0.01	−0.89	0.5	0.2	<0.05	1.50
Stipagrostis plumosa	93.1	6.6	<0.001	13.11	2.6	0.6	<0.05	3.33
Cyperus conglomeratus	8.3	24.2	<0.01	−0.66	1.7	5.0	<0.01	−0.66
Cynodon dactylon	2.0	1.0	<0.05	1.00	0.5	0.3	<0.05	0.67
Fumaria parviflora	6.0	11.0	<0.01	−0.45	0.4	1.1	<0.05	−0.64
Poa annua	1.0	1.0	>0.05	0.00	0.6	0.4	<0.05	0.50
Citrullus colocynthis	1.0	1.7	<0.05	−0.41	0.5	0.7	<0.05	−0.29
Solanum nigrum	4.0	20.0	<0.01	−0.80	0.2	1.2	<0.05	−0.83
Calotropis procera	1.4	7.3	<0.01	−0.81	0.6	1.7	<0.05	−0.65
Others	154.9	17.0			35.6	4.4		
Total	355.4	165.8			60.7	21.4		

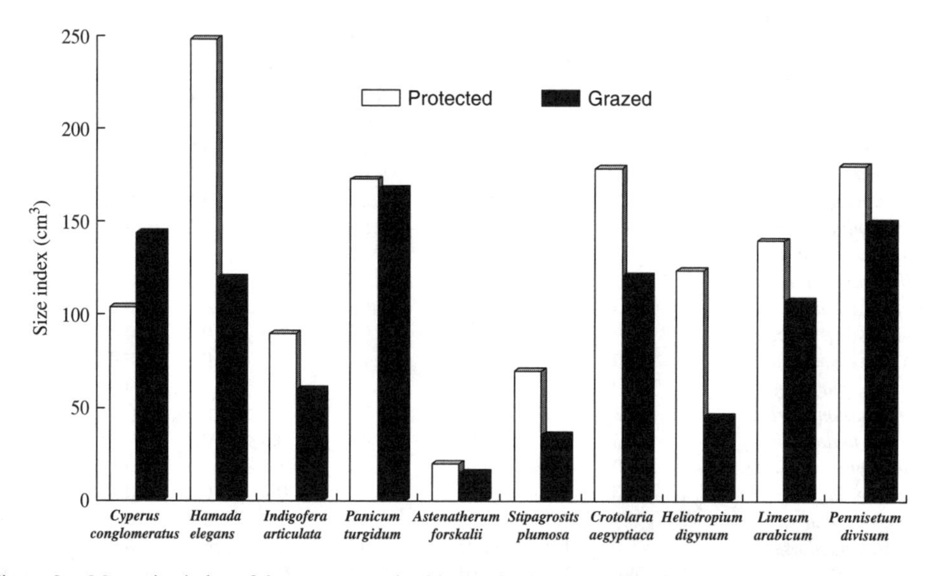

Figure 2. Mean size index of the common palatable species in protected and grazed sites.

4.4 *Effect of grazing on soil characters*

Some variables had significantly higher values outside than inside the protected sites such as phosphorus in Biat, potassium in Hili and Biat (Table 5), calcium, magnesium and sodium in Hili, Biat and Madam. On the other hand, no variables had the reverse.

Table 5. Soil characters (mean ± standard deviation) in both protected and grazed sites of the three studied locations.

	Hili			Biat			Madam		
Species	Pr	Gr	P-value	Pr	Gr	P-value	Pr	Gr	P-value
EC (mS/cm)	1.0 ± 0.2	1.2 ± 0.2	>0.05	1.0 ± 0.3	1.3 ± 0.2	>0.05	1.2 ± 0.7	1.8 ± 0.8	>0.05
pH	7.6 ± 0.3	7.3 ± 2.2	>0.05	8.0 ± 1.1	8.1 ± 1.0	>0.05	7.6 ± 1.2	8.1 ± 1.0	>0.05
Organic matter	2.5 ± 1.8	3.0 ± 1.3	>0.05	2.7 ± 1.2	2.6 ± 1.2	>0.05	3.2 ± 1.3	3.3 ± 2.5	>0.05
Sand %	91.2 ± 4.5	87.7 ± 4.1	>0.05	85 ± 5.7	84.7 ± 6	>0.05	84.1 ± 7.1	83.0 ± 2.0	>0.05
Silt + Clay	8.8 ± 0.2	12.3 ± 0.2	>0.05	14 ± 0.3	15 ± 0.7	>0.05	15.9 ± 0.1	17.0 ± 1.2	>0.05
N	1.2 ± 0.2	1.3 ± 0.1	>0.05	1.7 ± 0.2	1.8 ± 0.7	>0.05	1.2 ± 0.8	1.3 ± 0.1	>0.05
P	0.6 ± 0.1	0.7 ± 0.1	>0.05	0.5 ± 0.1	0.9 ± 0.2	<0.05	0.9 ± 0.2	1.0 ± 0.2	>0.05
K meq/l	4.7 ± 2.1	7.5 ± 3.2	<0.05	5.2 ± 1.7	7.0 ± 2.2	<0.05	5.2 ± 2.5	8.7 ± 1.7	>0.05
Ca	50.7 ± 3.0	107 ± 4.0	<0.05	56 ± 3.8	95 ± 2.2	<0.05	97.0 ± 2.1	125.0 ± 1.7	<0.01
Mg	67.1 ± 12.2	83.7 ± 7.7	<0.05	63 ± 5.2	92 ± 1.2	<0.05	72.5 ± 7.8	105.5 ± 15.2	<0.05
Na	113.2 ± 17.1	181 ± 18.2	<0.01	83 ± 9.7	121 ± 15	<0.05	97.7 ± 22.3	132.1 ± 37.0	<0.05

Pr: protected sites and Gr: grazed sites. P-values are according to one-way ANOVA

Table 6. The above – ground standing crop phytomass (kg dry weight/ha) of the common perennial species in the protected and grazed sites.

	Hili			Biat			Madam		
Species	Pr	Gr	P-value	Pr	Gr	P-value	Pr	Gr	P-value
Astenatherum forskalii	3.5	0.90	<0.05	7.5	1.5	<0.05	3.5	1.2	>0.05
Crotolaria aegyptiaca	75.0	17.5	<0.05	25.4	14.3	<0.05	25.0	18.9	<0.05
Cyperus conglomeratus	25.0	35.0	<0.05	25.0	35.0	<0.05	27.5	35.5	<0.05
Indigophera articulata	33.0	15.8	<0.05	25.0	16.5	<0.05	28.5	12.4	<0.05
Panicum turgidum	25.5	6.5	<0.05	47.5	9.5	<0.05	28.5	2.5	<0.05
Stipagrostis plumosa	75.0	15.8	<0.05	53.0	14.8	<0.05	94.0	24.8	<0.05
Acacia tortilis ssp. tortilis	240.0	215.0	<0.05	185.0	150.0	<0.05	220.0	195.0	<0.05
Prosopis cineraria	320.0	290.0	<0.05	285.0	250.0	<0.05	350.0	325.0	<0.05
Haloxylon salicornicum	220.0	55.5	<0.001	220.0	42.2	<0.001			
Fagonia indica	25.4	35.0	<0.05						
Limeum arabicum	50.0	20.0	<0.01						
Heliotropium digynum	32.0	35.2	<0.05						
Others	12.5	8.5	<0.01	19.5	12.4	<0.01	22.5	12.4	<0.01
Total	1136.9	750.7		892.9	546.2		799.5	627.7	

Pr: protected sites and Gr: grazed sites. P-values are according to one-way ANOVA

4.5 *Effect of grazing on standing-crop phytomass at different location*

The total above ground standing-crop phytomass is higher in the protected (1136.9 kg/ha) than in the free grazing (750.7 kg/ha) sites (Table 6). Some species have significant higher phytomass inside (e.g. *Prosopis cineraria, Acacia tortilis ssp. tortilis* and *Haloxylon salicornicum*); while some others have the reverse (e.g. *Cyperus conglomeratus*, *Fagonia indica* and *Heliotropium digynum*). The total above ground standing-crop phytomass is higher in the protected (892.9 kg/ha) than in the free grazing (546.2 kg/ha) sites. Some species have significant higher phyto-mass inside (e.g. *Prosopis cineraria, Haloxylon salicornicum* and *Acacia tortilis ssp. tortilis*); while some others have the reverse (e.g. *Cyperus conglomeratus*).

The total above ground standing-crop phytomass is much higher in the protected (799.5 kg/ha) than in the free grazing (627.7 kg/ha) sites. Some species have significant higher phytomass inside (e.g. *Prosopis cineraria* and *Acacia tortilis ssp. tortilis*); while some others have the reverse (e.g. *Cyperus conglomeratus*).

4.6 *Animal consumption*

It is notable that *Prosopis cineraria* provides the highest contribution to the daily animal diet in Hili, Biat and Madam (582 and 468.5 and 335.6 g/ha/day, respectively) (Table 7). *Acacia tortilis* has the second highest contribution to the daily diet for grazing animals in Hili and Madam (436 and 211 g/ha/day), while *Haloxylon salicornicum* has the second highest contribution in Biat (361.6 g/ha/day). The total daily consumption is about 2051 gm/ha/day in Hili, 1435 gm/ha/day in Biat and 745.1 gm/ha/day Madam).

The total annual consumption (Table 8) of all perennial species in Hili is about 748.6 kg/ha, of which about 212.4, 159.1 and 145.8 kg/ha from *Prosopis cineraria, Acacia tortilis ssp. Tortilis* and *Haloxylon salicornicum*, respectively. In Biat, the total annual consumption is 524 kg/ha, of which *Prosopis cineraria* contributes 171 kg/ha, while *Haloxylon salicornicum* contributes 132 kg/ha. In Madam, the total annual consumption is 272 kg/ha, of which *Prosopis cineraria* and *Acacia tortilis ssp. tortilis* 122.5 and 77 kg/ha. Annual consumption is much higher than production in Hili (269%) and Biat (260 %), while in Madam it is comparable to the production (117%). In Hili, Biat and Madam the annual production was estimates as 277.7, 202.1 and 232.2 kg/ha, respectively.

Table 7. Daily animal consumption (gm/ha/day) of the common perennial species in the protected and grazed sites in the three studied locations.

Species	Hili	Biat	Madam
Centropodia forsskaolii	4.0	12.3	3.4
Crotalaria aegyptiaca	134.5	41.8	24.0
Cyperus conglomeratus	43.0	41.1	26.4
Indigofera articulata	58.0	41.1	27.3
Panicum turgidum	44.0	78.1	27.3
Stipagrostis plumosa	134.5	87.1	90.1
Acacia tortilis ssp. tortilis	436.0	304.1	211.0
Prosopis cineraria	582.0	468.5	335.6
Haloxylon salicornicum	399.5	361.6	
Fagonia indica	44.0		
Limeum arabicum	89.0		
Heliotropium digynum	56.0		
Total	2051.0	1435.7	745.1

Table 8. Contribution of the different species to the total annual consumption (kg/ha) by domestic animals in the grazed sites in the three studied locations.

Species	Hili	Biat	Madam
Centropodia forsskaolii	1.5	4.5	1.2
Crotalaria aegyptiaca	49.1	15.2	8.8
Cyperus conglomeratus	15.7	15.0	9.6
Indigofera articulata	21.2	15.0	10.0
Panicum turgidum	16.1	28.5	10.0
Stipagrostis plumosa	49.1	31.8	32.9
Acacia tortilis ssp. tortilis	159.1	111.0	77.0
Prosopis cineraria	212.4	171.0	122.5
Haloxylon salicornicum	145.8	132.0	
Fagonia indica	16.1		
Limeum arabicum	32.5		
Heliotropium digynum	20.4		
Total	748.6	524.0	272.0

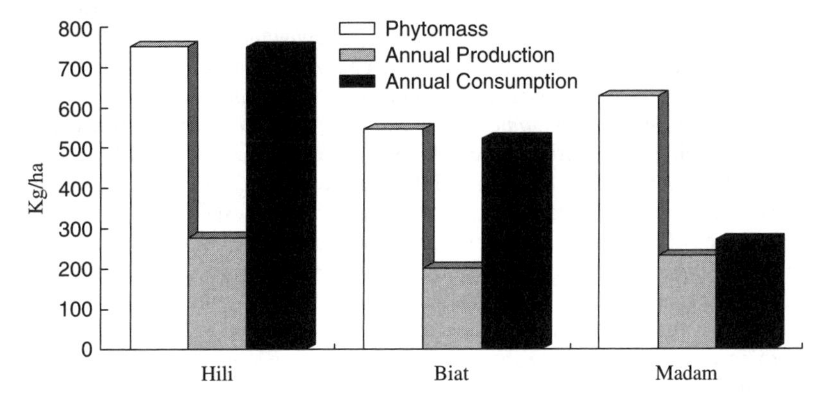

Figure 3. Total above – ground phytomass, calculated estimation of production and consumption by domestic animals (kg/ha) in grazed sites in the three studied locations.

It is possible to compare these estimates with the present consumption and also with the dry matter requirements which is estimated as 3 kg dry matter/animal/day (Le Houerou 1970 and El-Kady 1987). Calculated carrying capacity in Hili, Biat and Madam was estimated as 0.25, 0.18 and 0.21 head/ha, respectively. The relationship between the total above ground standing-crop phytomass, annual production and annual consumption by domestic animals is summarized in Figure 3.

5 DISCUSSION

The greater perennial species richness in the quadrates protected from intense grazing in the present study, supports the suggestion by Thalen (1979) that heavy grazing over an extended pe protected from grazing on desert rangelands increased after only one or two seasons. The difference between percentage covers on grazed and protected quadrates is much greater on sand substrata than on gravel substrata (Oatham et al. 1995a, b). One mechanism for increased production on sand substrata may be through the greater availability of water stored in the dunes to plants (Agnew 1988 & Bower 1982).

Defoliation in the over-grazed range ecosystems relates directly to decrease phytomass of most species, the present study demonstrates that the total phytomass increases due to protection or controlled grazing in UAE desert. The net primary production was estimated as approximately 712 kg/ha in the free grazing area, and 1046 kg/ha in the protected plots (0.37 of standing crop phytomass) as quoted by Shaltout & El-Ghareeb (1985). Some species exhibited a negative response to protection, and productivity of most species was more pronounced in the plots with controlled grazing after an initial period of protection. In fact, grazing both destroys and improves rangeland ecosystems. Defoliation is not the only effect that grazing animals exert on these ecosystems. Light nibbling and removal of standing dead shoots by these animals usually promotes vigor and growth of defoliated plants (Pearson 1965).

Shaltout et al. (1986) reported that continuous over-grazing and continuous protection at west Mediterranean desert of Egypt both have deleterious effects on vegetation. Protection leads to an initial increase in density of vegetation and deprives the ecosystem from the deposition of dung and urine of grazing animals. The stress created by the proximity of neighbors may be absorbed in an increased mortality risk for the whole plant or their parts, reduced reproductive output, growth rate and delayed maturity and reproduction.

Among the Thalen (1979) found that *Hammada elegans* (Bunge) Batch (=*Haloxylon salicornicum* (Moq.) Bunge ex Boiss.) one of the species recorded in the present study to be an important source of fodder for livestock all through the year on the Iraqi desert rangelands. The perennial grass *Stipagrostis plumosa* was found to be an important forage plant for livestock and gazelles in the Wahiba sands of Oman, especially after rain (Munton 1988). *Zygophyllum hamiense* was found to be only suitable for camels but in only small amounts as it caused ‘scour’ (Munton 1988). Munton (1988) suggested that the local dominance of *Zygophyllum hamiense* in the Wahiba sands may be due to overgrazing and elimination of more palatable species. According to the discriminant analysis, the *Heliotropium kotschyi, Salsola baryosma, Panicum turgidum, Zygophyllum hamiense, Stipagrostis plumosa, Fagonia indica* and *Haloxylon salicornicum* (on sand substrata) were classified as being characteristic of quadrates inside enclosure. It could be interpreted that these species were preferentially grazed in Wahiba sands of Oman. Munton (1988) suggested that *Heliotropium kotschyi* was important to livestock (such as goats) especially in dry periods, however that dominance of *Heliotropium kotschyi* in some areas may be the reuse selective grazing of other more palatable species. In the present study, the perennial grass *Panicum turgidum* found to be an important fodder species in the sand formations habitat for camels and goats especially after rain.

Thalen (1979) in his studies in Iraq, reported that *Hammada elegans* (=*Haloxylon salicornicum*), when protected from grazing showed an initial increase in percentage cover, then stopped growing and did not produce seed. He concluded that *Hamada elegans* need certain amount of grazing to remain vigorous. He went on to suggest that heavy grazing over a prolonged period would result in large areas of monospecific stands of species.

The increase of soil salinity outside in relation to inside the protected sites, which was observed in the present study, was reported also by Thalen (1979) in neighbouring Iraq. The vegetation in the free grazing area is sparse and the soil is usually exposed to wind and the direct heating effect of the sun. In the present study, the high nutrient content particularly of N in the free grazing area, may be related to the effect of grazing animals (Heady 1975). Consequently grazing pastures are richer in N than non-grazed. This finding has been reported by some authors (Ayyad and El-Kady 1982, Abulfatih et al. 1989, Ter Heerdt et al. 1991 & Hajar 1993). Thus the role of grazing animals in the redistribution of nutrients via their excreta in ecosystem productivity is significant. According to Till (1981), the amount of nutrients redistributed through this process is frequently of the same order as those applied in fertilizers (Shaltout et al. 1996).

Grazing animals select daily rations from available plants. Forage selectivity results from a highly complex interaction among three sets of variables operating over time: the plants being eaten, the animals doing the grazing, and the environment of both (Heady 1975). Each animal lives, grows, and reproduces on the food it selects, so information on the selective feeding of animals is an important ingredient to the success of management of range ecosystems. It is an established fact that sheep and goats are selective grazers (Arnold 1964a, b), and that the relationship between populations of these animals and their food resources is especially complex in an environment where herbage production is highly seasonal.

Vegetation is relatively abundant in relation to stocking rate; animals express their preference more freely than in deteriorated ranges. In grazed plots of the present study ten perennial species constituted the major part of sheep and goats diet: *Centropodia forsskaolii*, *Crotalaria aegyptiaca, Cyperus conglomeratus, Fagonia indica, Haloxylon salicornicum, Heliotropium digynum, Indigofera articulata, Limeum arabicum, Panicum turgidum* and *Stipagrostis plumosa*). Similar results are reported by Arnold (1964a) and Van Dyne & Heady (1965).

The amount consumed plant parts by domestic animals, in the present study, formed approximately 269% of the net primary production in Hili, about 260% of the net primary production in Biat. These deteriorated range ecosystems should be restored as soon as possible, the range managers should apply the grazing rotation technique in the two sites of Hili and Biat, after an initial protection of several years. This could be achieved through moving of grazing animals from one pasture to another on a scheduled basis. If the grazing period is short and the pastures few, each pasture is alternately grazed and non-grazed several times during a grazing season (Heady 1975).

6 CONCLUSION

1 Protection against grazing increase species richness and diversity of rangelands when compared with grazing.
2 Rangelands in UAE were subjected to over-grazing and their productivity was much lower than carrying capacity.
3 Stocking rated must be lowered to meet carrying capacities in most of location.
4 Rangelands should be protected from grazing pressure in the future for compensation and rehabilitation.
5 Soil of grazed areas was subjected to erosion and degradation as a result of vegetation loss.
6 Most of important range plants were threatened.

REFERENCES

Abulfatih, H.A., Emara, A.H. & El-Hashish, A. 1989. Influence of grazing on vegetation and soil of Asir highlands in Southwestern Saudi Arabia. *Arab Gulf Journal of Scientific Research* 7: 69–78.

Agnew, C. 1988. Soil hydrology of the Wahiba sands. *Journal of Oman Studies, Special Report* 3: 191–200.

Allen, S., Grimshaw, H.M., Parkinson, J.A. & Quarmby, C. 1974. *Chemical Analysis of Ecological Materials*. Blackwell Scientific Publications. London. 368pp.

Arnold, G.W. 1964a. Some principles in the investigations of selective grazing. *Proc. Australian Society of Animal Production* 5: 258–268.
Arnold, G.W. 1964b. Some principles in the investigations of selective grazing. *Proc. Aust. Soc. Anim. Production*. 5: 268–271.
Ayyad, M.A. & EL-Kady, H.F. 1982. Effect of protection and controlled grazing on the vegetation of Mediterranean desert ecosystems in Northern Egypt. *Vegetatio* 49: 129–139.
Barrs, R.M.T. 2002. Rangeland utilization assessment and modeling for grazing and fire management. *Journal of Environmental Management* 64(4): 377–386.
Bower, J.E. 1982. The plant ecology of inland dunes in Western North America. *Journal of Arid Environments*. 5: 199–220.
Canfield, R. 1941. Application of the line-intercept method in sampling range vegetation. *Journal of Forestry*. 39: 388–393.
Dean, W.C.J., M.T. Hoffman, M.E. Meadows & Milton, S.J. 1995. Desertification in the semi-arid Karoo, South Africa: review and reassessment. *Journal of Arid Environments* 30: 247–264.
Dunwiddie, P.W. 1997. Long-term effects of sheep grazing on sand plain vegetation. *Natural Area Journal* 17: 261–264.
El-Juhany, L.I., Aref, I.M. & El-Wakeel, A.O. 2002. Evaluation of above-ground phytomass and stem volume of three Casurina species grown in the central region of Saudi Arabia. *Emirates Journal of Agricultural Sciences* 14: 8–13.
El-Kady, H.F. 1980. *Effect of Grazing Pressures and Certain Ecological Parameters on Some Fodder Plants of the Mediterranean Coast of Egypt*. M.Sc. Thesis, Fac. Sci. Tanta Univ., Tanta. pp 97.
El-Kady, H.F. 1987. *A study of Range Ecosystems of the Western Mediterranean Coast of Egypt*. Ph.D. Thesis, Technical Univ., Berlin, pp 131.
El-Keblawy, A.A. 2003. Evaluation of the potentiality of using artificial forests as conservation sites for native flora of the UAE. *The Fourth Annual Research Conference of UAE University*. SCI: 44–48.
El-Keblawy, A.A. 2004. Effect of protection from grazing on species diversity, abundance and productivity in two regions of Abu-Dhabi Emirate, UAE. In: A.S. Alsharhan, W.W. Wood, A. Goudie, K.W. Glennie & E.M. Abdellatif (eds.) *Desertification in the Third Millennium*: 233–242. Rotterdam: Balkema.
Embabi, N.S., Yahia, M.A. & Al-Sharhan, A.S. 1993. *The National Atlas of UAE*. UAE University. Al-Ain.
Hajar, A.S.M. 1993. A comparative ecological study on the vegetation of the protected and grazed parts of Hema Sabihah, in Al-Baha region, South Western Saudi Arabia. *Arab Gulf Journal of Scientific Research* 11: 258–80.
Hill, M.J. & Braaten, R. 2003. A scenario calculator for effects of grazing land management on carbon stocks in Australian rangelands. Environmental Modeling and Software 18(7): 627–644.
Heneidy, S.Z. 1986. *A study of the Nutrient Content and Nutritive Value of Range Plants at Omayed, Egypt*. M.Sc. Thesis, Alex. Univ. 83pp.
Heneidy, S.Z. 1996. Palatability and nutritive values of some common plant species from the Aqaba Gulf area of Sinai, Egypt. *Journal of Arid Environments* 34: 115–123.
Heneidy, S.Z. 2002. Browsing and nutritive value of range species in Matruh area, a coastal Mediterranean region, Egypt. *Ecologia Mediterranea* 28(2): 39–49.
Holechek, J.L., Piper, R.D. & Herbel, C.H. 2001. Range Management, Principles and Practices. Prentice-Hall Inc., New Jersey. pp 587.
Janssen, M.A. and Anderies, J.M. 2004. Robust strategies for managing rangelands with multiple stable attractors. *Journal of Environmental Economics and Management* 47(1): 140–162.
Kassas, M.1995. Desertification: a general review. *Journal of Arid Environments* 30: 115–128.
Le Houérou, H.N. 1970. North Africa: Past, Present, Future, In: *Arid Lands in Transition* (ed. H.E. Dregene). pp.227–278. American Association for Advancement of Science, Washington, D.C.
Le Houérou, H.N. 1972a. An assessment of the primary and secondary production of the arid grazing lands of North Africa. In: L.E. Rodin (ed), *The Ecophysiologlcal Foundation of Ecosystem Productivity in Arid Zones*. Nauka. Leningrad. pp. 168–178.
Le Houérou, H.N. 1972b. Continental aspects of shrub distribution, utilization and potentials: Africa the Mediterranean region. In: McKell. C.M. Blaisdell, J.P. and Gordin, J.P. (eds.) *Wild lands Shrubs, Their Biology and Utilization*. USDA, US for Sero. Gen. Tech. Report. Int. l. Ogaden, Utah.
Le Houérou, H.N. 1975. The natural pastures of North Africa: types, production, productivity and development. In: *Proceedings of International Symposium on Range Inventory and Mapping in Tropical Africa*, Bamako, and Addis Ababa. ILCA.
Le Houérou, H.N. 1977. The rangelands of the Sahel. *Range Manage.* 33(1): 41–46.
Le Houérou, H.N. 1980. Chemical composition and nutritive value of browse in tropical West Africa. In: Le Houérou, M.N. (ed) *Browse in Africa*: 261–289. ILCA: Addis Ababa.

Mandaville, J.P. 1990. *Flora of Eastern Saudi Arabia.* Kegan Paul International, Riyadh.
Migahid, A.M. 1990. *Flora of Saudi Arabia.* 3 Vols. Riyadh University.
Ministry of Agriculture and Fisheries in UAE. 1996. *Annual Statistical Bulletin.* Karim Press, Dubai. (in Arabic).
Mousa, M.T. 2005. *Ecological Study on some Desert Rangelands.* Ph.D. Thesis. Tanta University.
Munton, P. 1988. Vegetation and forage availability in the sands. *Journal of Oman Studies, Special Report* No. 3: 241–250.
Oatham, M.P., Nicholls, M.K. & Swingland, I.R. 1995a. Manipulation of vegetation communities on the Abu Dhabi rangelands. I. The effects of irrigation and release from long term grazing. *Biodiversity and Conservation* 4: 696–709.
Oatham, M.P., Nicholls, M.K. & Swingland, I.R. 1995b. Manipulation of vegetation communities on the Abu Dhabi rangelands. II. The effects of top soiling and drop irrigation and release from long term grazing. *Biodiversity and Conservation* 4: 710–718.
Pearson, L.C. 1965. Primary production in grazing and ungrazed desert communities of eastem Idaho. *Ecology* 46: 278–286.
Shaltout, K.H. & El-Ghareeb, R. 1985. Effect of protection on the phytomass and primary production of ecosystems of the western Mediterranean desert of Egypt. I. Ecosystem of non-saline depressions. *Bulletin Faculty of Science, Alexandria Univ.* 25: 109–31.
Shaltout, K.H., El-Ghareeb. R. & Sharaf El-Din, A. 1986. Effect of protection on the phytomass and primary production of ecosystems of the western Mediterranean desert of Egypt. II. Ecosystem of coastal sand dunes. *Delta journal of Science* 10: 221–41.
Shaltout, K.H., El-Halwany, E.F. & El-Kady, H.F. 1996. Consequences of protection from grazing on diversity and abundance of the coastal low land vegetation in Eastern Saudi Arabia. *Biodiversity and Conservation* 5: 27–36.
Ter Heerdt, G.N.J., Bakker, J.P. & De Leeuw, J. 1991. Seasonal and spatial variation in living and dead plant material in a grazed grassland as related to plant species diversity. *Journal of Applied Eco*logy 28: 120–27.
Thalen, D.C.P. 1979. *Ecology and Utilization of Desert Rangelands in Iraq.* The Hague: Dr W. Junk Publishers. pp 428.
Till, A.R. 1981. Cycling of plant nutrients in pastures. In: *Grazing Animals.* (F. Morley. Ed.) Amsterdam, Elsevier Publishing Company.
Van Dyne, G.M. & Heady, H.F. 1965. Botanical composition of sheep and cattle diets on a mature annual range. *Hilgardia* 36: 465–492.
Whittaker, R.H. 1972. Evolution and measurement of species diversity. *Taxon* 21: 213–251.
Wilson, M.V. & Shmida, A. 1984. Measuring beta diversity with presence – absence data. *Journal of Eco*logy 72: 1055–1064.

Reclaiming the Desert: Towards a Sustainable Environment in Arid Lands – Mohamed (ed.)
© 2006 Taylor & Francis Group, London, ISBN 0 415 41128 9

Seed storage affects the competitive advantage of weeds: The case of *Portulaca oleracea* L.

T. Ksiksi, A. El-Keblawy & F. AL-Hammadi
Department of Biology, Faculty of Science, UAE University

ABSTRACT: Crop-weed competition causes major damage to land productivity because of its competitive edge over farm crops. The presence of weeds has been reported to reduce growth of a wide range of commercially grown crops. It is certainly true for UAE farmers who suffer from the competitive abilities of weeds, such as *Portulaca oleracea* L., in their farms. Hence, the effect of storage (stored vs fresh), seed burial depth, solarization duration (using Polyethylene plastic sheeting) of *P. oleracea*, a major weed in UAE farms, was investigated.

Percent germination of *P. oleracea* seeds was lowest for stored seeds (33.9 and 36.9% for stored and fresh seeds; respectively at $P < 0.05$). The results of the germination rate followed the same trend as percent germination. Seeds buried at 2.5 cm had a significantly lower germination percent than those buried at 15 cm ($P < 0.05$).

The analysis of variance also showed a significant storage by depth and depth by duration interactions ($P < 0.05$). Germination percent was consistently lower for seeds buried at 2.5 cm. This highlights the efficacy of soil solarization in ridding crops of unwanted weeds, such as *P. oleracea*. Moreover, seed germination percent, for both burial depths, declined as the solarization duration increased, which highlights that the duration of application of Polyethylene plastic sheeting plays an important role in the success of minimizing the impact of weeds. For germination rate, there was a significant depth by duration interaction ($P < 0.05$). Germination rate of both fresh and stored seeds was consistently lower for 2.5 cm burial depth than for 15 cm depth. In short, seeds of *P. oleracea* underwent detrimental effects from storage, burial depth and solarization duration. We posit that competition from *P. oleracea* may be greatly diminished if soil solarization becomes an integral part of weed control that is practiced early enough while *P. oleracea* seeds are still at the top soil. Solarization durations for as much as 30 days may significantly lessen the impact of *P. oleracea* competition of crops. Accumulating *P. oleracea* seeds over many seasons would make it harder for UAE farmers to minimize its competition with their crop species.

1 INTRODUCTION

Crop-weed competition causes major damage to land productivity and poses serious environmental threats. The presence of weeds has been reported to reduce growth of a wide range of commercially grown crops (Knowe et al. 1985; Nelson et al. 1985). Excluding environmental variables, yield losses in many crops are caused mainly by competition from weeds. The presence of weeds has also been shown to reduce soil water availability in juvenile stands of various tree species such as *Pinus elliottii* (Baker, 1973) and *P. radiata* (Sands and Nambiar, 1984), through increasing rates of water loss.

Portulaca oleracea is a species which occurs world wide and which has escaped from cultivation in many parts of the world. It is generally regarded a weed but, none the less, it is an Australian native plant, where it is causing major damage. It is a prostrate herb with fleshy, reddish stems and thick, succulent leaves which are oval shaped and about 25 mm long. Small yellow flowers occur in the leaf bases. In inland areas dense colonies of the plant appear after rain. The main feature of the plant which raises its status above that of a common weed is its edibility. The species was well known to the early settlers who often used the juicy leaves in salads and, cooked, as a substitute for spinach. The seeds are also edible and are usually ground and baked into a damper. *P. oleracea* is not often cultivated as an ornamental species due to its invasive tendencies. Even though *P. oleracea* is considered a nasty weed, some soil-borne phytopathogenic fungus, which are responsible for a root rot disease, were inhibited by extracts of *P. oleracea* (Mizutani et al. 1998).

Fortunately, many control tools have been developed to minimize the effect of *Portulaca oleracea* on agricultural lands. Soil solarization, for instance, is a natural hydrothermal process of disinfesting soils from plant pests that is accomplished through passive solar heating (Stapleton, 2000). Soil solarization is a relatively new method for controlling soil-borne pathogens. It is achieved manually by covering tilled and irrigated soils with continuous transparent polyethylene sheeting for specific periods of the growing season. The soil is heated by solar radiation. In production agriculture, like in our case, the principal use of solarization is probably in conjunction with green house grown crops (Stapleton, 1995). In the present study, the effects of storage (fresh vs stored), seed burial depth and solarization duration on *Portulaca oleracea* seed germination were studied.

2 MATERIALS AND METHODS

The experiment was conducted during the summer from June to July 2003 on a private farm near Al-Ain, UAE (Latitude N24°44, Longitude E55°46 and Altitude 306 m above sea level). The farm soil is sandy loam. Seeds of *Portulaca oleracea* were collected from different farms around Al-Ain city in early June 2003 (referred to fresh seeds). Stored seeds were collected back toward the end of 1999, stored in paper bags at room temperature.

Seeds were placed in 4 by 6 cm mesh bags. The soil was prepared for seeding and plots were set-up (24 individual plots – 20 × 2 m each). Drip irrigation was used throughout the trial period. The experiment was conducted in a completely randomized design with storage (fresh vs stored), depth of seed burial (2.5 and 15 cm) and duration of soils solarization – using polyethylene plastic sheeting – (15, 30 and 45 days) as factors (each replicated 4 times).

Germination rate was estimated using a modified Timson index of germination velocity (Khan and Ungar, 1984). Greater values imply faster rate of germination.

A three-way analysis of variance was used to test effects of the main factors and their interactions on the final germination percentage and germination rate. The germination percentages were arcsine transformed to meet the assumptions of ANOVA. All statistical methods were performed using SYSTAT (SYSTAT Inc, 2004).

3 RESULTS AND DISCUSSION

Percent germination of *P. oleracea* seeds was lowest for stored seeds (33.9 and 36.9% for stored and fresh seeds; respectively at $P < 0.05$). The results of the germination rate followed the same trend as percent germination (Figure 1). Seeds buried at 2.5 cm had a significantly lower germination percent than those buried at 15 cm ($P < 0.05$).

The analysis of variance also showed a significant storage by depth and depth by duration interactions ($P < 0.05$). Germination percent was consistently lower for seeds buried at 2.5 cm (Figure 1). This highlights the efficacy of soil solarization in ridding crops of unwanted weeds, such as *P. oleracea*. Moreover, seed germination percent, for both burial depths, declined as the solarization duration increased. In a study conducted by La Mondia and Brodie (1984) to evaluate the effectiveness of solarization, clear plastic sheeting reduced natural soil populations by 96 to 99% to a depth of 4 inches (10 cm) and significantly reduced the survival of encysted juvenites buried 6 inches (15 cm) deep, which highlights that the duration of application of Polyethylene plastic sheeting plays an important role in the success of minimizing the impact of weeds.

For germination rate, there was a significant depth by duration interaction ($P < 0.05$). Germination rate of both fresh and stored seeds was consistently lower for 2.5 cm burial depth than for 15 cm depth (Figure 1).

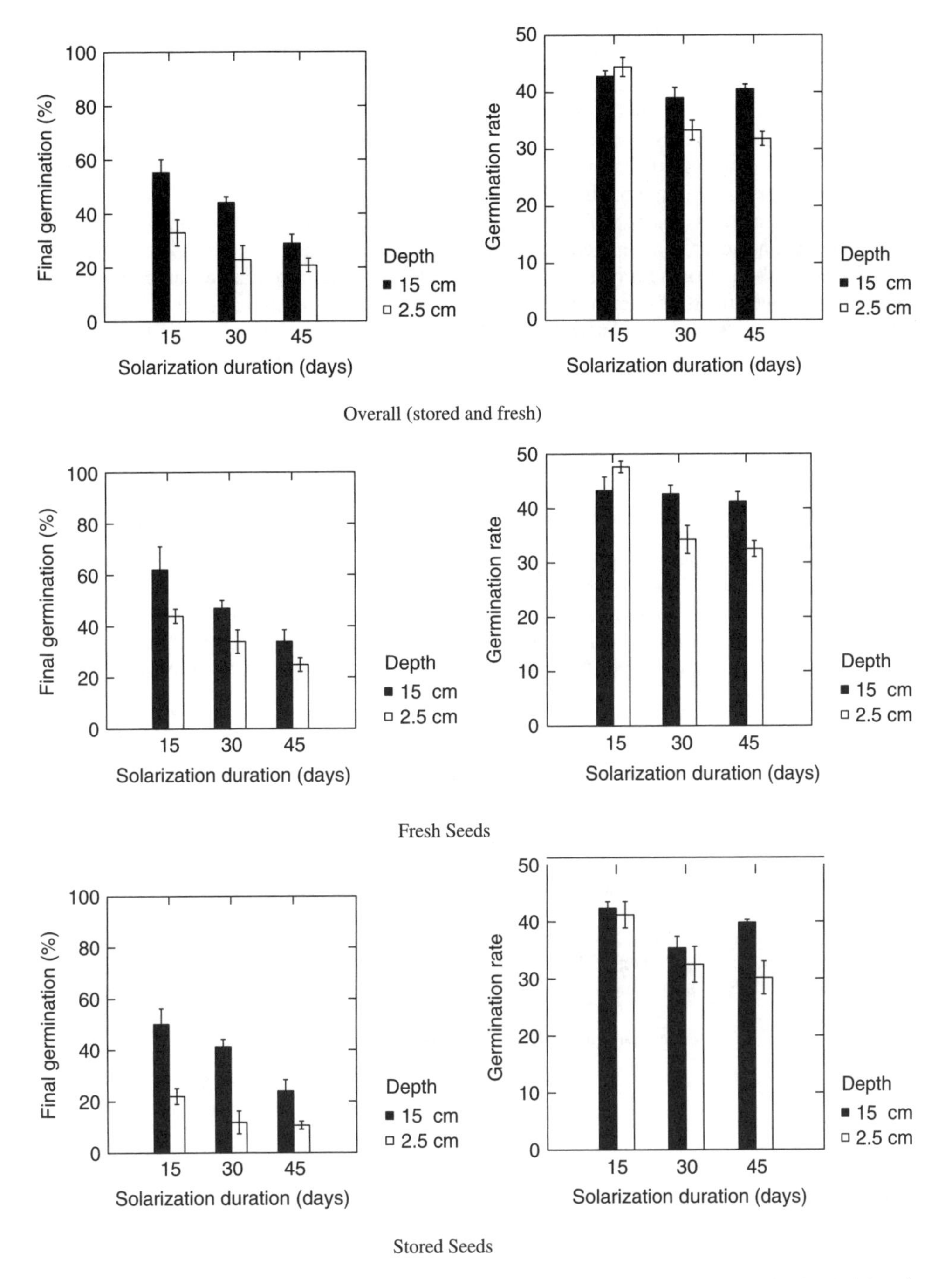

Figure 1. Effect of seed burial, burial depth and duration of soil solarization on final germination (Right column) and germination rate (Left column) of *Portulaca oleracea* L., a major weed in the United Arab Emirates.

In short, seeds of *P. oleracea* underwent detrimental effects from storage, burial depth and solarization duration. We posit that competition from *P. oleracea* may be greatly reduced if soil solarization becomes an integral part of weed control that is practiced early enough while *P. oleracea* seeds are still at the top soil. Solarization durations for as much as 30 days may significantly lessen the impact of *P. oleracea* competition on crops. Accumulating *P. oleracea* seeds over many seasons would make it harder for UAE farmers to minimize its competition with their crop species.

The above results show the importance of soil solarization in controlling weeds. The benefits of such control are many fold. Firstly, soil solarization improves crop production through minimizing competition for resources by weed species such as *P. oleracea*. Secondly, soil solarization minimizes the environmental effects of pesticide use. Thirdly, soil solarization improves financial return for farmers through minimizing the cost involved in weed control.

Balancing the need to control weeds in agricultural systems against the environmental hazards of pesticide use, however, is an immense challenge for researchers and farmers alike. The destruction of beneficial organisms such as arbuscular mycorrhizal fungi also may occur, which in turn reduce the positive effects of solarization. The infectivity of fungi was monitored by Katan and De Vay (1991) before and after solarization using a greenhouse biossay with Sorghum bicolor. Solarization apparently reduced fungi in soil indirectly by reducing weed populations that maintained infective propagules over the winter (Katan and De Vay 1991).

REFERENCES

Baker, J.B. 1973. Intensive cultural practices increase growth of juvenile slash pine in Florida sandhills. Forest Science 19: 197–202.

Katan, J. & DeVay, J.E. 1991: Soil solarization, CRC Press, Boca Raton, USA, 267 pages.

Khan, M.A. & Ungar, I.A. 1984. The effect of salinity and temperature on the germination of polymorphic seeds and growth of *Atriplex triangularis* Willd. *American Journal of Botany* 71: 481–489.

Knowe, S.A., Nelson, L.R., Gjerstad, D.H., Zutter, B.R., Minogue, P.J. & Dukes, J.H. 1985. Four-year growth and development of planted loblolly pine on sites with competition control. Southern Journal of Applied Forestry 9: 11–15.

LaMondia, J.A., & Brodie, B.B. 1984. Control of Globodera rostochiensis by solar heat. Plant Disease. 68: 474–476.

Mizutania, M., Hashidokoa, Y. & Taharaa, S. 1998. Factors responsible for inhibiting the motility of zoospores of the phytopathogenic fungus Aphanomyces cochlioides isolated from the non-host plant Portulaca oleracea. FEBS Letters 438: 236–240.

Nelson, L.R., Zutter, B.R. & Gjerstad, D.H., 1985. Planted longleaf pine seedlings espond to herbaceous weed control using herbicides. Southern Journal of Applied Forestry 9 (4): 236–240.

Sands, R. & Nambiar, E.K.S. 1984. Water relations of *Pinus radiata* in competition with weeds. Can. J. Forest Res. 14: 233–237.

Stapleton, J.J. & Devay, J.E. 1986. Soil solarization: a non-chemical approach for management of plant pathogens and pests. Crop Protection 5: 190–198.

Stapleton, J.J. 2000. Soil solarization in various agricultural production systems. Crop Protection 19: 837–841.

SYSTAT Inc. 2004. SYSTAT software Inc. Richmond, CA 94805 pp. 294.

Reclaiming the Desert: Towards a Sustainable Environment in Arid Lands – Mohamed (ed.)
© 2006 Taylor & Francis Group, London, ISBN 0 415 41128 9

Ecological sustainable gardens

K. Mohiuddin
United Arab Emirates University, Al Ain, Abu Dhabi, United Arab Emirates

ABSTRACT: Gardens are integral part of any urban development as they provide many environmental benefits. They must enhance the ecological benefits and be in harmony with the natural ecosystem. Native species adapt to the ecosystem of the region and maintain a sense of aesthetics, which is particular to the local region. So, care must be taken not to destroy the very qualities or resources that attracted to the site. Al Ain is the Garden City of UAE. This study is a review of the existing gardens in general with focus on Jebel Hafeet and its immediate environs (Mabazzara and Ain Al Fayeda) to determine the carrying capacity of this area as one combined ecological sustainable development. A GIS model was used to carryout the spatial analysis and to identify environmental impacts on the ecology of this area and beyond its boundaries for ecological sustainable development.

1 INTRODUCTION

Gardens are integral part of any development. They makes significant contributions to the quality of life. In addition to recreation and aesthetic values, they provide many environmental benefits and has positive effects on tourism. (John, 2003)

1.1 *Al Ain, The garden city of UAE*

1.1.1 *Location*

Al Ain is largest city of Abu Dhabi emirate and is located in the South East of UAE. It has five oases and the city takes its name from its largest oasis (Phil, 2001). The historic oases of Al Ain, Al Qattara, Al Hili, Al Mutaredh and Al Jimi provide an tranquil and impressive cultural environment with the net work of falajs, sand dunes and craggy ridges surrounds Al Ain (Fig. 1).

1.1.2 *Tourism philosophy*

The presence of oases and Late shaikh Zayed Bin Sultan Al Nahyans love for greenery are the reasons to transform Al Ain into Garden City of gulf. As tourism is world wide growing industry, Government recognizes the tourism potential and want to maintain the image of Al Ain as garden city to market as a tourist destination.

1.1.3 *Physical characteristic*

Jabel Hafeet, Ain Al Fayeda, Mubazzara, Zoo, Fun City and the restored architectural heritage are some of the features sufficient to establish Al Ain as a major tourist attraction in its own right. Jabel (mountain) Hafeet is unique in UAE. It is most prominent landscape feature rising majestically to a height of 1300 and visible from all areas of the Al Ain City. The irregular and unusual rock forms scenic quality. The 12-Km spectacular road from foot to summit with the series of hairpin bends is a contrast with the straight flat roads through the desert. Jabel Hafeet houses very rare and endangered species of animals and plants and protected under Federal Law No. 24 1999 for the protection and development of its environment.

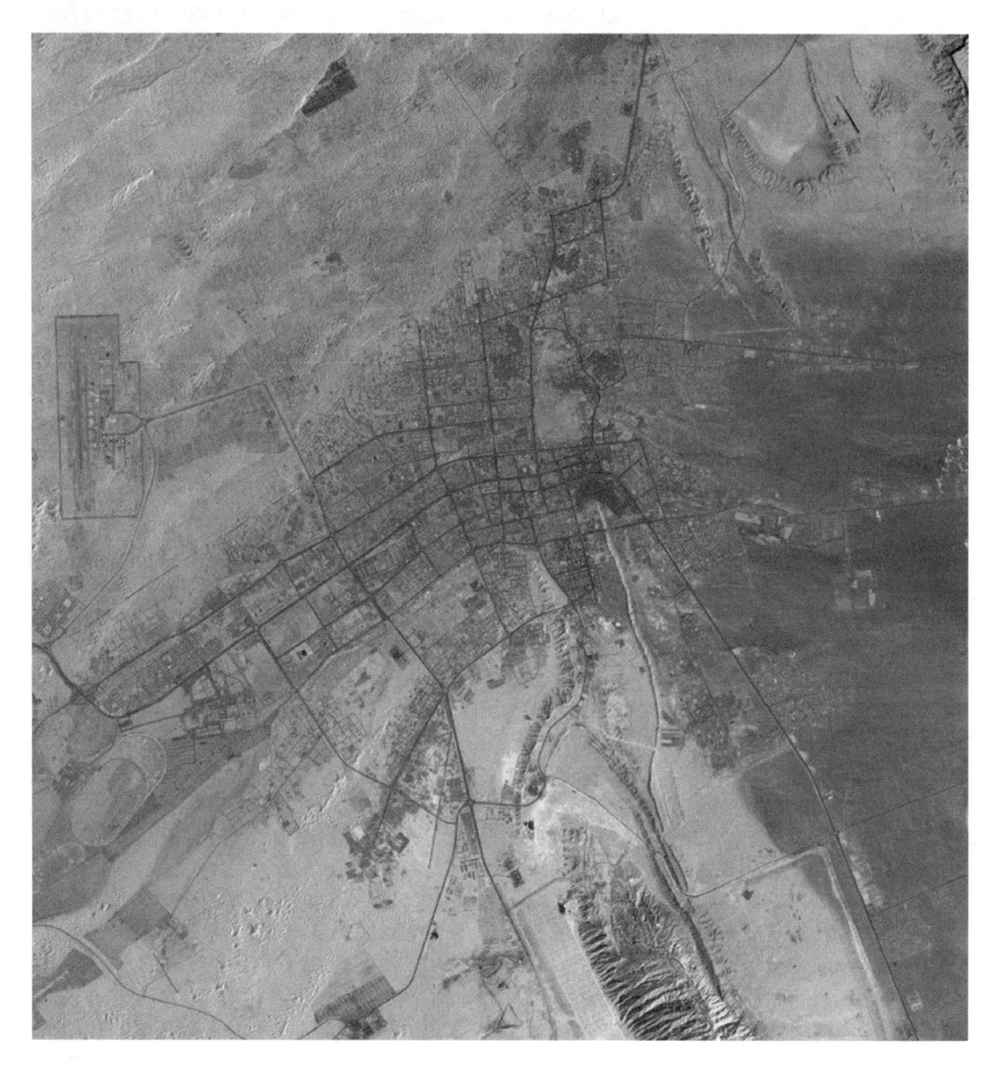

Figure 1. Satellite image of Al Ain (SPOT).

Ain Al Fayeda – The Ain (spring), is the reason for the existence of this family resort. It is a 30 year old development with varied ages of areas and elements around the artisan spring.

Mubazzara – The popular hot water springs are located in the foot of Jabel Hafeet.

1.1.4 *Climate*

Alain has bi-seasonal sub tropical arid climate, rain fall is infrequent and irregular, falling mainly in winter. In summer day temperatures exceeds 40 degrees C. and winters are much cooler dropping to as low as 4 degrees C. at night.

2 OVER REVIEW OF EXISTING GARDENS IN AL AIN

The gardens of Al Ain are typically with expansive tree and shrub plantation, walkways, grass areas, and ornamental fountains which do not respect the hot and dry climate. Gardens are not in accordance with the Master Plan 1985–2000 and many appear to have been conceived in the past

Figure 2. Digital Map of Al Ain showing the existing gardens.

without any clear strategy as the city was not grown so. Gardens are based on older design concepts without the consideration about the size, location, historic aspects or natural conservation. They have not been able to remain relevant to the present needs and lost the appeal. The continued fall in the water level and the availability of continued irrigation water to the existing unnecessary too large gardens is of grave concern.

Following is the map of Al Ain (Fig. 2) and list of existing gardens. They cover around 120 hectors in varying sizes and the distribution of green areas are mostly uneven.

1. Ain Al Fayeda (*public*) Approx. Area 207294 sqm
2. Hili Archeological Gardens (*public*) Approx. Area 187000 sqm
3. Jahli Garden (*women and children*) Approx. Area 56000 sqm
4. Public Gardens (*public*) Approx. Area 104000 sqm
5. Airport Garden – (*public*) Approx. Area 95000 sqm
6. Al Slemi Garden – (*public*) Approx. Area 67000 sqm
7. Al Towayya Garden(*public*) Approx. Area 99000 sqm
8. Al Salamat Garden (*family*) Approx. Area 85000 sqm
9. Al Oha Garden (*family*) Approx. Area 104950 sqm
10. Al Maqam Garden (*women and children*) Approx. Area 79000 sqm
11. Al Mutaradh Garden (*public*) Approx. Area 21000 sqm
12. Al Baladyia. Garden (*public*) Approx. Area 99000 sqm
13. Al Falaj Al Hazzaa Garden (*women and children*) Approx. Area 31000 sqm
14. Al Zahker Garden (*women and children*) Approx. Area 32000 sqm
15. Al Madheef Garden (*women and children*) Approx. Area 2063 sqm
16. Al Basra Garden (*women and children*) Approx. Area 55000 sqm

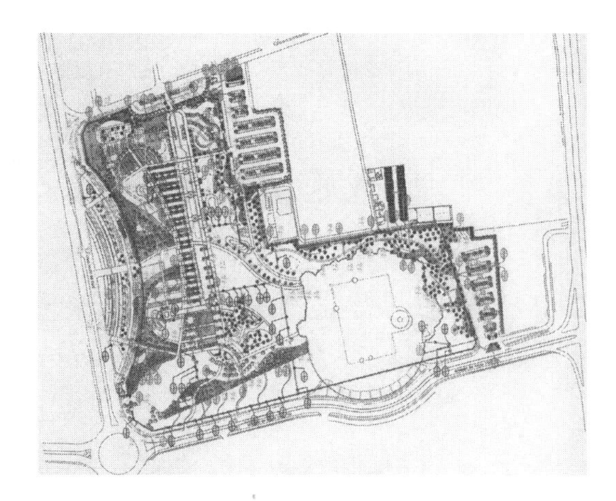

Figure 3. One combined garden as per new design.

17. Al Maraijeb Garden (*public*) Approx. Area 34000 sqm
18. Al Yahar Garden (*public*) Approx. Area 66500 sqm
19. Al Abusamra Garden (*public*) Approx. Area 22800 sqm
20. Al New Shuwaib Garden (*family*) Approx. Area 35000 sqm
21. Al Wagan 1 Garden (*women and children*) Approx. Area 7500 sqm
22. Al Wagan 2 Garden (*family*) Approx. Area 11700 sqm
23. Al Swehan Garden (*public*) Approx. Area 64000 sqm
24. Al Ramah Garden (*public*) Approx. Area 29000 sqm
25. Al Hayer Garden (*public*) Approx. Area 10000 sq m
26. Al Khazna Garden (*public*) Approx. Area 24000 sqm
27. Al Faqa old Garden (*public*) Approx. Area 30000 sqm
28. Al Faqaa new Garden (*public*) Approx. Area 33000 sqm
29. Al Quoa Garden (*family*) Approx. Area 44000 sqm
30. Al Yahar south Garden (*public*) Approx. Area 40000 sqm

3 EMERGING FUTURE NEEDS

After a thorough review of existing gardens a strategy should be develop for the conservation and restoration of habitat, plant life ecology. The exploitation of scenically or historical important sites should be based on creating appropriate ambience or setting and area of ecological interest to be respected.

The role and function of the garden has to be analyzed before the redesign. The catchment area concept needs to be related directly to the urban structure and cultural factors of Al Ain. The designs should comply with the socio-economical changing conditions, new tourism trends and future needs of all sectors and ages in the society. In the past one of the motives behind public parks in Al Ain was to provide rest and relaxation for workforce that had strenuous manual job. Presently about 50 percent of UAE Citizens population is under 15 years and their enthusiasm towards private garden is limited, so their own need should be considered.

3.1 *Changing trends*

An attempt is made to merge two gardens into one, as Jahli Garden and the work is in progress as per new strategy. Previously these were two adjacent gardens Jahli Park (women and children) and Public Gardens (family) with out any access to each other from inside. The architectural heritage of Jahli Fort is more prominent in the new design and has potential to attract regional tourists as

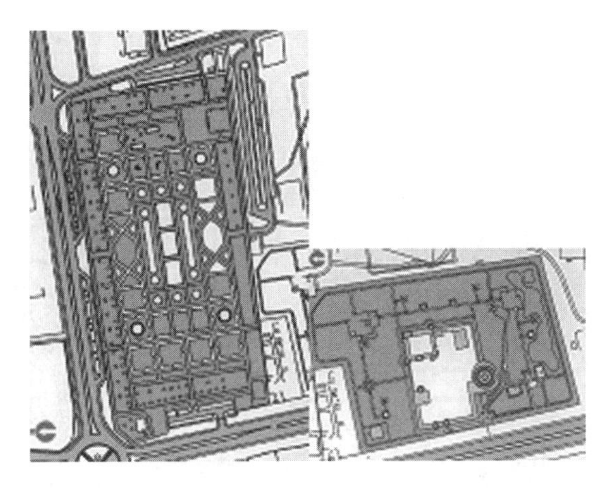

Figure 4. Jahli & Public Garden were two gardens.

Heritage Park. The larger size, functional landscape and closeness to the city center will make it a focal park of Al Ain. The previous parking problem has also been addressed.

4 TOWARDS NEW STRATEGY

Focus of my study is to analyze the area of Jabel Hafeet with its immediate environs for one combined development to ensure the long term conservation of the environment and ecology and still developing overall mountain setting as a major tourist attraction in the UAE.

4.1 *Objective*

Any development should be in harmony with its natural ecosystem. Care must be taken not to destroy the very qualities or resources which attracted to the site. It is essential to ensure that facilities will complement the physical characteristics of Jabel Hafeet, its topography, native flora and fauna, prevent dune destabilization. The objective is to provide guide lines for development of a GIS to carry out the spatial analysis to determine the carrying capacity for ecological sustainable development. of Jabel Hafeet and surroundings. Using the determined suitability develop a recreation and management plan that enhances the natural and ecological integrity. Further to provide possible remedies of problems associated with the deterioration of ecological heritage and threats that continues to face by adopting ecosystem.

5 ECOLOGICAL ANALYSIS OF THE OVERALL MOUNTAIN SETTING

Shaikh Sultan Bin Tahnoun Al Nahyan said "the mountain of Jabel Hafeet looms large over Al Ain and is an ever-present reminder of the forces of nature and of our own relative insignificance in the grand scheme of things. Jabel Hafeet supports a number of rare and unusual plants and animals besides its invaluable and irreplaceable archeological heritage." Sultan (2004). (Fig. 5).

5.1 *Physical characteristics*

Jabel Hafeet is composed of tertiary sedimentary lime stone. A major land mark rising sharply some 1240 m from the surrounding plain. the main mountain massif measures approx. 17 km $\times$ 3 km oriented on a NNW to SSE axis (Drew, 2004). It is visible from up to 50 km away on a clear day.

Figure 5. The majestic Jabel Hafeet.

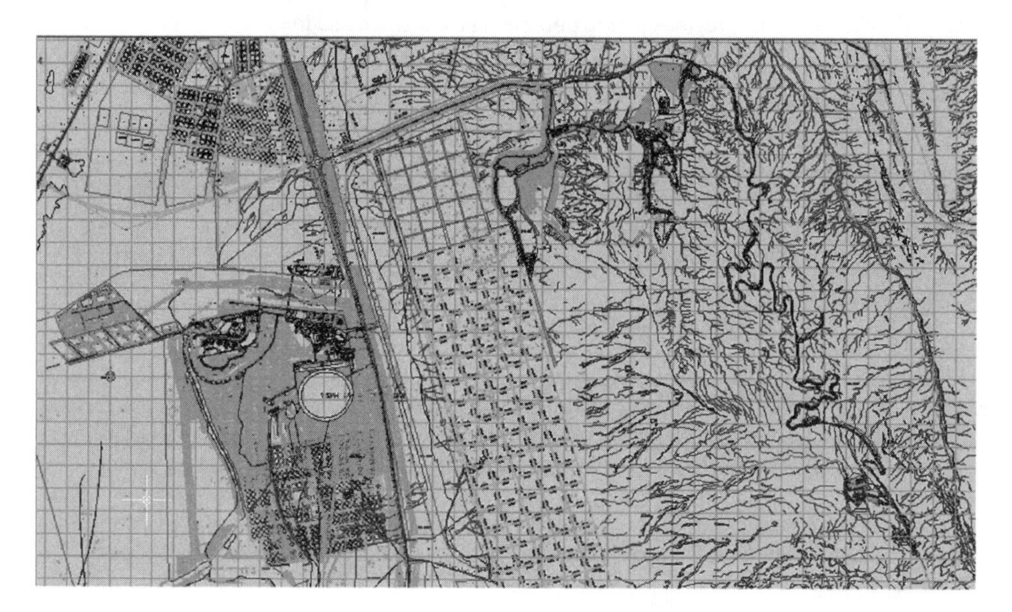

Figure 6. Showing Ain Al Fayeda & Mubazzara in the foot of Jabel Hafeet.

Although from a distance it might appear as a gigantic monolith, it is in fact deeply incised and contains a vast array of wadis (valleys) down. Tarabat is the largest and known as the hidden Wadi.

Ain Al Fayeda The nearest spring to be influenced by the hydrostatic head with in the Jabel Hafeet is Ain Al Fayeda. The garden was developed to exploit the Ain (spring) but the action of water is missing from the theme. It has strong manicured park image and a myriad of facilities and function and very popular as a family resort.

Mubazzara hot rich water Sping or Ain al Waal. Located in the foot of Jabel Hafeet and one of the unique principle tourist attractions of Al Ain. The hot springs were discovered in 1996. The water at Mubazzara is result of scouring and dissolution of pot holes within the lime stone of the wadi bed, and have no hydrological connection with the ground water. Three of the wells provide hot water into pools and other nine supply water into the artificial lake of Mubazzara (Fig. 6).

5.2 *Microclimate*

The lower part of the mountain would experience a similar climate of Al Ain, mean temperature are likely to drop by about one degree celsius for every 300 m increase in altitude, thus on average, the top of mountain will be four degrees cooler than bottom. Wadis are always cooler and generally more humid than areas exposed to direct sun and so provide a much more amenable range of microhabitats.

5.3 *Ecology, historical review*

Though the area of Jabel Hafeet is only 0.0025% of the Abu Dhabi land area, but it about 40% (ca160) of 390 species exists(Gary Sa Sa, 2004). So from the botanic point of view also it is the most important area. On the lower slopes of wadis, where the water is entrapped the vegetation with native trees are evident. Wadi beds host relatively lush vegetation and are floristically extremely rich.

The most important of the 50 families represented on Jabel Hafeet in terms of number of species are *Poaceae* (20 species) *and Asteraceae* (19 species). Other species rich families include *Asclepiadaceae, Brassicaceae, Fabaceae and Zygophylacea.*

5.3.1 *Characterisation of some prominent perennials*

- *Acidocarpus Orientalis* is a small evergreen tree, up to 6 m tall with conspicous large leathery leavesand in the UAE it is known only on Jabel Hafeet. The main population is at the head of Tarabat wadi, where shady conditions prevail for most part of the day.
- *Dodoneae Viscosa* is an evergreen shrub up to 4 m in height with willow like leaves.
- *Ficus Johannis ssp. Johannis* is a small tree that can attain a height of about 8 m.
- *Lycium Shawii* is a small spiny shrub up to 2 m tall with flowers present through out the year, extensively used in herbal medicine.
- *Moringa perregrina* reaches a height of 10 m the thin penulous branches are usually leafless, whitish pink.
- *Periploca Aphylla* is a medium sized, widely distributed shrub with complex flower structure.
- *Zizphus Spina-Chiristi* is one of the larger native tree source of herbal medicine in UAE up to 10 m in height. The tree is used for ornamental purposes with edible fruits.

5.3.2 *Mammals fauna*

Eighteen species of Mammals was believed to be living on or close to Jabel Hafeet but in 2003 ERWADA positively identified just these following eight species as living and breeding on Jabel Hafeet (Chris Lu Dr, 2004).

1. *Felis Cattus,* Feral cat
2. *Vulpes Cana Blandford*, Blandford fox
3. *Hemitragus Jayakari*, Arabian Tahr
4. *Acomys Dimidiaatus*, Egyptian Spiny Mouse
5. *Gerbiillus Gagner,* Wagner Gerbil
6. *Procavia Papensis*, Rock Hyrrax
7. *Rhino Pomamuscatellum*, Muscat Mouse – Tailed Bat
8. *Rousettus Aegygyptiacus,* Egyptian Fruitbat

5.3.3 *Insect fauna*

The wadis maintain a surprisingly diverse fauna, vibrant assembly of butterflies, moths, beetles, grasshoppers and other insects.

5.4 *Ecology at present, adaptable life*

Unfortunately the adaptable feral species such as red vented and white cheeked Bulbuls are increasing at the expense of native yellow-vented Bulbul. Some of the UAE rarest native birds appear to be extinct like Lapper Faced Vulture, Egyptian Vulture, Barbary Falcon and Trumpeter Finchand.

Figure 7. The adapted low-growing plant *Sesuvium sp. green* covers on mountain side at Mubazzarah.

The year round use of rotary sprinklers keeps the low-growing plant *Sesuvium sp. green* on mountain side at Mubazzarah (Fig. 7). This is attracting the house Mouse, as they can digest it. The availability of water dramatically increased also bats and feral cats posing a major threat to native reptiles, rodents and passerine birds (Chris Lu Dr, 2004).

Unfortunately the recent development causing undesirable impact on ecology as a result House Mouse could replace the Spiny Mouse and Red Fox to Blandfords.

5.5 *Archeological sites, mezyad fort and carins*

This area is most significant because it contains the oldest known archeological sites of UAE. Mezyad Fort is on the South East of Jabel Hafeet and has historical importance. There are many Carins (tombs) some have been reconstructed. The Carins are generally circular, above ground structures made of unworked stone. The sizes varies from 8 m to 4 m radius and are a maximum 3 m in height. (Robert Wa Ya, 2004).

5.6 *Cement Works and block factories*

Cement Works and block factories exists in the vicinity of Jabel Hafeet. The chimneys are most visually determental elements which is not compatible with the locational settings.

6 CONSERVATION AND RESTORATION OF PLANT LIFE

6.1 *Landscape development*

Natural landscape values may be easier to maintain if facilities are carefully dispersed. Native species adapt to the ecosystem and the region and maintain a sense of aesthetics and culture which is particular to the local region.

Plants species which establish themselves with some water during initial establishment period only and which naturally survive on moisture from dew fall should be planted Plants species such as *Euphorbia larica* and *Acocia Torills* are unpalatable for goats.

6.1.1 *Adoptive strategies of native vascular plants*

The native Plants has an mechanism serve to delay the onset of internal drought, perennail species like, drought-dedicious, dwarf shrubs, simoly loose their leaves during the hot season in order to reduce transpirational water loss, others develop minute leaves which are rapidly shed and the green stems are primarily responsible for photosynthesis. A number of species develop long rooting system that enable them to access water stored deep below the surface. Some store large amount of water in their stems, this enables them to survive for long periods without additional moisture.

6.1.2 *Land scape restoration*

The overriding principle of landscape restoration work is for the use of local natural materials including stone, gravel, sand and local indigenous flora. The use of design elements which blend with the natural landscape will ensure the man made aspects will not distract from the appeal of the Jabel Hafeet. Flora requires rehabilitation to reverse the inevitable process of desertification and to stimulate and increase in flora and fauna population and diversity.

6.1.3 *Repair of damaged areas*

The largest area requiring revegetation is to immediate south of cement factory. This area requires deep ripping, spreading of gravel a surface mulch and seeding during the winter months with local indigenous plants species.

7 STRATEGIC REVIEW

Typically, city strategies are reviewed every 20 years, and the vision for the area of interest is reexamined, With the advert of GIS and related technologies the city strategy can be reviewed and the vision for the area of interest can be reexamined with speed. (Khalid, 2003). GIS provides a useful tool for the collection, analysis and presentation of key data layers. Detailed inventories of natural and cultural features can be prepared to carryout the spatial analysis of the project area and to test the alternatives. GIS would automate the overlay process and the combination of homogenous layers of information to determine the site suitability for development and assist in the formulation of sustainable development spatial data layers (Eesri, 2001). The initial step in the strategic planning process is to gather primary and secondary data before planning and designing a project. This means collecting information on the environmental features of the site and its surroundings including flora and fauna; archaeology; geology; and meteorology. Identify impacts that the development may have on the ecology of the site and beyond its boundaries. Measure the quantities of environmental resources (water, energy, materials). Fundamental questions must be taken in consideration about soil erosion, vegetation and habitat protection, endangered species, wildlife movement, water quality, air pollution, and social implications.

8 APPLICATION OF THE STRATEGY

Any new development should be subjected to thorough environmental impact assessment (EIA) to ensure they do not compromise habitats and have minimal visual impact. Careful management of garden will ensure both wild life and landscape are sustained for future generations. Following are the few examples to process an EIA with GIS and related technologies to attain environmental sustainable development.

8.1 *Management strategy of Visual character and ambience*

Promote bio-diversity and to conserve the overall scenic quality.

8.1.1 *Architecture*

The building forms should be piece of sculpture, low profile and of shape, color and pattern with the harmony of the location and design should blend in with the rocky terrain. The appearance of the fence should be sympathetic to the natural landscape and it would need to be sited to avoid places where it would unreasonably interfere with the view. 3D model rendered in true color and

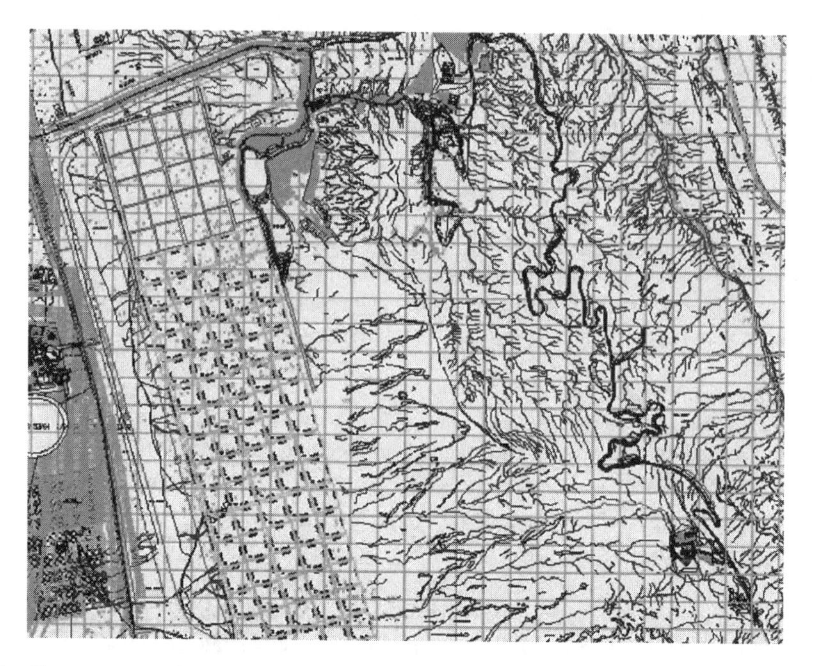

Figure 8. Showing the farms which do not regard the topography.

texture of proposed development can be viewed from any desired height and angle from Jabel Hafeet virtually, for visual impact assessment (Khalid, 2002).

8.1.2 *Date plantation*

95 farms measuring 183 × 183 m has been laid out as date palm plantation on the flat land between the Jabel Hafeet and Ain Al Fayeda. Each farm is fenced and separated from each other by a grid of access roadways. Palms are planted in rows, 4 of date palms, *Phoenix Doctilifera* and two of *Zizyphus Spina-Christi.* These farms are laid out with no regard to the topography (Fig. 8) or natural features of the landscape. The visual impact is seen in the satellite image and this plantation is not compatible to the natural its settings. The growth of vegetation on mountain resulting the recent rainfall can be measured and compare with the one which is prior to 2002 (when goats were allowed to roam freely).

8.2 *Management strategy of ecological sensitive areas*

Wadis are always cooler and generally more humid than areas exposed to direct sun so they provide a much more amenable range of microhabitats. The majority of plants occurs in the wadis and is rich in diversity of fauna. So the wadis and the dune areas are environmentally sensitive. They can be located, and mapped along with areas of high use areas.

8.3 *Management strategy of water resources*

The integration of the remotely sensed Natural Resource Information System, which consists of databases soil, geology, geomorphology, metrology and groundwater model in GIS can effectively be utilized for Water Resource Management (Yousuf, 2002). A natural dome of fossil ground water exists within the mountain, but at too low a level for there to be any natural springs at the foot of the mountain. The nearest spring to be influenced by the hydrostatic head with in the Jabel Hafeet is Ain Al Fayeda.

12 wells have been sunk at the foot of Jabel Hafeet. Three of the wells provide hot water and other nine supply water into the artificial lake of Mubazzara. Measure whether the unregulated dumping of variety of waste materials across extensive areas may contaminate the surrounding

Figure 9. The Ariel photo of Jabel Hafeet showing the spectacular road and Mubazzara in the footings.

ground water or not. The impact on surrounding ground water by dumping of variety of waste materials across extensive areas can be assessed.

Any of the existing dams hold enough water to maintain a permanent water supply for the wild life can be determine by GIS.

9 CONCLUSION

Its scenery, archeology, fauna and flora require to be protected from incompatible development. The development should not be at cost of the fragile natural habitat that is unique to the mountain (Fig. 9). Every possible step to be taken to restore the damaged landscape, rehabilitation of its habitats and increase bio diversity.

10 RECOMMENDATIONS

All entertainment facilities to be diverted to Ain Al Fayeda and Jebel Hafeet should be left for conservation. The link between Jebel Hafeet and Ain Al Fayeda will create many opportunities for a substantial coordinated leisure environment within the imposing setting of Jebel Hafeet. Monorail and cable cars are the most environment friendly means of transportation. Sustainability can be attain by an effective buffer zone around the area to ensure that such an important wild life asset can expand rather continue to shrink. The majestic character of Jabel Hafeet should be maintain, as it benefits the bigger picture by retaining the natural scenery as viewed from most approach roads. Over development would compromise this special feeling of isolation and nature power. The more the surrounding land changes from its original character and becomes clutter with development, the less imposing and less impressive the Jebel Hafeet becomes. The building forms should be piece of sculpture, low profile and of shape, color and pattern with the harmony of the location. Landscape should be planned to be simple without irrigated planting. No special features which in any way distract from the dramatic scenery of Jabel Hafeet should be incorporated.

ACKNOWLEDGEMENT

I would like to thank the following authorities of Al Ain Town Planning & Survey Department, Municipalities & Agricultural Department, and Gardens Department, who provided the valuable information for my study. Engineer Talal Salmani, Engineer Salma Saeed Al Maammari, Engineer Bakriaa Abdul Raheem, and Engineer Tariq Al Zarouni.

REFERENCES

Chris Lu Dr (2004); The Mammals of Jabel Hafeet – Jabel Hafeet, a natural history 169 – Emirates National History Group – Al Ghurair printing & publishing house.
Drew (2004); The Reptiles of Jabel Hafeet – Jabel Hafeet a natural history 149 – Emirates National History Group – Al Ghurair printing & publishing house.
Erdas (2004); www.erdas.com/virtualGis/3Dcapabilities
Gary Sa Sa (2004); Flora & Vegetation of Jabel Hafeet – Jabel Hafeet, a natural history 65, 66, 67 – Emirates National History Group – Al Ghurair printing & publishing house.
John (2003).r.; www.svr_ato.com/sustainabledesign techniques
Karen C. H.; GIS for landscape architects (1999) 38–39, 52–53. Esri press
Khalid (2002); 3D virtual environments; MEA 2002 proceeding
Khalid (2003); Virtual Urban environments – SUPS 6 proceedings
Micheal Br Ho (2004); Insects of Jabel hafeet – Jabel Hafeet a natural history – Emirates National History Group – Al Ghurair printing & publishing house. 94
Phil I. (2001); Alain oasis-AlAin Municipality 1–1.
Ranav S (2001); www.esri.com. Urban and regional planning lessons
Robert Wa Ya (2004); The Archeology of Jabel Hafeet, Jabel Hafeet, a natural history – Emirates National History Group – Al Ghurair printing & publishing house. 48, 50
Sultan (2004); preface – Jabel Hafeet a natural history – Emirates National History Group – Al Ghurair printing & publishing house. 8
Yousuf (2001); Simulation model for irrigation management 12, 13 SUPS 6 proceedings.

Energy

Reclaiming the Desert: Towards a Sustainable Environment in Arid Lands – Mohamed (ed.)
© 2006 Taylor & Francis Group, London, ISBN 0 415 41128 9

The energy sustainability accelerator: An informed protocol for participatory sustainable energy planning in Sudan

S.E. Badri
Ahfad University for Women, Omdurman, Sudan

ABSTRACT: This paper examined the emerging role of active participation in planning for sustainability in the context of the energy sector in Sudan. It explored the value of integrating participatory planning and education for sustainability. The Energy Sustainability Accelerator (ESA), an informed protocol for energy participatory planning was developed by the author and tested within the research process. The initial impetus for developing ESA stemmed from the realisations of: (1) changing the models and nature of how traditional energy research handle the relationships of the socio-economic and environmental impacts of energy planning with future vision and strategies to achieve the required change in energy planning, and (2) traditional methodologies usually fail to deliver sustainable solutions to similar research problems, on the other hand an action paradigm is more relevant in being explicit about more active involvement in the formation, undertaking and execution of real-world solutions.

The ESA process adopted creative involvement of stakeholders in energy planning within a framework that brings experts together with other stakeholders including legislators and members of the National Assembly. They were invited to participate in creating long-term vision of sustainability in energy planning in Sudan. The process involved provision of leadership skills, technical support, and planning and design expertise to guide the participants and ensure that they are fully informed of all relevant ground-rules, considerations, constraints, and potential options. The research tested ESA at the decision-making and stakeholders' level, using an action-based, interpretive methodology. Interviews were conducted with participants before and after the process, providing data about changes in understanding resulting from their experience. The research pointed to the importance of understanding participatory planning as a societal process that is both engaging and meaningful. It demonstrated the benefits of an iterative process in which planning at one level of scale informs, and is informed by the other level. The research had the two major outcomes; (1) in-depth exploration of the theoretical basis of informed participatory approach, as exemplified by ESA, and (2) contribution to practice through investigating ESA's potential to meet key challenges of Sustainable Energy Development.

1 INTRODUCTION

1.1 *Energy planning and sustainability*

It is widely agreed that to implement 'planning for sustainability' is a difficult task, still there are several methodologies that can be applied to develop sustainable solutions. In addition the author takes the view that attempting to solve problems associated with energy planning before they are applied in practice is a sensible precaution, and posits that participatory processes will be necessary to do so successfully.

Sustainability lies in the interplay of maintaining environmental quality, and promoting economic vitality and social equity. Environmental and developmental concerns raised during recent years, together with the challenge of climate change mitigation, represent the pressing items of achieving Sustainable Energy Development (SED) through finding new approaches and strategies to conciliate these issues to the energy services to shift from unsustainable patterns to a more sustainable through good technical, environmental and socio-economic prospects, bearing in mind the importance of appropriate and equitable measures for implementing SED.

1.2 *SED principles and directives*

Key principles and directives of SED are based on the recommendations of different international fora such as The World Summit on Sustainable Development (WSSD) in 2002 and The Year 2001 Session of the United Nations Commission on Sustainable Development (CSD-9). They both called for capacity building, institutional strengthening, support to technical research, financial assistance, and increased cooperation and participation at all sectoral levels. Ambitious and broad-ranging application of SED would require integrating environmental and sustainability awareness into all levels of policy making and early stages of decision making in energy planning. This offers an objective-led framework, setting standards in terms of outcomes that should be achieved, laying out detailed prescriptions of actions to be undertaken, and offering opportunity for improving the way that energy is managed. Locally, there are many benefits to be gained from implementing innovative SED principles and directives in energy planning in Sudan.

2 SUDAN ENERGY

2.1 *Overview*

Situated in Northeastern Africa, Sudan covers a land area of about 2.5 million square kilometers. Of this area, 34% is classified as desert, 20% as shrub/semi-desert, 38% as woodland/forest, 7% as agricultural and 1% as swamp/wetland. In spite of the recent development in the oil sector, Sudan energy consumption is still characterized by high reliance on traditional fuels. Of the estimated total energy consumption of 10 million TOE in 2000, firewood, charcoal and other forms of biomass constitute the overwhelming majority of the fuels, while oil and small percentage of electricity satisfy commercial energy consumption. Because of the high reliance on traditional fuels, it mounts to about four times higher than commercial energy. Sectorally, industrial, transport and residential consumers remain the largest customers of electricity, concentrated in the major urban centers.

Sudan's primary environmental concerns center on the over-exploitation of its forests and marginal agricultural lands. The ten-year (1992–2002) Comprehensive National Strategy (CNS) identifies the pursuit of environmental protection as priority actions related to climate change policy. Sudan has signed and ratified the UNFCCC on 19 November 1993, signifying its commitment to the global environment and limitation of GHG emissions. These include protection and development of the rural environment, rehabilitation/preservation of ecosystems for sustainable development, promotion of the use of renewable energy resources and enhancement of environmental awareness among concerned groups. Annual average solar radiation is very high throughout the country, ranging from 436 W/m^2 in the south to 640 W/m^2 in the north. However, this potential remains largely untapped. Solar energy began receiving Government support in Sudan since 1970 when the Institute of Solar Energy (ISE) was established in 1974. ISE later in 1990, was renamed the Energy Research Institute (ERI), as part of the Ministry of Higher Education and Scientific Research. ERI developed competence in the field of renewable energy applications and research, pursuing active programmes in a variety of areas including solar, wind and biomass energy. Solar energy programmes are prominent among the technology-based renewable energy programmes of ERI, including solar thermal technology (hot water systems, cookers, dryers, solar passive architecture etc.), solar photovoltaic technology (lanterns, fixed systems, pump sets) as well as information dissemination, marketing, standardisation of products and research and development (R&D). In general solar energy programmes concentrated on the two applications below

- Solar thermal technologies, where the heat produced is used to operate devices for heating, cooling, drying, water purification and power generation. Those devices are suitable for use by rural communities including solar hot water heaters, solar cookers and solar drier
- PV systems which convert sunlight into electricity including applications such as lighting, pumping, communication and refrigeration.

2.2 *Renewable energy initiatives*

Several initiatives related to energy in general, and PV systems in particular, have been undertaken in Sudan since the 1970's, for example.

- The UNDP Global Pumping Project to test and monitor the performance of different types of PV pumps (1978–81)
- USAID technical assistance projects in the period 1980–86. These are the Energy Planning and Management Project, in collaboration with the Ministry of Energy and Mining (MEM)
- The UNICEF/WHO Extended Immunization Programmes (1990–94)
- UNDP technical assistance project on Sustainable Rural Energy Development SRED (2002).

2.3 *The UNDP-SUDAN medium sized project SUD/98/G36 (PV Project)*

This PV Project is officially known as the Barrier Removal to Secure PV Market Penetration in Semi-Urban and Rural Sudan Project. UNDP-SUDAN is the implementing agency for the Global Environment Facility (GEF) for the PV Project. It conforms to the environmental objectives stated in the Country Strategy Note (CSN), derived from the 10-year Comprehensive National Strategy (CNS) 1992–2002 for Sudan. CSN identifies the pursuit of environmental protection as integrally related to its goals of increasing the prosperity of the population. A number of priority actions outlined in CSN are related, either directly or indirectly, to climate change. These include protection and development of rural environment rehabilitation/preservation of ecosystems, for sustainable development, promotion of the use of renewable energy resources and enhancement of environmental awareness among decision-makers and concerned. The PV Project objectives are as follows:

- To meet the suppressed and growing demand for electric energy in rural and semi-urban Sudan through reliable, domestic PV systems as a substitute for fossil based generating units
- Market penetration of PV technology in rural and semi urban Sudan
- Sustainable technical, institutional and financial infrastructure created to support market penetration efforts.

The target beneficiaries of the project include

- Sudan's rural and semi-urban community, namely households, businesses, service institutions, and industries, having access to information about, and incentives to use PV systems, thus meeting their suppressed demand for electricity
- The Sudanese people in the targeted areas, who were to benefit from the economic stimulus created by supplying/improving the electricity needs
- The local and international private sector and NGOs involved in the business of consultation, design, installation, financing and management of energy in general and PV systems in particular
- The Ministry of Energy and Mining as well as stakeholders involved in the project, which will benefit from capacity building programs
- The Government of Sudan, which will benefit from institutional strengthening, as well as saving on investment and expenditure in fuel for oil-fired thermal plant
- The global community, which indirectly would benefit from reduced potential emission of greenhouse gases.

3 ADOPTING THE INSTITUTIONAL/POLICY ACTIVITY OF THE PV PROJECT AS THE CASE STUDY OF THE RESEARCH

3.1 *The institutional/policy activity of the PV project*

The PV project objectives focuses on the institutional/policy side, undertaken with the collaboration of the concerned stakeholders, for an overall assessment of policies, legislation and regulations to identify the barrier removal strategies to be recommended to the government for adoption in support of PV programmes. Those barriers are as follows

- Weak formal energy institutions at the state level
- Incapacitated private sector institutions
- Poor organizational and outreach structure at the local levels
- Lack of a national policy strategy for the promotion/use of PV systems.

3.2 *The research partners*

The preparation of any development framework will require skills in negotiation with a variety of the different parties involved so as to achieve consistency and integration within that framework. Never the less the mechanisms that the PV Project adopted to reach the energy policy-makers did not include community involvement, focusing on more systematic approaches inclusive of forums, workshops, and working with NGOs and networks. The PV Project mechanisms had identified areas and issues of strategic importance in the energy policy context, but had not engaged public or stakeholder participation in developing ideas for the needed change, which was not sufficient in getting vital commitment/breakthrough in the so far traditional processes. Accomplishing the above twin goals required active collaboration from concerned parties, such as the research underhand. The author, being already involved with the concerned parties used what is known as mediation techniques in participatory planning contexts to convince three major actors of the system under study to come together to achieve the following:

- Establish a partnership between the research, the PV Project, the Energy Committee of the National Assembly, and Ahfad University for Women
- Adopt the institutional/policy activity of the PV Project as the most suitable context for a case study for the research

The action research paradigm, adopted in the research process, aims to contribute both to the practical concerns of people in an immediate problematic situation and to further the goals of social science simultaneously. This dual commitment also governed the formation of the above partnership.

4 THE ESA PROCESS

4.1 *The research process*

The research process was developed with a view that classical methods in planning is not going to deliver sufficient changes to move towards sustainability, and that new models are necessary to make such change. Attention to the processes of capacity building and communication facilitates dialogue leading to the development of sustainable plans. It facilitates investigating energy planning systems in Sudan from the perspective of SED handled by the author through participatory process in an educational set-up, applying the specially designed toolkit ESA.

The process was structured around the 10 stages of the ESA, linking its various components and exploring in detail key ideas that emerged from the process at each stage, as shown in Table 1. Application of the stages of ESA, below. The stages provided different protocols that can be used at multiple levels, such that several different groups can use the same framework simultaneously,

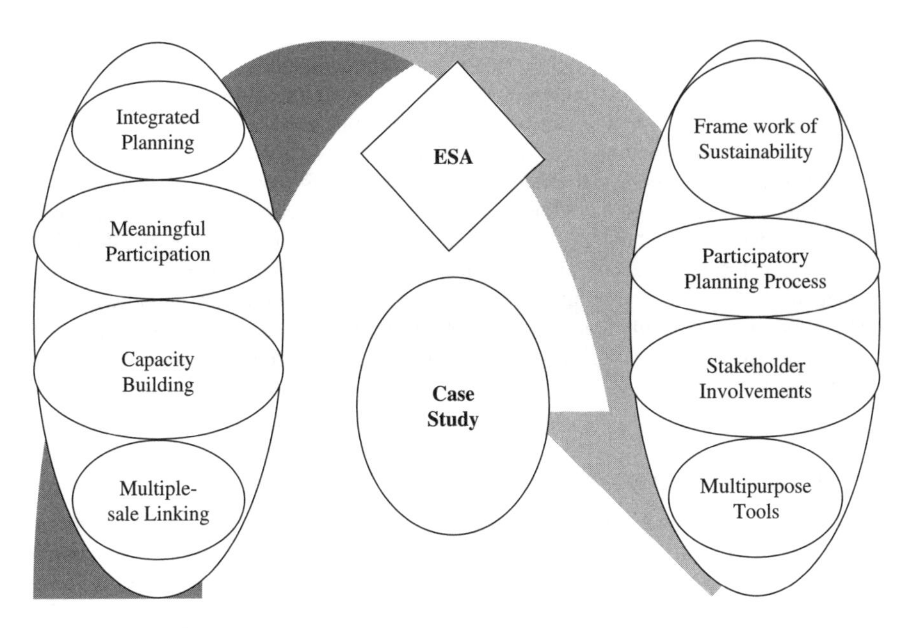

Figure 1. The research process.

Table 1. Application of the stages of ESA.

Core participants sector	Organisation
Creativity	Brainstorming future possibilities Social learning between different stakeholders and parliament
Context	Building a picture of existing assets Dialogue and goal setting
Sustainability	Analysis of resource against sustainability criteria Developing indicators, Analyzing systems
Limits and solutions	Analysis of limits and problems, developing solutions Selecting innovations, Creating strategies, and Making agreements
Values and goals	developing shared vision and goals and allow for stages of learning
Filtering ideas	Testing ideas against goals, sustainability criteria and current limits
Integrated decision Making	Deciding priorities for action
Action planning	Prioritising actions and develop strategy to implement plans
Implementation	Implement plan and allow for stages of execution
Review	Review from the implementation process

and are able to communicate easily with each other. The objectives of the such parallel process were as follows:

- Engages participation by involving the stakeholders, decision-makers, and legislators in the design of options
- Stakeholder goals and knowledge are clarified and developed into more visible decision making framework
- Decision making process encourages focus on common goals and values
- Teaches skills of testing ideas against participants' goals and sustainability principles
- Interacting dialogue between stakeholders at different levels of scale.

4.2 *The research process participants*

Stakeholder identification and recruitment was carried out within existing contacts, networks built by the PV Project over its couple of years of operation targeting energy sector members. Invitations were sent to those contacts, putting in mind that there were no resources for additional outreach to hard-to-reach stakeholders and social groups. Nevertheless it was possible to inform wide range of stakeholders about the process, including the private sector, as shown in Table 2. Participating Stakeholders, below. Each cycle of the Action research process had its own category of participants, mainly selected by the author according to the requirements and objectives of the cycle. All of the core participants of the second and third cycles were interviewed both before and after the process. The total number of participants was considered quite acceptable. There was considerable range of people who attended the workshops in addition to the core participants, and contributed to the discussions in these workshops.

4.3 *Applying the research process*

The research framework brought experts together with lay people that required skilful facilitation, and was seen as a collaborative event which allowed people from different backgrounds to work together at both decision-making and stakeholders' levels of scale. The order of the ESA steps followed was given different emphases to suit the context and time available. The process aimed at building the capacity of the participants in developing sustainability criteria in decision making, through focus on social capital, environmental integrity, and economic vitality. The order of the ESA steps followed was given different emphases to suit the context and time available. The process aimed at building the capacity of the participants in developing sustainability criteria in decision making, through focus on social capital, environmental integrity, and economic vitality, hence building level by level the basic steps in the process of SED, as follows

- Developing indicators
- Analyzing systems
- Selecting innovations
- Creating strategies
- Making agreements.

Table 2. Participants.

Core participants sector	Organisation	Number	Role
Partners	PV Project	10	members
	ECNA	20	members
	AUW	5	staff
NGOs	SECS	4	members
	Fredrich Ebert	2	members
	BBSAWS	2	members
Public sector	MEM	4	staff
	MF	4	staff
	MST	2	staff
	ME	10	staff
	ERI	2	staff
Private sector	SOLARMAN	2	staff
	HAJAR	2	staff
	EDGE	1	staff
Total	14	80	

The process was held in two all-day workshops, each took a total of 5 hrs. Both workshops were facilitated by the author assisted by the research partners as stand alone events to provide opportunities to engage participation from additional stakeholders.

4.4 *The first workshop*

The first workshop was held at Ahfad University for Women and took the form of educational framework of sustainability at the stakeholders' level. The participants included members of the National Assembly, NGOs, academia, and public and private sector organisations. The ultimate objective of the workshop was to encourage all stakeholders to work together towards achieving wide scale results in sectoral energy planning, which significantly benefited the task of promoting SED. The workshop was addressed by one key figure in energy planning, Prof. ElTayeb Idris, the general secretary of the Ministry of Science and Technology, a veteran renewable energy expert in Sudan and the ex-head of the Energy Research Institute. He built clear and explicit picture of Sudan energy assets using presentation aids as a stimulus for dialogue. Sustainability principles were delivered to the participants as part of the process, as well as examples and case studies of sustainable technologies. That presentation illustrated possibilities for the participants to get encouraged to share knowledge through discussion. The author acted as facilitator, while the partners of the research process; the PV Project and the National assembly members were key workshop mobilisers. The participants chose to build trust with each other, and to create permanent linkages between them, realising the opportunity of that gathering in promoting future links and joint action necessary for a successful venture. The major outcome of the workshop and the whole process was the initiative presented by the participants to establish a special framework in the form of a national forum for sustainable energy planning, with different energy stakeholders' membership including parliament members. The forum aims to assist in building sustainable energy future for the country. The participants worked immediately on realising this initiative and selected a steering committee for the forum. The group was named the "Sudanese Parliamentary-Civic Society Group for Sustainable Energy Development" (SPCGSED). The objectives set by the imitative included the following:

- Public awareness, education and dissemination of sustainable energy principles
- Influencing national and local policies towards energy service provision to poor and marginalized communities and removing barriers to energy conservation and energy efficiency
- Promotion of legislative measures in the field of energy covering dissemination of know-how and best practices involving all stakeholders
- To link Global, National and Local energy issues, adopting a transparent, participatory, national-driven approach to energy planning
- Cooperation with sustainable energy international community; thus ensuring linkages with related environmental protection, regional policy, and cooperation agencies
- Help provide sustainable energy services to households, rural schools and health centers, community services, water pumping and micro-enterprises
- Promote grassroots energy initiatives that protect and conserve the environment while at the same time generating sustainable livelihood opportunities.

4.5 *The second workshop*

The second workshop, targeted decision-makers, and was held in the National Assembly. The workshop was addressed by His Excellency the Assembly Speaker and the Minister of Electricity, and was attended by representatives of the UN, EU and the major embassies in Khartoum, in addition to the invited participants, who included the research partners, wide representation of energy stakeholders, and the assembly members already involved in the process.

The workshop started with speeches delivered by the above dignitaries, the head of the Energy Committee of the National Assembly, PV Project Officer, and the author, followed by long sessions of open discussions, jointly facilitated by the author, the PV Project Officer, and the head of the Energy Committee. Dialogue was encouraged amongst the participants to realise SED envisioning process. The participants focused on the role of the government; however, the approach gave greater weight to other actors such as the private sector, academic and research institutes, and civil society bodies including community groups.

Stakeholder goals were clarified and developed into decision making framework that recommended number of policy and institutional measures. The participants proposed to take into account the importance of identifying national energy objectives, strategies and priorities to the development of stakeholders' capacity to respond to and manage the current changes in energy planning requirements, through availing relevant information and technical advice nationally and locally. They also recommended broadening the approach of energy planning in Sudan to become more sustainable, strategic and dynamic, through applying the following resolutions:

- Full support SPCGSED to help stakeholder to assume their role in energy planning as a lever in meeting broader livability and sustainability goals
- The generous initiative presented by the Minister of Electricity to immediately draw an official project proposal of solar electrification of 1000 Sudanese villages to be presented to the immediate fiscal plan
- Drawing legislative manuscripts in the field of sustainable energy to be introduced by the Energy Committee, supported by the National Assembly members/participants, covering promotion and adoption of SED for open vote in the National Assembly's immediate session.

5 FINDINGS OF THE RESEARCH PROCESS

5.1 *The analysis of the participants experience*

The analyses of the participants' experience of the planning processes was done, using data from in-depth before and after interviews and the participants' observations, paying close attention to the experience as a learning one. The theoretical construct applied to the analysis of both the before and after surveys targeting the participants of the process was based on the Theory of Reasoned Action (TRA). As the name implies, TRA is based on the assumption that human beings usually behave in a sensible manner; that they take account of available information implicitly or explicitly consider the implications of their actions. TRA postulates that in order to change behaviour it is necessary to change either the pertaining attitudes and/or subjective norms by changing the corresponding underlying beliefs through an educational process.

The stages of ESA were explored through qualitative analysis of participants' experience. Results from this analysis have been positive, indicating a marked increase in participants' understanding of sustainability, and their capacity to design creative solutions to move towards the goal of sustainable development. Interviews with participants indicated that they were made more aware of the importance of underlying processes in shaping and influencing energy plans. It was important that elements of the environment, social capital, and the economy were considered, as this helped participants to broaden their perceptions, often looking outside their area of expertise and interest. They felt that they developed skills in forging links amongst these different areas, becoming more able to develop solutions.

5.2 *Assessment interpretation*

TRA was tested using results of the analysis of participants' experience to prove that the ESA educational process undergone by the participants positively affected their attitudes towards sustainability in energy planning. Interviews with participants indicated that they were made more aware of the importance of the underlying processes in shaping and influencing energy plans. Table 3. Assessment of participants' attitudinal change before and after the process, below presents an assessment of participants' behavioral ability towards attitudinal change before and after the process.

Table 3. Participants.

Training perceptions *(B – A > 0)	Number of participants scoring
Being aware of The Planning Process	
the objectives of this process	34
my objectives for being involved in this process	30
my organisations' objectives for being involved	28
the measurements of success for this process	18
the difference of this process to my past experiences	37
Being aware of The Partnership	
my role in the partnership	37
what excites me about the partnership	30
ability to make a change	32
the advantages my organization has gained	26
the contributions my organization has made to the partnership	13
Being aware of Participatory Planning	
the extent my experience of participatory planning	23
the extent my experience of communication with stakeholders	22
the advantages and disadvantages of participatory planning	31
my using different participatory tools	28
Being aware Planning for SED	
the principles, challenges, options and strategies of SED	37
the extent of encouraging planning for SED in my organization	26
my organization strategies for SED	21
my organization plans for SED update over time	19
SED strategies are translated to activities and levels of scale	15
local and grassroots activities inform the strategic view of SED	9
Being aware Sustainability	
understanding sustainability	26
my abilities of planning for sustainability	21
abilities of my organization of planning for sustainability	14
tools used for achieving sustainability	23
my experience in this field	26
incorporate information into planning	13
the decision making principles	17

*(B – A), (B minus A): change compared with initial pre-trained ability
A: perceptions before training, pre-trained ability (1 to 5).
B: perceptions after training, post-trained ability (1 to 5).

6 THE SUDANESE PARLIAMENTARY-CIVIC SOCIETY GROUP FOR SUSTAINABLE ENERGY (SPCGSE)

6.1 *The society as the major outcome of the research process*

The Sudanese Parliamentary-Civic Society Group for Sustainable Energy (SPCGSE) was established in 2003 to act as the first of its kind national energy forum which links individuals and groups from all over Sudan. SPCGSE can be described as follows

6.2 *Vision*

SPCGSE stated in its provisional mandate that it strives to strengthen the role in sustainable energy development in Sudan through advocacy, action, information exchange, and provision of training and supporting research. SPCGSE is committed to bringing affordable, energy-efficient products and services to the energy marketplace in Sudan. SPCGSE also works to establish a formal and comprehensive structure to pursue sustainable energy transformation, providing information on projects, reports, links, and publications.

6.3 *Mission*

The mission of the group is excellence in promoting community involvement and informed decision making and the formulation and communication of scientific knowledge that promote effective decision-making supportive of SED at all levels of the society.

6.4 *Membership*

Membership includes all energy stakeholders concerned with promotion of sustainable energy development in Sudan.

6.5 *Objectives*

Objectives include the following

- promoting community involvement and informed decision making in energy planning
- Promote education and research in SED
- Influence energy strategies and actions in order to enable deployment of SED Creating Strategies
- To act as a lobby group targeting the energy agenda of the National Assembly
- Contact and partner with others who are working on, or are interested in starting, energy projects in their community
- Organise training courses and workshops for activists and workers and energy professionals
- Develop a national database of community energy projects – see what others have tried and achieved, and share experiences
- Organise annual conferences on information about sources of grant funding support from a panel of experts for supporting SED
- Publish regular newsletters and provide information about new pilot energy projects, annotated bibliography of key references containing detailed design guidelines and other practical aids for community energy planning, and resource directory that lists information sources and technical resources including organizations, internet sites, modeling tools, consultants and businesses dealing with various aspects of energy planning.

REFERENCES

Akanksha Chaurey, J., Gururaja, Malini Ranganathan, and Babu, Y.D. A multi-stakeholders approach towards new partnerships for concrete actions. Tata Energy Research Institute, New Delhi. Pacific and Asian Journal of Energy 11(2): 141–150. 2001.

Brack, D., Calder, F., Dolum, M. From Rio to Johannesburg: The Earth Summit and Rio +10. Briefing Paper New Series, No. 19. Royal Institute of International Affairs, London. March 2001.

Capra, F. The turning point: science, society, and the rising culture, New York. 1982.

Churchill, Anthony. Overview to 'Financing the Global Energy Sector – the Task Ahead', Journal of the World Energy Council, December pp. 6–8. 1997.

Císcar, Juan Carlos. IPTS Photovoltaic Technology and Rural Electrification in Developing Countries: The Socio-Economic Dimension. The IPTS Reports. 1999.

Claussen, Eileen. Climate Change Solutions: A Science and Policy Agenda, Pew Center on Global Climate Change. December 6, 2004.

Connell, and Kubisch. A Theory of Change Approach to Evaluation. Washington, D.C., The Aspen Institute. January 5, 2004.

CSD IX on Energy and Energy-Related Aspects, Agenda 21, Chapter 7 & 9. 2001.

de Geus. Planning as Learning. Vol. 38, pp. 1–3. and The University of Strathclyde 1998.

Dyson, M. and Gregory, J. Solar Development Corporation: A new Approach for Photovoltaics. Paper presented at the 14th European Photovoltaic Solar Energy Conference and Exhibition. Barcelona, June/July 1997.
Ecoal. The road to sustainable development. Vol. 38, pp. 1–3. de Montfort University. June 2001.
ElKhalifa, Ali. The role of private investment in promoting applications of solar energy technologies in Sudan. Conference Paper. (1994).
Elliott, D. Renewable Energy and Sustainable Futures. CTS Special Sustainable Futures' Issue, pp. 261–274. April/May 2000.
Flavin, C. and O'Meara, M. Financing Solar Electricity. The off-the-grid solution goes global. World Watch, Vol. 10 (3). May/June 1997.
Goldenberg, J., Johansson, T.B., Reddy, A.K.N., Williams, R.H. Energy for the New Millennium. Ambio, Vol. 30, No. 6, pp. 330–337. September 2001.
Gore, W. Assessment of the Socio-economic and Environmental Impacts of the PV Applications of the Project Barrier Removal To Secure PV Market Penetration in Semi-Urban. Sudan SUD/99/G35/A/1G/99, MEM, UNDP, GEF. 2000.
Hanna, K.S. The paradox of participation and the hidden role of information: a case study. Journal of the American Planning Association, 66(4), pp. 398–410. 2000.
Hassan, H. Wardi. Social and Environmental Impacts of Community Based Solar Energy Interventions. UNDP report. 1997.
ITDG. Power to the People: sustainable energy for the world's poor forum for action on energy and poverty reduction. Johannesburg. 2002.
Jesper Munksgaard and Anders Larsen. A Sustainable Energy Policy. Cambridge University Press. August 1999.
Kemmis, S. and McTaggart, R. Participatory Action Research. Handbook of Qualitative Research, Second Edition. N.K. Denzin and Y.S. Lincoln, Ed. Thousand Oaks, Sage Publications: 567–605. 2000.
Koch-Weser Ciao. Climate Change and Its Likely Impacts on Development, Address to the UN Framework Convention on Climate Change(COP3). Kyoto. December 8. 1997.
Musu. I, E. Ramieri, V. Cogo Sustainability Indicators: An Instrument for Agenda 21 Working Paper 01.98 Fondazione ENI Enrico Mattei Venice. 1998. http://www.feem.it/web/index.html
Tippett, L.H.C. 1969. Participatory Planning in River Catchment, London. Oxford University Press.
United Nations. Economic and Social Council Commission on Sustainable Development Report on the Ninth session. Economic and Social Council Official Records. Supplement No. 9. 2001.

ABBREVIATIONS

AUW	Ahfad University for Women
CNS	Comprehensive National Strategy
ECNA	The Energy Committee of the National Assembly
ERI	Energy Research Institute
ESA	The Energy Sustainability Accelerator
GEF	Global Environment Facility
ME	Ministry of Electricity
MEM	Ministry of Energy and Mining
MST	Ministry of Science and Technology
MET	Ministry of Environment and Tourism
NGO	Non-Governmental Organization
PV	Photovoltaic
PV Project	The UNDP project Barrier Removal to Secure Photo-Voltaic Market Penetration in Semi-Urban Sudan
RSED	Rural Solar Energy Development Project
SECS	Sudanese Environment Conservation Society
SED	Sustainable Energy Development
SRED	Sustainable Rural Energy Development
TOE	Tons of Oil Equivalent
TPR	Tripartite Review
UNDP	United Nations Development Programme
UNFCCC	United Nations Framework Convention on Climate Change

Reclaiming the Desert: Towards a Sustainable Environment in Arid Lands – Mohamed (ed.)
© 2006 Taylor & Francis Group, London, ISBN 0 415 41128 9

Improving the energy and resource efficiency by imparting relevant technical education and skill training for achieving sustainable cities

Z. Ali
DDG (Tech), NISTE, Islamabad, Pakistan

ABSTRACT: The three major players for achieving sustainable cities are economy, environment and society. Instead of working on the improvement of each of the three spheres separately, a result-oriented approach is to work for interlinking the economy, environment and society. Actions to improve conditions in a sustainable community take these interlinks into consideration. Sustainable environment depends on sustainable development, which in turn linked to sustainable construction practices. Engineers, Architects and Construction Managers can improve the energy and resource efficiency of commercial and residential buildings by the use of practices for water conservation, better indoor quality, use of day light, energy efficiency for which we need to impart relevant technical education and skill training. Without the emphasis of these aspects in technical education and skill training the approach towards sustainable cities is rather going to be incomplete. First starting from awareness among higher secondary school classes and later inclusion of these aspects into engineering and technical education and in skill training in vocational institutes will provide the desired results.

1 INTRODUCTION

Efficient use of resources i.e construction materials, energy and water is the key for a sustainable city. Engineers, Architects and Technicians needs to improve the energy and resource efficiency of buildings by the use of practices for energy efficiency, water conservation, better indoor quality, use of day light, energy for which we have to impart relevant technical education and skill training. Without the emphasis of these aspects in technical education and skill training the approach towards sustainable cities is rather going to be incomplete. Starting from awareness among higher secondary school classes and then inclusion of these aspects into engineering/technical education and in skill training in vocational institutes will provide the desired results.

1.1 *Sustainable city*

For a Sustainable City the achievements in physical development, economic and social sectors are made to last. This means that a sustainable city utilizes its supply of resources in an efficient manner and follows a development path in which the present progress does not takes place at the expense of future generations. An equilibrium has to be maintained between different issues in order to achieve an across the board development. Instead of tackling the issues separately, a result-oriented approach needs to be adopted for interlinking the economy, environment and society. Sustainable Communities have the advantages of good quality of life, job opportunities, natural open spaces, green parks as lungs of the city, clean air, quality drinking water, reduced waste, equality, lower crime and a sense of participation.

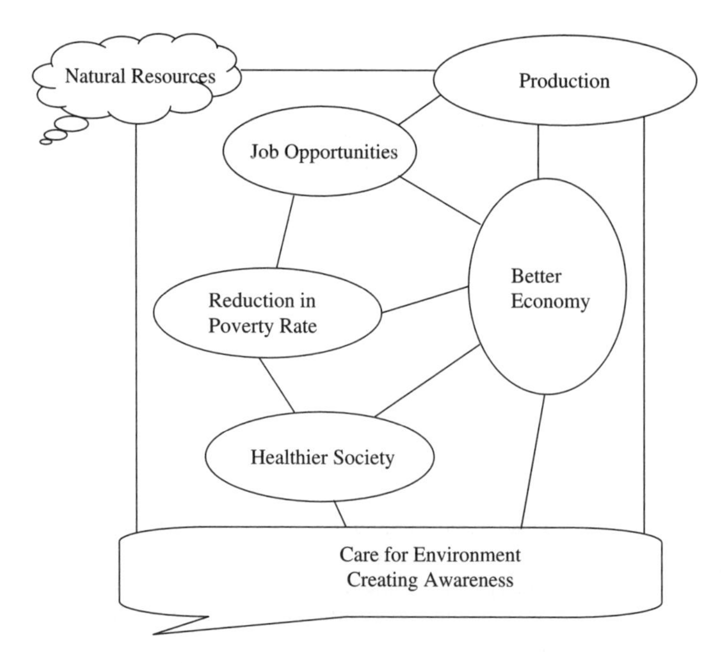

Figure 1. Interlinks between Environmental Care, Economy & Society.

1.2 *Interlinks in economy, environment and society*

Natural resources provide materials for production, which creates job opportunities and improvement in economy. Decline in job opportunities create poverty, which in turn affects society. Materials used for production, outdoor/indoor air quality, water quality affect health which have an effect on workers productivity. Health problems due to outdoor/indoor air quality, exposure to toxic materials on construction site/production units/offices/residences, have an effect on productivity and create a burden on health budget. Low productivity affect economy where as we need a better economy for job opportunities for healthier sustainable cities. Thus for sustainable cities, we need to keep in mind these interlinks in order to adopt an integrated approach. (Figure 1).

2 BASIC CONCEPTS

The basic concepts of energy efficiency, resources efficiency, water conservation, indoor and outdoor air quality, use of indigenous materials are:

2.1 *Energy efficiency*

Energy Efficiency can be achieved by using materials and systems that meet the following criteria:

- Materials, components, and systems that help reduce energy consumption in buildings and facilities.
- Planning and designing of a building system utilizing maximum daylight.
- Orientation and placement of building/house, keeping in view the sun cycle to maximize natural heating and cooling efficiency.
- Conservation of energy through proper insulation.
- Increased insulation for exterior walls and roofs.

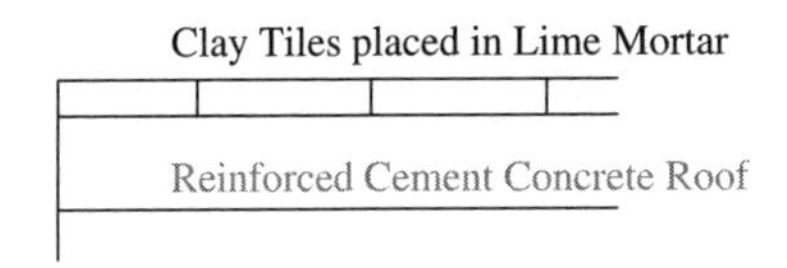

Figure 2. Clay tiles with lime mortar.

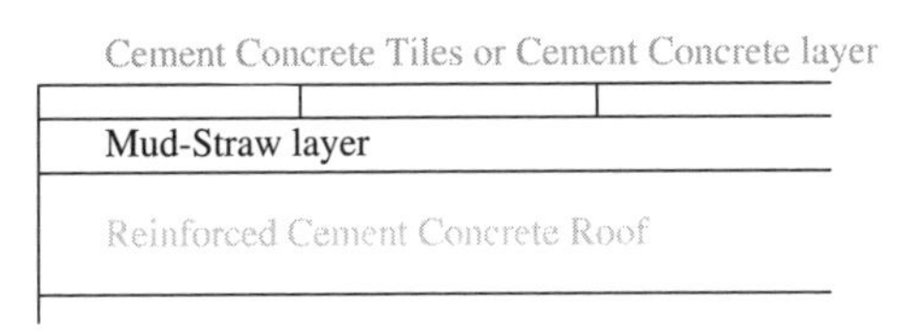

Figure 3. Mud straw as insulation.

- Locally available insulation materials and shades for reducing heat so as to minimize energy required for air conditioning. The examples of locally available insulation materials are:
  - Clay Tiles placed with lime mortar on concrete roofs as roof insulation (this reduces the amount of energy consumption for air conditioning) (Figure 2).
  - A layer of mud-straw on top of which a thin layer of cement mortar or cement concrete tiles as roof insulation. (This helps in reducing heat which in turn minimize energy requirement for air conditioning) Figure-3 (The Author had utilized both these techniques)

Further more, sustainable buildings use sensible design and planning to minimize energy usage over the lifetime of a project. Day lighting systems use strategically placed windows for easy access of light into buildings, substantially reducing the need for artificial light. Sensible building envelope design maximizes solar heating. These simple features can reduce building energy requirements by as much as 30%.

2.2 *Resource efficiency*

It can be accomplished by using materials that meet the following criteria:

- Recycled content: Products with identifiable recycled content.
- Natural or renewable: Materials obtained/harvested from sustainable managed sources.
- Resource efficient manufacturing process: Products manufactured with resource-efficient processes including reduced energy consumption.
- Reusable or recyclable: Select materials that can be easily dismantled and reused or recycled at the end of their useful life.
- Recycled or recyclable product packaging: Products enclosed in recycled content or recyclable packaging.
- Durable: Materials that are longer lasting or are comparable to conventional products with long life expectancies.

Minimizing construction waste: An important factor in resource efficiency is the construction waste. Construction waste should be minimized and Contractors should develop plans to minimize construction waste. Targets should be set with the aim of minimizing waste production on each project.

2.3 *Water conservation*

It can be obtained by using materials and systems that meet the following criteria:

- Systems and methods that contribute in reducing water consumption in buildings and conserve water in landscaped areas.

- Installing such devices in over-head tanks, which can minimize over flow of water from the over-head tank. (This will conserve energy as well since the unwanted electricity used in pumping water will be saved.)
- Use of sprinklers for watering lawn.
- Using such plants for landscape requiring less amount of water.

It can also be obtained by:

- Avoiding wastage of water on excessive car washing in front of houses. (A practice quite common in our cities.)
- Opening and regulating taps efficiently while using for water in houses and offices.
- Repairing of water systems in time to avoid wastage of water.

2.4 *Indoor air quality*

In the last several years, a growing body of scientific evidence has indicated that the air within homes and other buildings can be more seriously polluted than the outdoor air in even the largest and most industrialized cities. Other research indicates that people spend approximately 90 percent of their time indoors. Thus, for many people, the risks to health may be greater due to exposure to air pollution indoors than outdoors. (CPSC Document # 450)

Indoor Air Quality can be improved and enhanced by using materials that meet the following criteria:

- Low or non-toxic: Materials that emit few or no carcinogens, reproductive toxicants, or irritants or produce lesser irritating emissions.
- Minimal chemical emissions: Products that have minimum chemical emissions.
- Moisture resistant: Products and systems that resist moisture and minimize or limit the growth of biological contaminants in buildings.
- Healthfully maintained: Materials, components, and systems that require only simple or non-toxic methods of cleaning.

The quality of the indoor environment has a large impact on the people who use buildings. In many industrialized areas, people spend 90% of their time indoors, and research indicates that indoor air quality, exposure to natural sunlight, and the use of non-allergenic materials can improve the health and productivity of a building's occupants (Bartlett and Baldwin 1994).

Pollutants in our indoor environment can increase the risk of illness. Several studies by EPA and independent scientific panels have consistently ranked indoor air pollution as an important environmental health problem. While most buildings do not have severe indoor air quality problems, even well maintained building can sometimes experience episodes of poor indoor air quality (EPA 1997).

2.5 *Outdoor air quality*

Outdoor air quality also affect health of urban city population when polluted by vehicles and in some cases by industrial units inside cities or adjacent to city limits. Outdoor air polluted by vehicular and other emission from industrial units is generally referred as air pollution. Air pollution affect health, which in turn have an effect over productivity of workers and health budget.

According to a study the air quality in six major cities in Pakistan is deteriorating rapidly. The number of vehicles in those cities has increased by some 300 percent. Air pollution in Karachi, Islamabad, Rawalpindi and Lahore has reached serious proportions because of growing levels of CO_2 and SO_2. In Islamabad alone the number of registered vehicles was just 60,000 where as in March 2005, the number of vehicles was 340,000 according to an official of vehicle registration authority. Almost similar is the situation in other major cities, Karachi and Lahore.

2.6 *Indigenous construction materials*

Use of indigenous locally available material has manifold effects:

- Reduction in transportation cost. (This will conserve energy as well)
- Contribution in local economy. (Improvement in economy)
- Employment generation for local people. (Job opportunities)

3 IMPARTING RELEVANT TECHNICAL EDUCATION & SKILL TRAINING

"Building on more than 30 years of experience in environmental education, education for sustainable development must continue to highlight the importance of addressing the issues of natural resources (water, energy, agriculture) as part of the broader agenda of sustainable development. In particular, the links with societal and economic considerations will enable learners to adopt new behaviors in the protection of the world's natural resources, which are essential for human development and indeed survival. Humanity is dependent on the goods and services provided by ecosystems. Thus, the protection and restoration of the Earth's ecosystems is an important challenge." (UNESCO, Education for Sustainable Development, United Nation Decade 2005–2014). Education for sustainable development must find a central place across the full spectrum of educational endeavors if it is to provide the opportunity for all people to learn the values, behavior and lifestyles required for positive societal transformation (Ikeda 2004).

In view of the increasing awareness and need for sustainable cities, the concepts and information about achieving sustainable cities should now be propagated. Moreover, the basic information/ideas should also be included in general/science education, technical education and skill training. For example information about efficient use of resources, energy efficiency, water conservation, importance of water, indoor air quality, outdoor air quality, etc.

3.1 *Basic concepts at school level*

Basic concepts of efficient use of resources, efficient use of energy, realization of importance of water and water conservation should be included in social studies and science subjects in 9th & 10th (Secondary school certificate or Matriculation) and in trade skill subject of recently introduced scheme of Matric (Tech) in our country. It will be better, if the basic concepts be included in other streams like commerce, arts of secondary school certificate as we have to educate the society as well. The same should also be conveyed through seminars and other annual functions. Another aspect is to include poster projects about sustainable cities, sustainable buildings in 9th class or 10th class for creating awareness about sustainable buildings, green building materials and conservation of resources.

3.2 *The three streams*

After ten years of schooling i.e after passing 10th class i.e Secondary school certificate/Matriculation there are three streams for getting vocational, technical or engineering education, respectively (Figure-4).

3.2.1 *Vocational education*

The energy and resource efficiency concepts should be included in 12 to 18 months skill training/trade programs termed as vocational education offered at various vocational training institutes since the pass out from these vocational training institutes enter into trades that have a direct impact on environment.

3.2.2 *Technical education*

A good number of students opt for three years of diploma course in different technologies after 10th class. They work as technicians and form a vital link between engineers and skilled worker. For

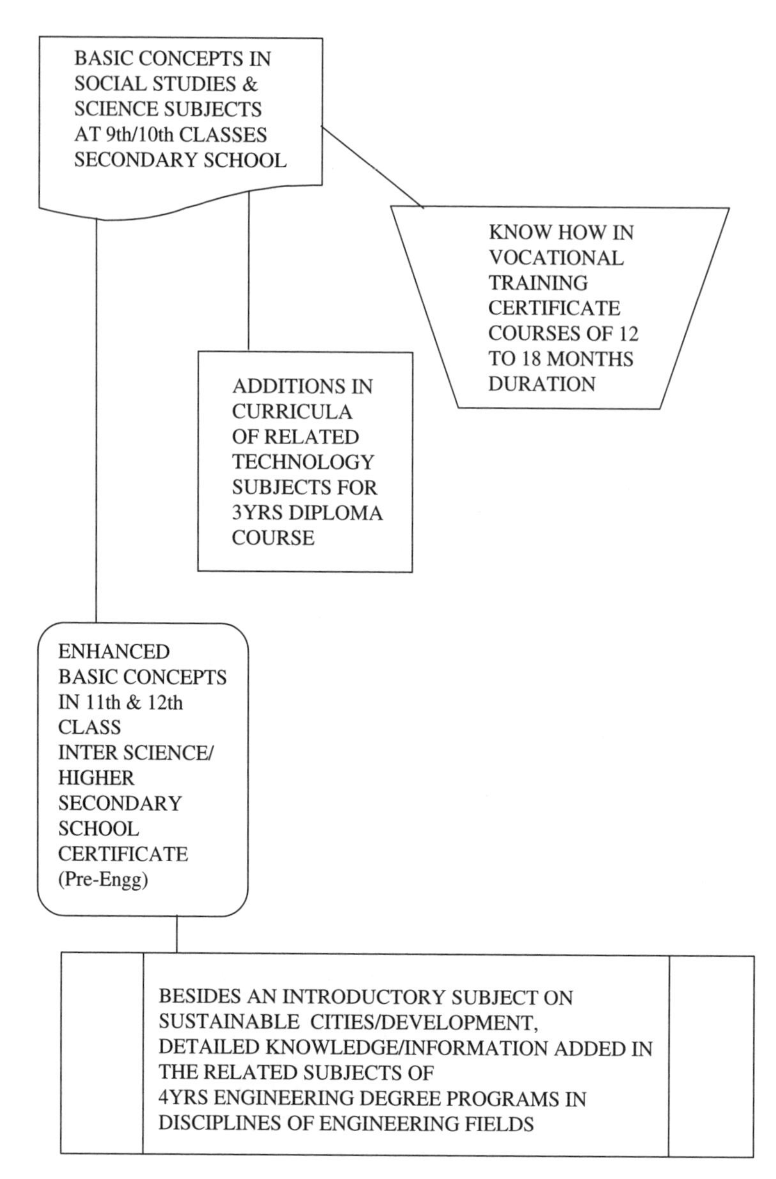

Figure 4. Vocational, technical and engineering streams.

sustainable cities, engineers, architects, technicians and contractors have to meet the increasing demand for energy and resource efficient buildings. As construction companies and contractors move to meet market demands, there will be an increasing need for technicians who have knowledge of sustainable building concepts. To meet this need, educators will have to develop new curricula and techniques. An initial approach will be to introduce these concepts in the related respective areas of buildings, materials, energy, etc.

3.2.3 *Intermediate/higher secondary certificate/college education*

For those who wish to go for engineering education at engineering universities, they first have to successfully complete 2 years Intermediate science (Pre-Engg) course at science colleges. Parallel to this other streams are Inter commerce, Inter arts, etc. An introductory course should be given to all streams (as awareness for society) and initial concepts introduced in Inter science (Pre-Engg) as they have to pursue engineering education.

3.2.4 *Engineering education*

After the entry requirement of two years college education, the four years engineering degree program at engineering universities starts. Besides an introductory subject on sustainable cities/sustainable development, an initial approach will be to introduce these concepts in the related respective areas of different engineering fields.

But ultimately, the most important area will be the development of "whole building" education and "whole city" education. Unlike conventional buildings, sustainable buildings utilize an integrated "whole building" approach that optimizes a building's energy efficiency, indoor environmental quality, and resource and material utilization. For example, an increase in the thermal efficiency of a building envelope will often reduce the tonnage of air conditioning required to cool the building. At the same time, increasing the thermal efficiency of a building often reduces fresh air intake, which in turn can lead to sick building problems.

Unfortunately, most of the engineering/technical education programs use a modular approach to education that is similar to the design and construction process of a building. On construction projects, designers, engineers, technicians and contractors perform their respective tasks without regard to the project as a whole. Similarly, technical education programs often provide well-conceived individual technical subjects, while failing to provide students with a complete understanding of how building systems are integrated. As such, there is a real need for the development of a "whole building" approach to engineering/technical education that will allow students to understand not only the parts of a building, but how the whole building operates as well. This will require more emphasis on how buildings are developed and designed, and how interdisciplinary teams can be used to maximize energy efficiency, reduce resource waste, and improve the environmental quality of the structures we construct. Then this approach can be transformed into a "whole city" approach to engineering/technical education.

In the beginning, keeping in mind the above mentioned concepts, following areas can be included in engineering/technical education:

Energy efficiency

- Information about materials, components, and systems that help reduce energy consumption in buildings and facilities.
- Additions in design courses of a building system utilizing maximum daylight.
- Proper insulation techniques for conservation of energy.
- Increased insulation for exterior walls and roofs.
- Know how about locally available insulation materials and shades for reducing heat so as to minimize energy required for air conditioning.

Materials efficiency

- Materials and products selection by evaluating several characteristics such as reused and recycled content, zero or low gassing of harmful air emissions, zero or low toxicity, locally harvested materials having sustainability, recyclable, durability, longevity, and locally produced. Such products promote resource conservation and efficiency.
- Use dimensional planning and other material efficiency strategies. These strategies reduce the amount of building materials needed and cut construction costs. For example, design rooms on 4-foot multiples to conform to standard-sized wallboard and plywood sheets. This will reduce wastage as well.
- Reuse and recycle construction and demolition materials. For example, using inert demolition materials as a base course for a parking lot keeps materials out of landfills and costs less.

Water efficiency

- Minimize wastewater by using ultra low-flush toilets, low-flow showerheads, and other water conserving fixtures.
- Use of re-circulating systems for centralized hot water distribution.
- Install point-of-use hot water heating systems for more distant locations.
- Use of state-of-the-art irrigation controllers and self-closing nozzles on hoses.

Indoor air quality

Recent studies reveal that buildings with good overall environmental quality can reduce the rate of respiratory disease, allergy, asthma, sick building symptoms, and enhance worker performance. Selection of construction materials and interior finish products with zero or low emissions to improve indoor air quality should be included in the curricula of technical education. Many building materials and maintenance products emit toxic gases, such as volatile organic compounds (VOC) and formaldehyde. These gases can have a detrimental impact on occupants' health and productivity.

4 CONCLUSIONS

1. The need of the hour is to introduce and include sustainable building concepts in engineering and technical education for having sustainable cities.
2. Some introduction should also be given in general and science education, so that city people who are going to use a house or a building can ask/demand for better indoor quality, energy efficiency, etc.
3. Further research in sustainable cities and sustainable development to be carried out for the better and efficient use of our resources keeping in mind healthier environment for urban population and future generations.
4. The three areas of economy, environment and society needs to be interlinked for a purpose oriented approach.
5. Use of local/indigenous building materials needs to be encouraged.
6. Research in local low cost environmental friendly building material to be carried out.

REFERENCES

Bartlett, P. & Baldwin, R. 1994. Assessing the Environmental Impact of Buildings in the UK. Building. Research Establishment. Proceedings of the US Green Building Conference 1994.

CPSC Document # 450, Consumer Product Safety Commission and Environmental Protection Agency.

EPA 1997. An Office Building Occupant's Guide to Indoor Air Quality, Indoor Air Publications, EPA.

Ikeda, D. 2004. Education for Sustainable Development, The Japan Times, Nov. 22, 2004.

UNESCO 2005–2014. Education for Sustainable Development, United Nation Decade.

Global warming

Reclaiming the Desert: Towards a Sustainable Environment in Arid Lands – Mohamed (ed.)
 ISBN 0 415 41128 9

Modeling GHG emissions reductions from different waste management strategies: Case study

M.A. Warith & A.K. Mohareb
Department of Civil Engineering, Ryerson University, Toronto, Canada

ABSTRACT: The threat of climate change caused by the emission of anthropogenic carbon has encouraged the international community to develop agreements such as the United Nations Framework Convention on Climate Change and the Kyoto Protocol. Upon ratification of the Kyoto Protocol, Canada committed, through the 2008–2012 commitment period, to reducing its greenhouse gas (GHG) emissions to 570 MT eCO_2, from 1990 levels of 610 MT eCO_2. The solid waste sector in Canada generated 24 MT eCO_2 in 2000, 23 MT eCO_2 of which were produced by landfill gas (LFG). The transport of waste likely generated a further 740 kT eCO_2 that are not accounted for in the waste sector by the National Greenhouse Gas Inventory. It is likely that waste transport emissions will increase as fewer landfills are sited and further source separation of wastes occurs. The benefits of source reduction, recycling, LFG capture for energy recovery, composting, anaerobic digestion and incineration are discussed. This paper presents an estimation of the GHG emissions from the solid waste sector using the waste disposal, recycling, and composting data from Ottawa, Canada for the year 1999, as well as the results of an audit of residential units performed the same year. This evaluation determined that, among the options examined, waste incineration, further source separ-ation of recyclables, and anaerobic digestion of organic wastes have the greatest benefits for reducing GHG emissions in the City of Ottawa's waste sector. Challenges surrounding the installation of incineration facilities in Canada suggest that improved diversion of recyclable materials and anaer-obic digestion of organic materials are the optimal options for the City of Ottawa to pursue.

1 INTRODUCTION

There are significant emissions reductions to be achieved from diverting solid waste from ultimate disposal. There are life cycle emissions reductions to be achieved through different waste management strategies. Some waste management options offer direct emissions reductions from the point of disposal (i.e. incineration, LFG recovery). This paper provides modeling scenarios for GHG emissions generated through several waste management options, using the City of Ottawa as a case study. The model employed in this study is the Integrated Waste Management Model (IWM) produced by the Environment and Plastics Research Institute and Corporations Supporting Recycling (EPRI and CSR, 2002).

1.1 *GHG emissions from the management of MSW*

In order to model the GHG emissions from the solid waste sector using the IWM model, a set of inputs is required. These include:

- Categorization of the solid waste stream (how much waste is generated, how much of each category of waste is sent for landfilling, recycling, etc.);
- Knowledge of the destination of the wastes, i.e. how much of each component of the solid waste stream is recycled, composted, incinerated, digested anaerobically, or landfilled;
- For recyclables, the distance from materials recovery facilities to markets;

Table 1. Waste disposal statistics for the Region of Ottawa-Carleton, 1999 (Solid Waste Division, 2000).

	Amount generated (T)
Residential solid waste	262,160
Sent for disposal to landfill	181,633
Waste sent for recycling	56,840
Sent from MRF to market	55,287
Paper sent from MRF to market	44,024
Plastics, glass and metals sent from MRF to market	11,263
Sent from MRF to disposal	1,553
Organic waste sent for composting	23,687
Compost produced	14,200
Organic waste diverted by backyard composting (estimated)	7,000

- For landfilled/incinerated wastes/digested wastes, the distance travelled to the site of waste treatment and disposal;
- For a landfill, the extent of LFG capture and ultimate use of LFG (flaring, conversion to energy); and
- For compostable materials, their distance to the composting site.

The city of Ottawa, Ontario, is Canada's capital, located at the confluence of the Gatineau River, the Rideau River and the Ottawa River, on the south shore of the Ottawa River. On Jan. 1, 2001, the amalgamated 11 former municipalities, combined to create the City of Ottawa. The City of Ottawa had a population of 774,000 in 2001, and is growing rapidly; the population is projected to be greater than 1 million by 2011 (City of Ottawa, 2003).

2 MEETING THE KYOTO PROTOCOL IN THE CANADIAN WASTE INDUSTRY

Solid waste management decisions affect the amount of GHG emissions being generated by the waste sector. Source reduction decreases the amount of materials being consumed, eliminating all of the related GHG emissions. Recycling enables the reduction of GHG emissions at many stages of the material life cycle, as it diverts materials from landfills (especially beneficial when diverting rapidly biodegradable materials) and reduces processing emissions (Pickin *et al.*, 2002; NOPP, 2002). Diversion of organic materials, for either composting or anaerobic digestion, removes a significant portion of readily degradable waste from landfills, preventing these wastes from degrading anaerobically and generating CH4. Incineration also prevents organic wastes from decomposing anaerobically by combusting these wastes to generate power or steam.

Of the 31% of waste that was diverted from the MSW stream, 22% of the total material was sent for recycling, and the remaining 9% was organic waste sent for composting. The City of Ottawa estimates that of the material sent for composting, 60% becomes compost product, while the other 40% is lost as moisture and CO_2 (Solid Waste Services Division, 2002). This contradicts Brunt *et al.* (1985), who state that 30–50% of organic materials that are composted result in a compost product. There were 70,000 backyard composters in the Region of Ottawa-Carleton in 1999, and the city estimates that these composters divert an average of 100 kg/unit/year. This reduces the amount of organic waste sent for disposal by 7,000 T (Solid Waste Division, 2000).

One of the required inputs for the model is the amount of waste generated in each specific category. The waste audit results are provided in Table 2.

The distance that recycled materials travel from MRFs to markets is a required input for the model. These values are provided in Table 3.

Finally, the emissions from the transport of waste to the MRFs and the landfill are necessary inputs for the model. Knowing the fraction of the City of Ottawa's waste that is dealt with by the

Table 2. Results of a waste audit performed in Ottawa during the autumn and winter of 1999 (IES, 2000).

Materials	Refuse collected total (kg) (1)	MRF total contamination (kg) (2)	Net recyclables total (kg) (3)	Gross recyclables total (kg) (4) = (2) − (3)	Total waste generation (kg) (5) = (1) + (4)
Other materials	827.80	25.00	0.00	25.00	852.80
Disposable diapers	152.60	13.70	0.00	13.70	166.30
Tires	16.90	0.00	0.00	0.00	16.90
Textiles	89.50	0.50	0.00	0.50	90.00
Construction and demolition	168.50	3.90	0.00	3.90	172.40
Household special waste	12.00	0.00	0.00	0.00	12.00
Unclassifiable items	388.30	6.90	0.00	6.90	395.20
Total	3704.60	88.35	2415.45	2503.80	6208.40

Table 3. Distance to markets for recyclable materials from Ottawa.

Material	Paper	Glass	Ferrous materials	Aluminum	Plastics
Distance to market (km)	50	0	550	800	500

Table 4. Estimated distances travelled by all waste handling vehicles, Region of Ottawa-Carleton.

Material handled	Distance travelled by vehicles handling residues (km)	Distance travelled per tonne of waste (km/T)
Garbage	1309300	7.2
Blue box	468200	38.7
Black box	747800	16.7
Leaf & Yard waste	383100	16.2

City's own fleet, the total km travelled by all garbage, recycling and organic collection vehicles in the City of Ottawa (both public vehicles and private vehicles) can be determined (Table 4).

All of the required information is now available to model the GHG emissions generated by waste in the City of Ottawa/Region of Ottawa-Carleton, and to model the potential of reducing emissions in Ottawa.

3 MODELING GHG EMISSIONS FROM THE WASTE SECTOR IN THE CITY OF OTTAWA

There are several models available to help determine the best options for reducing GHG emissions from waste in the City of Ottawa. This study employed the Integrated Waste Management Model for Municipalities, produced by Corporations Supporting Recycling (CSR) and the Environment and Plastics Research Institute (EPRI), and offered through the University of Waterloo (EPRI and CSR, 2002).

The modeling of GHG emission considered 8 scenarios:

1. The "do-nothing" approach (all waste is sent to landfill, no recycling occurs, no LFG capture occurs);
2. The current approach of the City of Ottawa (LFG is captured and flared, recycling and composting occur at present rates);
3. LFG capture system upgrade (LFG capture rate increases to 50%, energy conversion facilities are installed, at 30% energy conversion efficiency);

4. Improved diversion rates of materials currently being recycled, without improvements at any other point in the system (capturing 50% of the recyclable/compostable material that are currently not being captured);
5. Diversion of food waste (30% capture rate), with organics diverted sent for composing;
6. Diversion of food waste (30% capture rate), with organics diverted sent for anaerobic digestion;
7. Source reduction of recyclable materials by 10%;
8. Incineration of materials being sent to landfill, with energy recovery.

4 RESULTS FROM THE MODELING EXPERIMENTS

The results displayed in this section are indicative of the GHG emissions and energy consumption statistics that may be possible through the 8 scenarios analyzed. They should not be assumed to be exact, due to the assumptions necessary in the model and the actual conditions that are present in waste management in Ottawa.

The emissions generated from the base case (the current state of waste management in the City of Ottawa are displayed in Table 5. A summary of the emissions is demonstrated in Table 6. The emissions have been categorized as collection (or transport) emissions, landfilling emissions, recycling emissions (this category includes emissions from reprocessing recycled materials),

Table 5. GHG emissions from the base case (case 2) scenario.

Scenario	Management option	Mass of waste managed (T)	CO_2 emissions (T)	CH_4 emissions (T eCO_2)	N_2O emissions (T eCO_2)	GHG emissions (T eCO_2)
2 – current case	Collection	262,160	7,687	173	2,085	9,946
	Landfill	182,255	1,221	90,993	384	92,598
	Recycling	56,218	14,516	33	6,062	20,611
	Composting/ anaerobic digestion	23,687	0	0	0	0
	Energy from waste (incineration)	0	0	0	0	0
	Gross emissions	262,160	23,424	91,199	8,531	123,155
	Virgin material displacement credit	n/a	−45,622	−1,153	−11,507	−58,282
	Net emissions	262,160	−22,198	90,047	−2,975	64,873

Table 6. Gross and net GHG emissions from scenarios run in IWM model.

Scenario	Mass handled differently from Case 2 (T)	Gross GHG emissions (T eCO_2)	Virgin materials displacement credit (T eCO_2)	Net GHG emissions (T eCO_2)
1 – landfill all waste	79,905	212,517	0	212,517
2 – current case	0	123,155	−58,282	64,873
3 – upgrade LFG capture system	182,255	106,268	−58,282	47,987
4 – increase diversion of manufactured goods by 50%	22,611	116,695	−83,689	33,006
5 – diversion of organic waste (30%) – composting	27,028	102,606	−58,282	44,324
6 – diversion of organic waste (30%) – anaerobic digestion	27,028	107,454	−58,282	49,173
7 – source reduction of manufactured goods (10%)	8,495	118,137	−55,444	62,693
8 – incineration of waste	182,255	103,337	−58,281	45,056

composting or anaerobic digestion emissions (these two categories are combined as there are no cases where both composting and anaerobic digestion occur) and energy from waste emissions.

The reduction in GHG emission in all cases are presented in Table 7.

The results of the modeling are displayed in the following figures. Figure 1 demonstrates the gross emissions from each case. Figure 2 displays gross emissions from each case by emission source. Figure 3 displays the net emissions, with the crosshatched section demonstrating the emissions reductions achieved by recycling and source reduction in each scenario.

The cases that generate the fewest gross emissions are case 5 (diversion of food waste for composting) (102.6 kT eCO_2) and case 8 (incineration of refuse) (103.3 kT eCO_2), respectively. However,

Table 7. Reductions in GHG emissions and energy consumption for 8 cases compared with case 2.

Scenario	Mass handled differently from Case 2 (T)	Gross energy consumption reduction compared to base case (GJ)	Net energy consumption reduction compared to base case (GJ)	Gross GHG emissions reductions compared to base case (T eCO_2)	Net GHG emissions reductions compared to base case (T eCO_2)
1 – landfill all waste	262,160	869,420	−1,060,232	−89,362	−147,644
2 – current case	0	0	0	0	0
3 – upgrade LFG capture system	182,255	65,239	65,239	16,886	16,886
4 – increase diversion of manufactured goods by 50%	22,611	−244,641	487,417	6,459	31,867
5 – diversion of organic waste (30%) – composting	27,028	10,548	10,548	20,549	20,549
6 – diversion of organic waste (30%) – anaerobic digestion	27,028	26,852	26,852	15,700	15,700
7 – source reduction of manufactured goods (10%)	8,495	93,891	−99,365	5,018	2,180
8 – incineration of waste	182,255	1,386,164	1,386,164	19,818	19,818

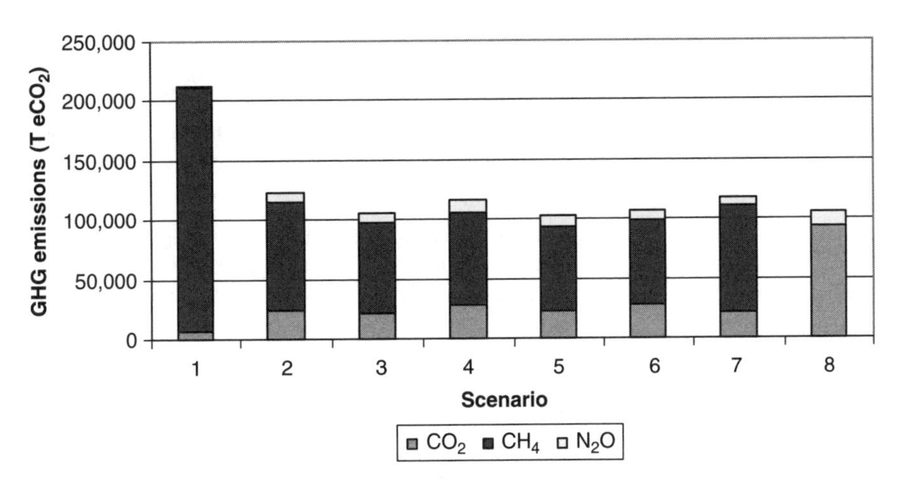

Figure 1. Gross GHG emissions from the Region of Ottawa-Carleton's waste sector for the eight waste management scenarios proposed, by greenhouse gas.

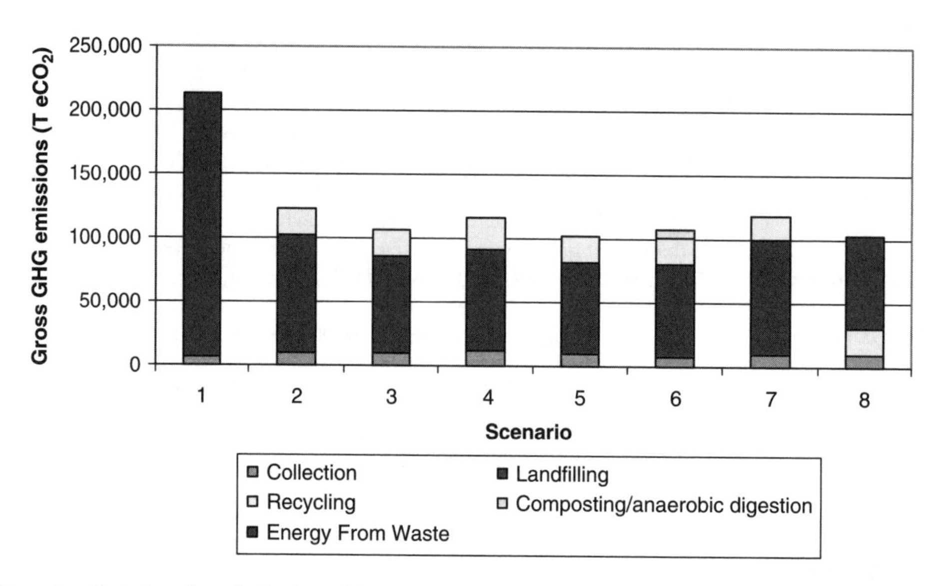

Figure 2. Emissions from the Region of Ottawa-Carleton's waste sector for the eight waste management scenarios proposed, by emission source.

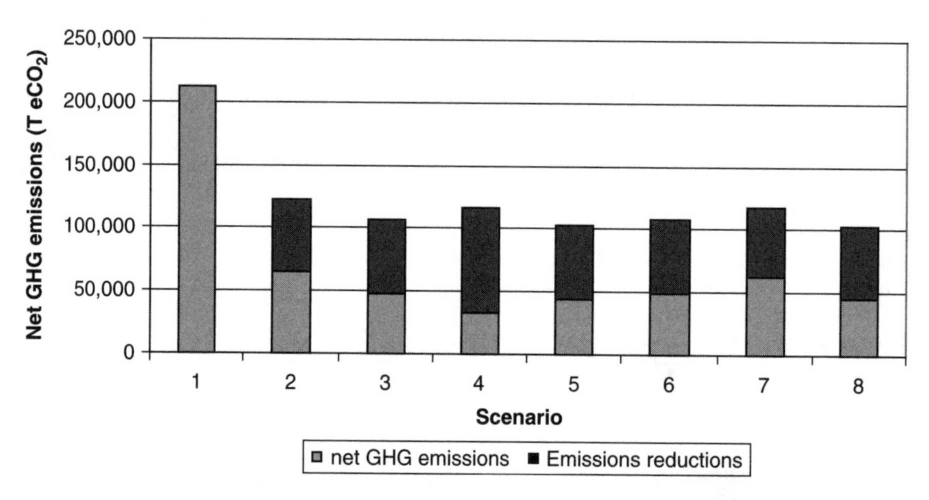

Figure 3. Gross (full bar) and net emissions (in solid blue) from the Region of Ottawa-Carleton's solid waste sector for the eight waste management scenarios proposed.

for net GHG emissions, the lowest amount was generated by scenario 4 (capturing 50% of the remaining recyclable material) (33 kT eCO_2), while scenario 5 (44.3 kT eCO_2) and scenario 8 (45 kT eCO_2) also emitting fewer GHGs than the other cases considered. It should be noted that these figures only consider the emissions from the generation of waste by consumers, handling of waste by collectors, and disposal of waste. The model does not consider all emissions throughout the life cycle of the materials. Surprisingly, the anaerobic digestion scenario does not demonstrate lower emissions than the composting scenario (emissions are 49.2 kT eCO_2), though displacing fossil fuel generated power, as would be the case in anaerobic digestion, should reduce emissions. This will be examined in the discussion on energy consumption of different waste management strategies.

The fossil fuel emissions that are displaced by the generation of power associated with some of the waste handling methods discussed, such as incineration, LFG capture for energy and anaerobic

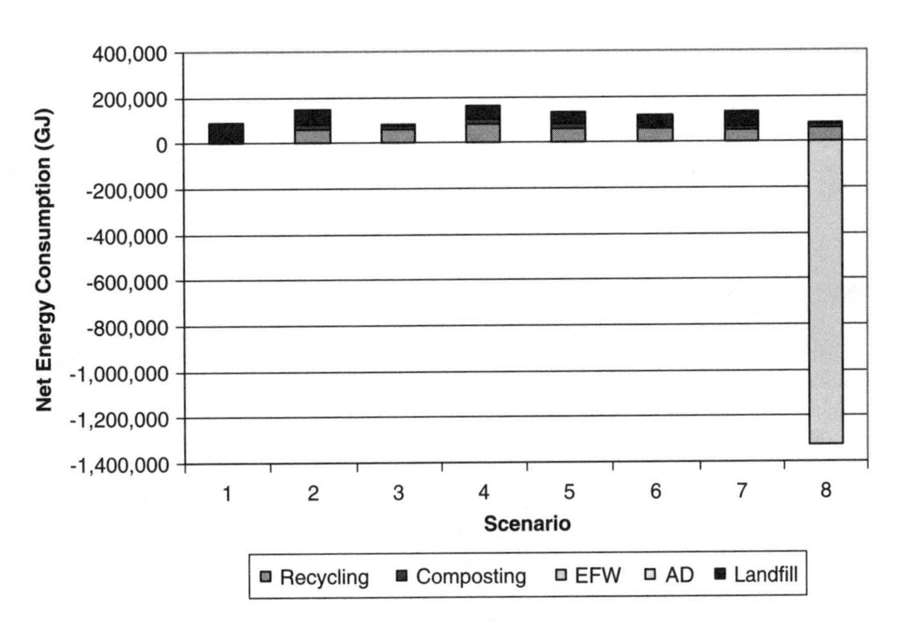

Figure 4. Net energy consumption for the Region of Ottawa-Carleton from the eight waste management strategies proposed, excluding energy consumed and avoided through the reprocessing of recycled materials.

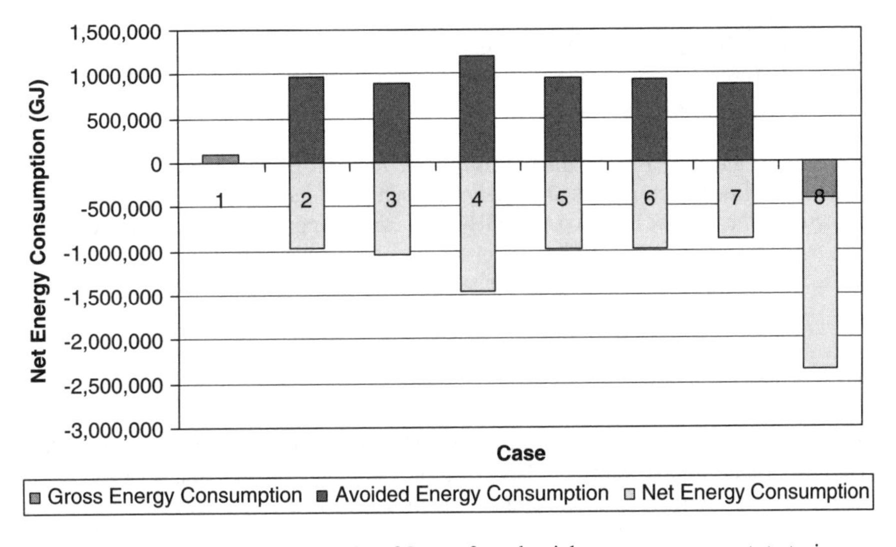

Figure 5. Net energy consumption for the city of Ottawa from the eight waste management strategies proposed.

digestion, are not incorporated into the figure. The net GHG emissions (waste emissions less emissions offset through recycling) from each scenario are demonstrated in Figure 3.

It is also interesting to determine how much energy is being consumed in each of the scenarios, as energy consumption is linked to greenhouse gas emissions. Most emissions recognized by the model result from energy consumption, including CO_2, CH_4 and N_2O emission. The exceptions to this are LFG emissions. Figure 4 demonstrates the energy consumed through the waste management cycle excluding energy consumed by material recycling and energy displaced by avoiding the use of virgin materials through recycling.

Figure 5 displays the energy consumed by waste management, including the energy consumed by processing recycled materials and the energy displaced by avoiding the production of virgin materials. The consideration of energy consumed through the reprocessing of recycled materials account for the much higher positive energy consumption totals in Figure 5. The energy consumed through reprocessing recycled materials was between 5 and 10 times greater than the energy consumed in waste management for scenarios 2 through 7, though this energy consumption is more than compensated for by avoided energy consumption through the production of materials from virgin feedstocks.

5 CONCLUSIONS

The modeling scenarios analyzed in this paper indicate that emissions are reduced the most through diverting 50% more of the waste that is currently recycled. Though this may seem a challenge, there already has been a noticeable increase in diversion rates among single-unit residences in Ottawa between 1997 and 1999; paper diversion increased from 67% to 81% (capturing 42% of what was being discarded) while container diversion increased from 51% to 65% (capturing 29% of what was being discarded) (Solid Waste Division, 2000). Minimum recycled content legislation will further promote material recycling, encouraging greater GHG emissions reductions. Incineration offers significant emissions reductions, but the political cost of this option may be too high. Source reduction reduced net emissions by only about 0.25 T eCO_2/T waste reduced, but this was partially because of the assumption in that scenario that only manufactured materials would be source reduced. Most LFG emissions come from the degradation of food and yard wastes; paper diversion has been successful in Ottawa, with 64% of paper being diverted at present. Since it was assumed that the capture rates of recyclable materials would remain consistent, this scenario witnessed a reduction in the amount of material being recycled. Source reduction has perhaps a much greater benefit that the model predicts. Anaerobic digestion has potential to reduce the emissions from the sector, but its greatest potential lies in reducing net energy consumption. A law banning the landfilling of organic material, such as the one implemented in Nova Scotia in 2000 (Nova Scotia Environment and Labour, 2003), would increase the amount of organic material being diverted, making anaerobic digestion a more feasible emissions and energy consumption reduction strategy. The benefits of composting lie in the low per-tonne cost for GHG reductions. Improved LFG capture reduced GHG emissions by 17,000 T eCO_2, as well as exporting energy. It appears that recycling and anaerobic digestion each offer significant GHG emissions and energy consumption reductions.

6 RECOMMENDATIONS

Canada's waste sector generated 24 MT eCO_2 in 2000, of which 23 MT eCO_2 were produced by the decomposition of organic wastes in landfills. The main options for Canada's solid waste sector to contribute to reducing emissions are through source reduction, recycling, LFG capture, composting, anaerobic digestion, and incineration. LFG capture, composting, anaerobic digestion and incineration will directly reduce emission from landfills, while source reduction and recycling will indirectly reduce emissions, perhaps to a greater measure, through displacing the processing of virgin materials. The IWM model was used to analyze the emissions from the waste sector. It was found that further diversion of materials through recycling has the greatest effect on reducing emissions, mostly through reducing production emissions through feeding recycled material instead of virgin material to production processes. Incineration follows with the next highest levels of GHG emissions reductions, followed by composting. Energy consumption is reduced the most through incineration with energy recovery, followed by recycling and improved LFG capture. The IWM underestimates the GHG emissions and energy consumption reductions on anaerobic digestion. It is possible that anaerobic digestion will reduce emissions and energy consumption significantly, but at 30% diversion of food wastes as well as current diversion rates of yard wastes, the emissions reductions will still rank behind diversion of an additional 50% of the recyclable material and incineration. Further efforts to divert recyclable materials will achieve the greatest amounts of emissions reductions in the Canadian solid waste sector.

ACKNOWLEDGEMENTS

This paper could not have been completed without the help of Ralph Torrie of Torrie-Smith Associates, Kala Harris at the City of Ottawa, Frédéric Tremblay at la Ville de Gatineau, and Alain David of Environment Canada, whose wisdom, as well as their references and files have proven invaluable.

REFERENCES

Brunt, L.P., Dean, R.B. and Patrick, P.K. (1985). "Composting." In: *Solid Waste Management, selected topics*. Suess, M.J. (editor). World Health Organization, Geneva, Switzerland, pp. 70–75.

City of Ottawa (2003). *Demographic-Economic Facts*. Document #25-08, Development Services Department, Research & Technical Services Division, City of Ottawa, Ottawa, ON.

EPRI and CSR (Environment and Plastics Research Institute and Corporations Supporting Recycling) (2002). *Integrated Waste Management Model for Municipalities*. Microsoft Excel Model, University of Waterloo, Waterloo, ON. Available at: http://www.iwm-model.uwaterloo.ca/english.html.IES (Integrated Environmental Services) (2000).

IWSA (Integrated Waste Services Association) (2000). *About Waste-to-Energy – Clean, Reliable, Renewable Power*. IWSA, Washington, DC. Available at: http://www.wte.org/waste.html

NOPP (National Office of Pollution Prevention) (2003a). "Landfill Gas – CCI – Anaerobic Waste Digestion." Technical Bulletins, National Office of Pollution Prevention, Environment Canada, Ottawa, ON. Available at: http://www.ec.gc.ca/nopp/lfg/en/issue17.cfm

NOPP (National Office of Pollution Prevention) (2003b). "Landfill Gas – SUBBOR Pilot Waste Processing Project," Technical Bulletins, National Office of Pollution Prevention, Environment Canada, Ottawa, ON. Available at: http://www.ec.gc.ca/nopp/lfg/en/issue18.cfm.

Nova Scotia Environment and Labour (2003). *Status Report of Solid Waste Resource Management in Nova Scotia*. Department of Environment and Labour, Halifax, NS. Available at: http://www.gov.ns.ca/enla/pubs/status03.pdf

Solid Waste Services Division (2002). *Division Profile and Summary of Activities in 1999, 2000 & 2001 – Structure, Roles and Responsibilities, Accomplishments, Data and Activity Accounts*. Transportation, Utilities and Public Works, The City of Ottawa, Ottawa, ON.

Reclaiming the Desert: Towards a Sustainable Environment in Arid Lands – Mohamed (ed.)
© 2006 Taylor & Francis Group, London, ISBN 0 415 41128 9

Challenges of Japan's petroleum industry to prevent global warming

T. Nishikawa
Technology and Environmental Safety Department, Petroleum Association of Japan, Tokyo, Japan

ABSTRACT: The Kyoto Protocol went into effect on February 16, 2005 following Russia's ratification. More than seven years had passed since the Kyoto Protocol had been adopted at the third session of the Conference of Parties (COP) in December 1997 in Kyoto, Japan. According to this agreement, by 2010 Japan must reduce its emissions of CO_2 and other greenhouse gases by 6% compared to 1990 levels (requiring a reduction of 13.6% compared to actual figures from FY2002).

In April 2005 the Japanese government announced a Cabinet decision on a basic plan entitled the Kyoto Protocol Target Achievement Plan. This plan details the amounts of CO_2 emissions allowable for each of five sectors: the industrial sector, energy-conversion sector, commercial sector, residential sector, and transportation sector. The goal of the plan is to maintain CO_2 emissions at +0.6% of FY1990 levels (1.056 billion t-CO_2) in FY2010.

Home fuel in the oil refining industry is classified under the energy-conversion sector. In other types of industry, energy consumption from power plants in the electricity industry corresponds to this classification. Since 1997, major industries in the industrial sector and the energy-conversion sector have formulated and made public the Voluntary Action Plan on the Environment through the impetus of Nippon Keidanren. The Japanese government has also recognized their achievements in the Kyoto Protocol Target Achievement Plan. The plan aimed at achieving these goals by FY2010 calls on the industrial sector to steadily enforce their plan of voluntary action and to be thorough in energy management at factories and the like.

Through the Voluntary Action Plan, the oil refining industry set a goal to reduce the unit energy consumption in oil factories by 10% in FY2010 and has exceeded this goal. The aggregate amount of energy savings has been estimated at 2.5 million kl.

This paper covers the efforts of Japan's oil factories aimed at energy savings, including (1) results of the Follow-up to the Voluntary Action Plan on the Environment, (2) primary steps taken for saving energy, (3) shifts in the amount of CO_2 emissions, and (4) fluctuations in unit energy consumption, and the factors behind these efforts. The paper also details efforts related to reducing CO_2 emissions through the supply of environmentally compatible petroleum products. It is our hope that this report will effectively contribute to the rationalization of energy use and the production of environmentally compatible products by oil factories throughout the world in order to achieve the current goal of reducing greenhouse gases (GHG).

1 INTRODUCTION

In response to the entry into force of the Kyoto Protocol in February 2005, the Japanese government formulated the "Kyoto Protocol Target Attainment Plan." It presents Japan's commitment to reduce total greenhouse gas emissions 6% from the reference year by the first commitment period, and is the first step toward achievement of the ultimate goal of the United Nations Framework Convention on Climate Change. The plan includes measures for attaining the 2010 CO_2 reduction target in the industrial sector through steady promotion of the Keidanren Voluntary Action Plan

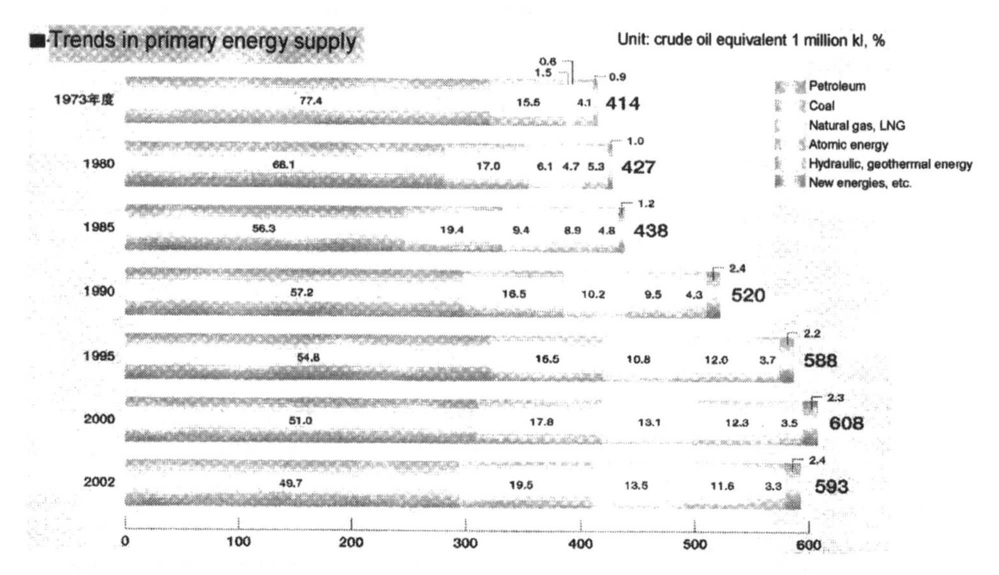

Figure 1. Trends in primary energy supply in Japan.
Source: Ministry of economy and industry, comprehensive energy statistics.

that has been implemented to date, and emphasizes the ever increasing significance of voluntary action plans in attaining the 2010 targets.

As shown Fig. 1, petroleum is a key energy accounting for 48.9% of Japan's primary energy consumption (FY2003), and the significance of its role is expected to continue.

As energy is the engine of Japan's economy, maintaining a balance between global warming countermeasures and energy supply could be considered a priority issue for the mid to long term.

The "Basic Energy Plan" approved by the Cabinet in October 2003 and the "Energy Demand Outlook for 2030 (report)" presented by the Comprehensive Resource Energy Survey Committee's Demand and Supply Group in March 2005 state the following:

(1) Petroleum will continue to be important in the future from the perspective of economy and convenience
(2) The energy demand and supply outlook for 2030 shows that "the share of petroleum will decrease, but petroleum will continue to be an essential energy source accounting for a share of about 40%."

Japan's petroleum industry is fulfilling its responsibility of securing a stable supply of oil, which is the essential energy supporting both the environment and economy. At the same time, it is working to supply clean petroleum products that have little burden on the environment and to disseminate effective utilization methods of petroleum as a precious, limited resource. Refineries, the source of petroleum products, are also making active efforts for environmental conservation and improvement of energy efficiency. As shown in Fig. 2, energy consumption in refineries corresponds to a mere 5% of domestic oil consumption. Nevertheless, the petroleum industry acknowledges energy conservation in refineries as a priority environmental measure, and has been making continuous efforts for energy conservation since the first oil shock of 1973.

The Petroleum Association of Japan formulated the "Petroleum Industry's Voluntary Action Plan for Global Environment Conservation" in February 1997. As a member of Keidanren, the association established a goal of reducing unit energy consumption in refineries as part of its efforts to conserve the global environment. Since then, it has been examining and disclosing its performance in the following six areas each year. The section below discusses the eighth follow-up results of fiscal year 2004 and past energy conservation measures in reference to "(1) Global warming prevention measures (energy conservation)."

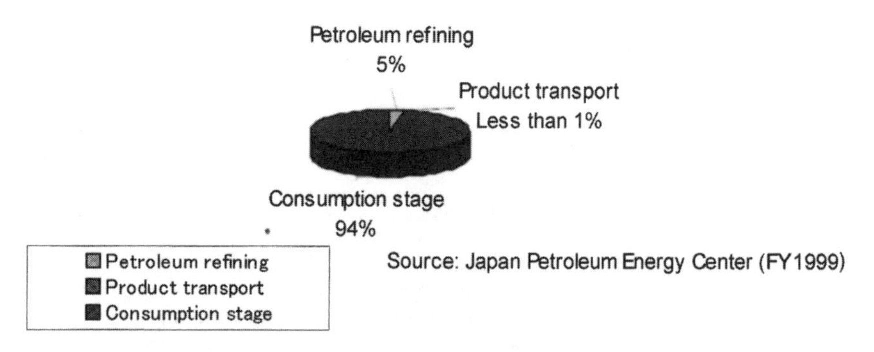

Figure 2. Ratio of energy used for petroleum refining to oil consumption.

(1) Global warming prevention measures (energy conservation)
(2) Waste control measures
(3) Creation of environmental management systems
(4) Overseas technical cooperation
(5) Marine environment conservation measures
(6) Promotion of publicity activities

2 GOALS AND ACHIEVEMENTS OF THE FY2004 "PETROLEUM INDUSTRY'S VOLUNTARY ACTION PLAN FOR GLOBAL ENVIRONMENT CONSERVATION"

2.1 *Targets (FY2010/FY1990)*

The Petroleum Association of Japan established the following targets in its voluntary action plan for the production, transport, and consumption sectors. Every year, it follows up on their performance, and continues to make efforts toward achievement of the targets.

Production sector	Unit energy consumption in refineries	10% reduction
Transport sector	Amount of fuel consumed for the transport of petroleum products	9% reduction
Oil consumption sector	Amount of energy conserved with the diffusion of petroleum cogeneration	1.4 million KL/year
Waste materials	Amount ultimately disposed	66.7% reduction

2.2 *Follow-up results*

Table 1 shows the follow-up results for fiscal year 2004. Each sector showed steady performance, and the production sector in particular already achieved its target owing to efforts that will be discussed later. Its future goal is to keep maintaining that level.

3 GLOBAL WARMING PREVENTION MEASURES

3.1 *Energy conservation efforts in refineries*

3.1.1 *Reduction targets*

The petroleum industry is promoting energy conservation based on the "unit energy consumption in refineries" adopted by refineries around the world as a tool for assessing the effects of energy conservation efforts. Its target for fiscal year 2010 is the achievement of a 10% reduction from the fiscal year 1990 level (Table 2).

Table 1. Follow-up results of the FY2004 voluntary action plan.

	FY2004 performance	Comparison with FY 1990 performance	Rate of achievement
United energy consumption in refineries	8.80	▲13.6%	136%
Fuel consumption in the transport sector	1.4 million KL	▲7.3%	73%
Energy conserved with the diffusion of petroleum cogeneration	1.06 million KL (2.65 million KW diffusion increase)	1.06 million KL (2.65 million KW diffusion increase)	78%
(Reference) CO_2 emission	43.54 million tons	10.51 million ton increase	–
Waste materials ultimately disposed	0.12 million tons	▲87.9%	132%

Table 2. Target of the voluntary action plan of refineries. (Unit: crude oil equivalent KL/crude oil distillation unit equivalent throughput 1,000KL.)

Subject matter	FY1990 (reference year)	FY2010 (target year)	Reduction target	Reference
Unit energy consumption in refineries	10.19	9.17	▲10%	Corresponds to approx. 1.29 million KL/year

Table 3. Trends in unit energy consumption in refineries. (Unit: crude oil equivalent energy consumption KL/crude oil distillation unit throughput 1,000 KL.)

Subject matter	FY1990 (reference year)	FY2003	FY2004	2004/1990
Unit energy consumption in refineries	10.19	8.87	8.80	86.4%

3.2 *Performance and outlook of unit energy consumption and energy consumption*

Unit energy consumption in refineries: As shown in Table 3 and Fig. 3, unit energy consumption in refineries was 8.80 in fiscal year 2004 owing to the promotion of energy conservation efforts. This corresponded to a 0.7% decrease from fiscal year 2003, a 13.6% decrease from fiscal year 1990, and a 136% achievement of the target.

Table 3 shows that unit energy consumption steadily decreased from fiscal year 1990 and basically leveled off in fiscal year 1999. Hereafter, refineries would need to employ new processes to improve the quality of petroleum products in response to changes in the demand structure, such as further demands for light oil and for sulfur-free gasoline and diesel fuel (below 10 ppm). This would cause an increase in energy consumption, but as long as the petroleum industry continues and makes additional energy conservation efforts to maintain the present unit energy consumption, it is sufficiently possible to attain the fiscal year 2010 targets.

Energy consumption in refineries: As shown in Table 4, energy consumption in fiscal year 2004 was 16.69 million KL (crude oil equivalent). This corresponded to a 29.7% increase from fiscal year 1990, but basically maintained the same level as fiscal year 2003, with only a slight decrease of 0.4%. Energy consumption in refineries has hardly changed since 1997.

Compared to when no measures were implemented, energy consumption decreased an estimated 2.5 million KL (crude oil equivalent) from fiscal year 1990.

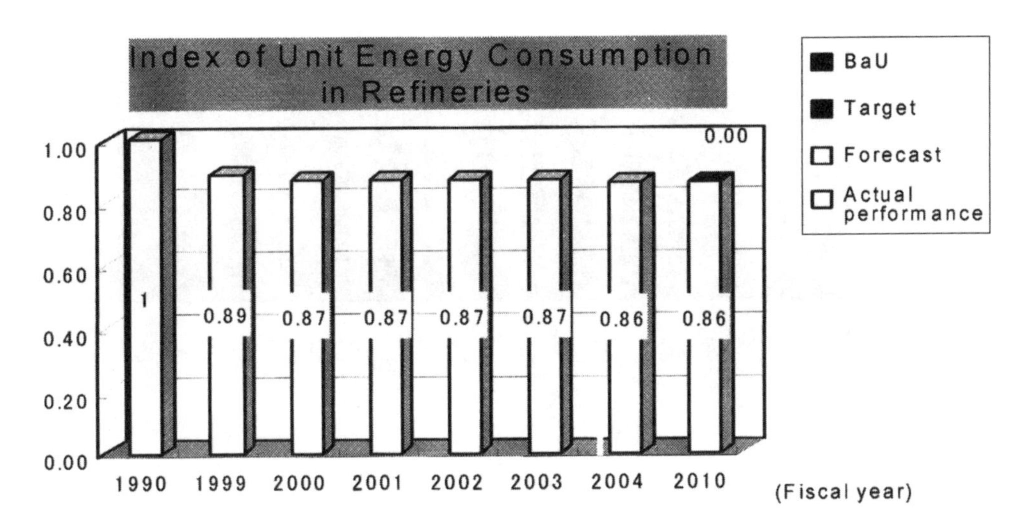

Figure 3. Trends in the index of unit energy consumption in refineries.

Table 4. Trends in energy consumption in refineries. (Crude oil equivalent million KL/year.)

Subject matter	FY1990	FY2003	FY2004	2004/1990
Energy consumption	12.87	16.75	16.69	29.7%

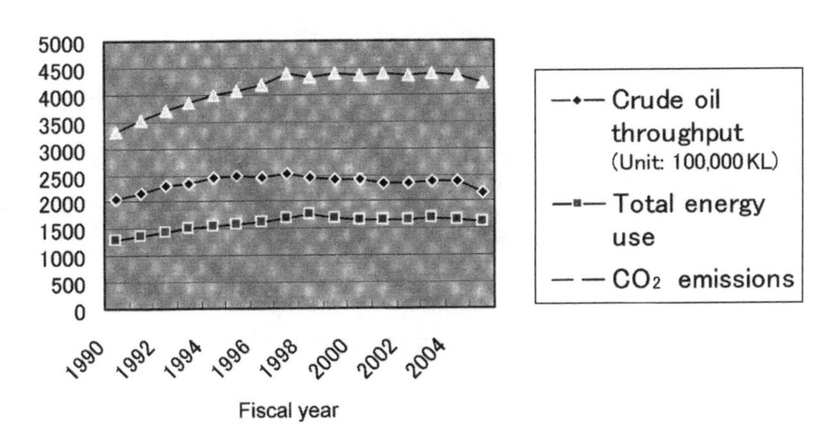

Figure 4. Trends in crude oil throughput, energy utilization, and CO_2 emissions.

Note: Reduction effect is equal to the difference between the value calculated based on the assumption that each year's unit energy consumption in refineries is the same as that of fiscal year 1990 and the value of actual performance.

When considering fiscal year 1990 as the reference year, crude oil throughput marked a 15% increase in fiscal year 2004, as shown in Fig. 4. This lead to a 30% increase in energy use.

Changes in the product demand structure: There was a shift in demand structure to light oils. Specifically, demand for heavy oil decreased due to a shift in fuels in industrial sectors and power

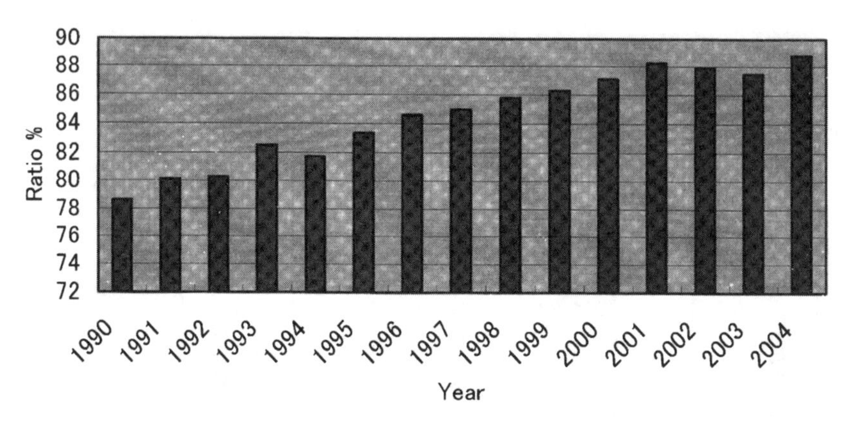

Figure 5. Light oil ratio.

plants, while the demand for automotive fuels such as gasoline and diesel fuel increased. As shown in Fig. 5, the ratio of light oil increased from 77% in fiscal year 1990 to 89% in fiscal year 2003.

In order to respond to these demand changes, refineries newly constructed and increased the operations of cracking units for converting heavy oil to gasoline and diesel fuel, which resulted in increased energy consumption.

Environment-friendly quality: In order to respond to demands for environment-friendly product quality, which called for lower benzene content in gasoline and lower sulfur content in gasoline and diesel fuel, refineries needed to newly construct and increase operations of benzene extraction units and gasoline and diesel fuel desulfurization units. This in turn increased energy consumption. Fig. 6 shows trends in the sulfur content of gasoline and diesel fuel.

Energy consumption in refineries in fiscal year 2004 marked an approximately 30% increase compared to fiscal year 1990. Fig. 7 shows an analysis of the effects of this increase.

Response to increased product demand	15%
Response to changes in product demand structure	24%
Response to demand for environment-friendly quality	8%
Total	47%
Reduction through energy conservation efforts	▲17%

4 PRESENT STATE OF GREENHOUSE GAS EMISSIONS

4.1 *CO_2 emissions from refineries*

CO_2 emissions from refineries totaled 43.54 million tons in fiscal year 2004, corresponding to a 31.8% increase over fiscal year 1990. This could be attributed to efforts to respond to the stable supply of petroleum products, changes in demand structure, and demand for environment-friendly quality mentioned above. This increase exceeded the CO_2 emission reduction effect of energy conservation efforts (improvement of unit emission) carried out by the petroleum industry (▲17%).

However, owing to the efforts of the petroleum industry, CO_2 emissions have basically leveled off in the past few years. Compared to when no measures were implemented, a reduction of approximately 6 million tons of CO_2/year was achieved.

As shown in Table 5 and Fig. 8, CO_2 emissions decreased 310,000 tons compared to the previous year (FY2003). Unit CO_2 emission was 22.9 CO_2-tons/1,000 KL in fiscal year 2004, corresponding to a 12.3% decrease from fiscal year 1990. In the past few years, it has leveled off.

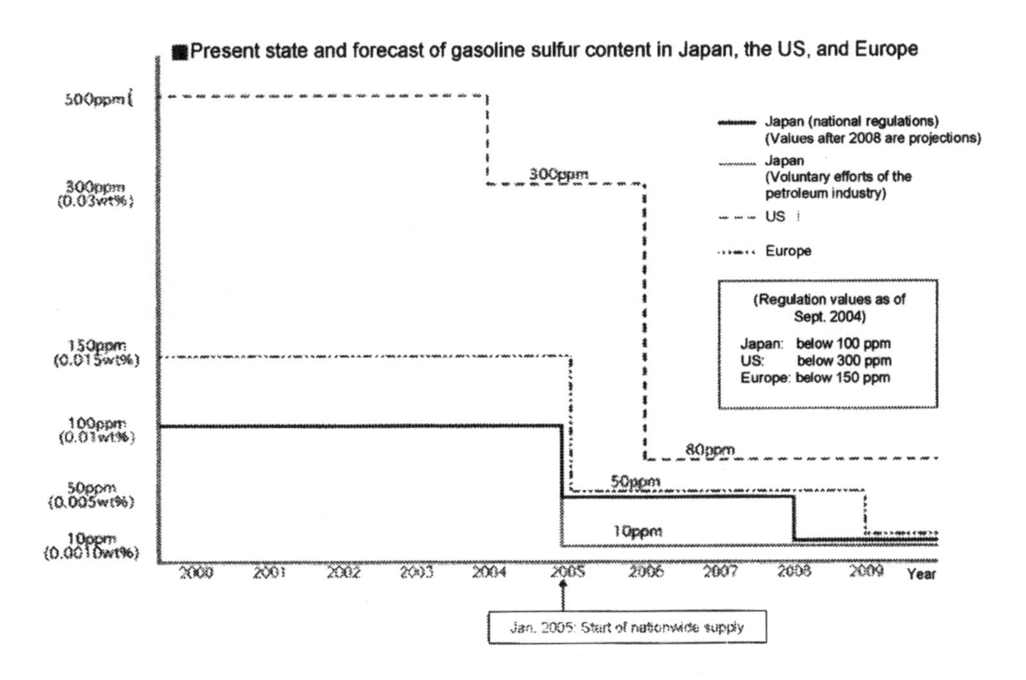

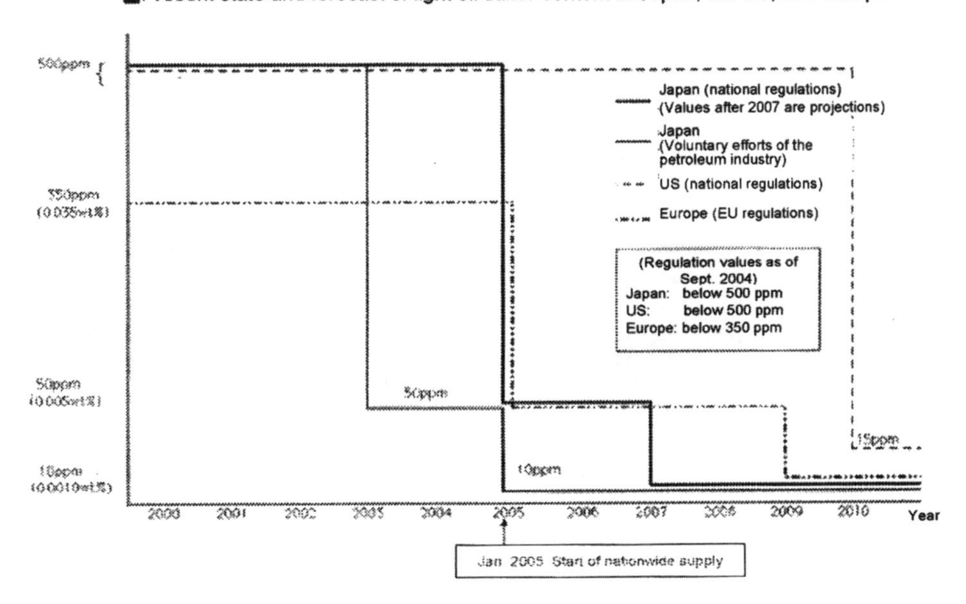

Figure 6. Trends in the sulfur content of automotive gasoline and diesel fuel in Japan, the US, and Europe.

Note: Unit CO_2 emission is the amount of CO_2 emissions per crude oil distillation unit equivalent throughput.

CO_2 reduction effect is equal to the amount of energy conservation calculated based on the assumption that each year's unit energy consumption in refineries is the same as that of fiscal year 1990 and converted to CO_2 values.

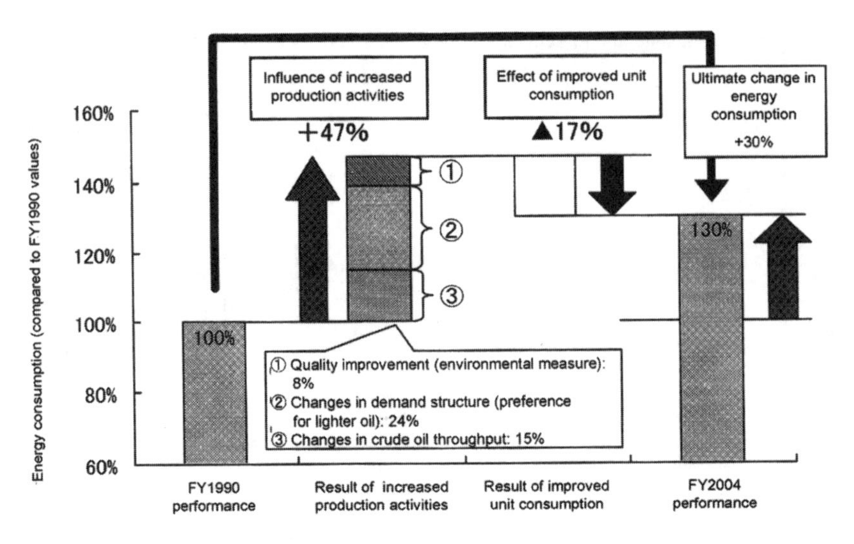

Figure 7. Quantitative analysis of factors involved in the increase in energy consumption.

Table 5. Comparison of CO_2 emissions and unit CO_2 emission in refineries.

Subject matter	FY990	FY2003	FY2004	2004/1990
Unit CO_2 emission (note)	26.1	23.2	22.9	87.7%
CO_2 emission (million ton)	33.00	43.85	43.54	131.8%

Unit CO_2 emission = (CO_2-tons/crude oil distillation unit equivalent throughput 1,000 KL)

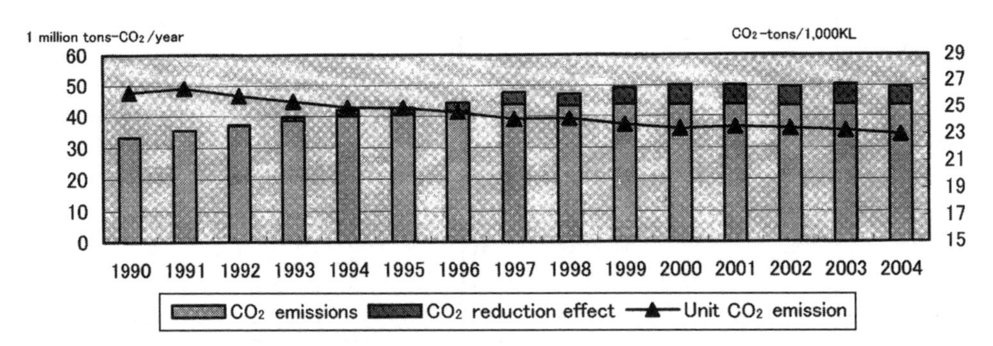

Figure 8. Trends in CO_2 emissions and unit CO_2 emission of refineries.

5 ENVIRONMENTAL MEASURES RELATED TO THE QUALITY OF PETROLEUM PRODUCTS

5.1 *Supply of environment-friendly products*

The petroleum industry is the world's leader in environmental measures that address the quality of automotive fuels such as gasoline and diesel fuel (see Fig. 9).

A measure to reduce the benzene content of gasoline to below 1 volume% was enforced from January 2000 in the effort to reduce the emissions of hazardous air pollutants from automotive exhaust gases and storage tanks.

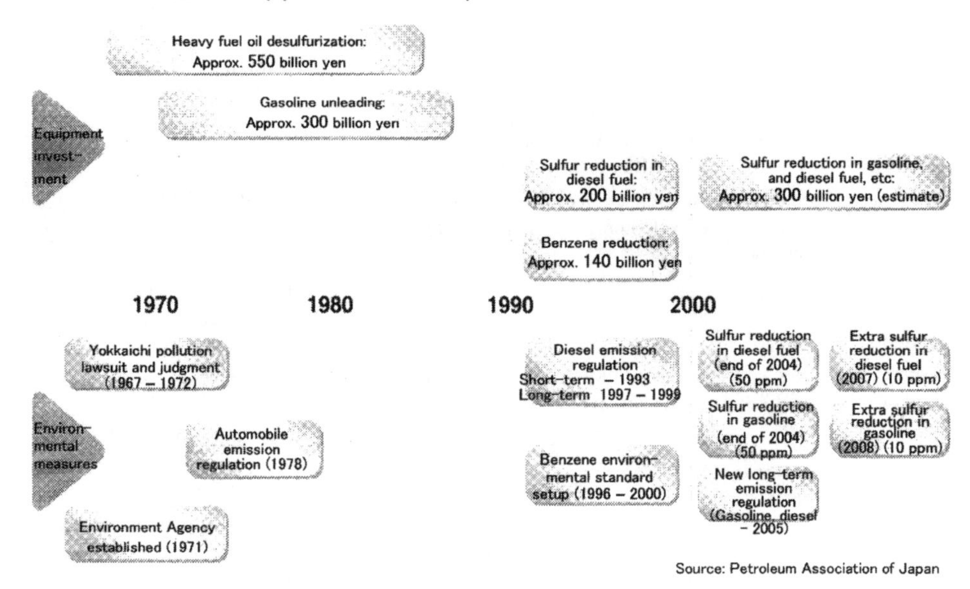

Figure 9. Trends in automotive fuel quality improvement measures and equipment investment in Japan.

As a measure to help reduce atmospheric hydrocarbons that are the primary cause of photochemical smog, summertime gasoline vapor pressure was lowered from 78 kPa and below to 72 kPa and below from 2001, and further down to 65 kPa and below from the summer of 2005.

In regard to diesel fuel, sulfur content was reduced from 0.5 mass% to 0.2 mass% in 1992, and from 0.2 mass% to 0.05 mass% in 1997, in order to reduce nitrogen oxides (NO_x) and particulates (PM) from the exhaust gases of diesel automobiles (trucks and buses).

In April 2003, prior to the enforcement of national regulations, the petroleum industry voluntarily moved up its schedule to reduce sulfur to 0.005 mass% (50 ppm) by about two years. Furthermore, from January 2005, it began the national supply of sulfur-free gasoline and diesel fuel containing less than 10 ppm of sulfur. This step was taken ahead of the scheduled national regulation that considered it appropriate to make sulfur-free gasoline and diesel fuel compulsory from fiscal year 2008 and 2007, respectively. It was a measure that set a precedent for the world.

As shown in Table 6, the elimination of sulfur from gasoline and diesel fuel is not only effective in achieving clean automotive gases as an air pollution measure. It could also maximize engine mileage performance when combined with an exhaust after treatment device that utilizes the properties of sulfur-free fuels, as well as improve mileage, and ultimately reduce CO_2 as a global warming countermeasure.

5.2 *Environment conservation measures and energy consumption in refineries*

The supply of environment-friendly products inevitably leads to an increase in energy consumption in refineries. The pursuit of energy conservation and energy efficiency continues to be a difficult challenge for the petroleum industry. For instance, as already mentioned, measures to reduce benzene from gasoline and sulfur from diesel fuel involved the construction of new refinery facilities and more energy consumption. In addition, the petroleum industry began the production of sulfur-free (less than 10 ppm) gasoline and diesel fuel from 2005. As will be discussed later, Japan's measures for energy conservation through the efficiency improvement of refinery operations are highly advanced measures compared to other countries. However, energy consumption in refineries is expected to increase hereafter in response to further demands for environment-friendly quality.

Table 6. Reasons why "Sulfur-free" fuels are environment-friendly: The effects of sulfur-free fuels on existing automobiles and automobiles with new type engines are as shown below.

		Gas emission countermeasures	Global warming countermeasures
Gasoline	Existing automobiles	SO_x reduction Improved NO_x reduction effect with longer life three-way catalysts	
	Automobiles with new type engines	NO_x reduction effect with usages of NO_x storage-reduction catalysts	Improved mileage with usage of direct-injection engines and lean-burn engines
			Fuel conservation effect with longer regeneration frequency of NO_x storage-reduction catalysts
Diesel (Light oil)	Existing automobiles	Improved PM reduction effect owing to prevention of DPF clogging	Fuel conservation effect with longer DPF regeneration frequency
		Improved NO_x reduction effect with longer life three-way catalysts	Fuel conservation effect with longer regeneration frequency of NO_x storage-reduction catalysts
	Automobiles with new type engines	Improved PM reduction effect owing to prevention of DPF clogging	Fuel conservation effect with longer regeneration frequency of DPF storage-reduction catalysts
		NO_x reduction effect with usage of NO_x storage-reduction catalysts	Fuel conservation effect with longer regeneration frequency of NO_x storage-reduction catalysts

In this way, the increase in CO_2 emissions accompanying the increase in energy consumption in refineries is "an inevitable result" of environmental conservation. At the same time, however, the promotion of high-performance automotive engines and the diffusion of high-mileage, ultra-low exhaust emission automobiles could improve the mileage of automobiles and greatly contribute to the reduction of CO_2 emissions and the promotion of exhaust gas countermeasures in the transportation sector. In other words, it is possible for the petroleum industry to reduce CO_2 emissions throughout Japan.

6 ENERGY CONSERVATION MEASURES IN REFINERIES

Refineries are making the following energy conservation efforts to achieve the targets of the petroleum industry (see Tables 7 and 8). They are also implementing energy conservation measures that promote the efficient utilization of heat and energy in cooperation with plants of other business in the vicinity rather than by a single refinery alone, through the complex renaissance project and other initiatives.

Energy conservation measures implemented by refineries in fiscal year 2004, including measures related to both refining facilities and utility facilities, are expected to produce an energy conservation effect of about 230,000 KL per year (crude oil equivalent, design-based). Equipment investment amount is estimated to be about 8 billion yen per year.

7 INTERNATIONAL COMPARISON OF ENERGY EFFICIENCY IN REFINERIES

Energy consumption efficiency of Japan's refineries is equal to or higher than that of refineries in the United States and Europe. Solomon Associates (a consulting firm in the US) conducted a

Table 7. Measures related to refinery facilities.

Present state	Future response (incl. technical development)
Promoting heat recovery • Complete control over heat and cold insulation of vessels and pipes • Installation, cleaning, and replacement of furnace air preheaters • Installation of waste heat boilers • Installation of various heat exchangers • Recovery of flare gas	*Promoting heat recovery* • Improvement of performance of heat and cold insulation materials • Addition of various heat exchangers • Installation of heat pumps • Development of technology for the recovery of low temperature waste heat and other unused energies
Promoting high efficiency and optimization of facilities • Reduction of air in furnaces (low O_2) • Mutual heat utilization between refining equipment • Optimization of pump capacity (impeller cuts) • Installation of process turbines (recovery of pressure energy) • Promotion of computerized control • Review of operation management values	*Promoting high efficiency and optimization of facilities* • Achievement of high-efficiency furnaces and heat exchangers • Strengthening of process turbines • Advanced computerized control • Achievement of high-performance oil refining catalysts (operations at even lower temperatures) • Achievement of inverter-type large scale motors • Introduction of a single-step loading adjustment mechanism in compressors

Table 8. Measures related to utility facilities.

Present state	Future response (incl. technical development)
Promoting heat recovery • Repair and replacement of furnace walls and various heat insulation materials • Installation, cleaning, replacement of air preheaters • Installation of boiler water preheaters • Installation of steam superheaters • Installation of economizers (heat recovery from waste gas)	*Promoting heat recovery* • Upgrading and achieving high efficiency air preheaters and other heat recovery equipment • Heat sharing among plants
Promoting high efficiency and optimization of facilities • Reduction of boiler air • Achieving high efficiency in power generation steam turbines • Optimization of pump capacity (impeller cuts) • Promotion of computerized control • Review of operation managements values	*Promotion of high efficiency and optimization of facilities* • Promoting high efficiency and optimization of various facilities *Introduction of high efficiency power generation facilities* • Upgrading to high efficiency power generation facilities • Promoting the installation of cogeneration facilities
Introduction of high efficiency power generation facilities • Gas turbine cogeneration • Diesel engine cogeneration	

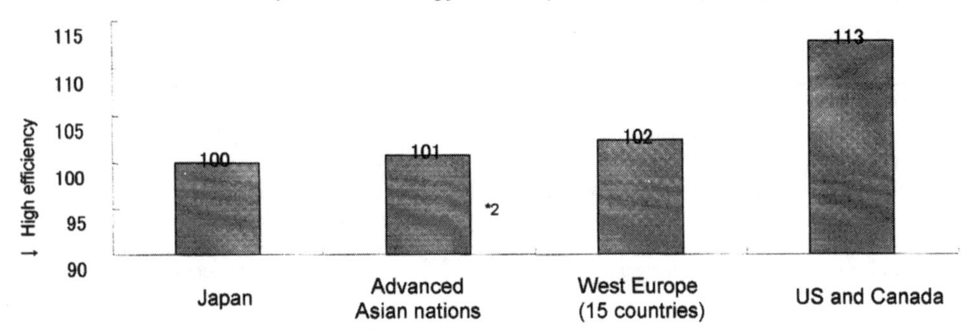

Figure 10. International comparison of energy consumption efficiency in refineries.
Created based on results of a survey conducted by Solomon Associates
*1 Solomon Associates' own index that uses equivalent throughput and has similar properties to unit energy consumption of refineries adopted in the voluntary action plan.
*2 Refers to Korea, Singapore, Malaysia, and Thailand, and excludes China.

comparative survey in 2002 based on its own energy consumption index (uses equivalent throughput; has similar properties to unit energy consumption of refineries adopted in the voluntary action plan; the lower the value, the higher the efficiency). According to the survey, when Japan was given a value of 100, the advanced Asian nations (Korea, Singapore, Malaysia, Thailand; excludes China) scored 101, Western Europe (15 countries), 102, and the United States and Canada, 113.

8 CONCLUSION

In its voluntary action plan for fiscal year 2010, Japan's oil refining industry established a goal of "reducing unit energy consumption in refineries by 10%." Owing to efforts that aimed to achieve the goal, it has produced results that exceed the goal. Cumulative energy conservation is estimated as approximately 3 million KL. When calculated from as far back as 1973, the year of the first oil shock, energy conservation effect amounts to a total of approximately 10 million KL. In this way, the oil industry worked to reduce CO_2 emissions through the steady implementation of energy conservation measures in refineries. At this rate, with continued efforts it should be sufficiently possible to achieve the fiscal year 2010 targets.

Japan's Kyoto Protocol Target Attainment Plan states that, in order to fulfill the commitment to achieve a 6% reduction over the value of 1990 which Japan pledged in the Kyoto Protocol, Japan must implement measures to achieve an additional emission reduction corresponding to 12% (approx. 148 million tons-CO_2 equivalent) as well as policies for their promotion in addition to those measures and policies that it are currently being implemented. The voluntary action plan of the Keidanren adopted by 34 major industries calls for a reduction in CO_2 emissions to below 0% from fiscal year 1990 to 2010. In these ways, with the steady implementation of voluntary action plans in the industrial sector, achievement of the targets looks promising. In regards the issue of CO_2 emission control in the private transportation sector, the shift to sulfur-free gasoline, diesel fuel, and other transportation fuels is expected to produce effective results in the sector.

Nevertheless, global scale efforts that transcend industries and country borders are indispensable to solving the global warming problem. Japan's petroleum industry hopes its energy conservation efforts could offer some point of reference to the environmental challenges of petroleum industry around the world.

Reference 2 Trends in energy consumption in refineries.

FY	Crude oil throughput (million KL) (1)	Crude oil distillation unit equivalent throughput (million KL)	Energy consumption (crude oil equivalent million KL)(2)	Cf (3)	Unit energy consumption in refineries (4)	CO_2 emission (million C-ton) <million CO_2-ton>	Light oil ratio (%)
1990	205.61	1,263	12.87	6.14	10.19	9.00 <33.00>	78.6
1991	216.81	1,336	13.77	6.16	10.31	9.63 <35.32>	80.1
1992	229.23	1,424	14.35	6.21	10.08	10.06 <36.87>	80.3
1993	234.77	1,518	15.07	6.47	9.93	10.53 <38.61>	82.5
1994	246.05	1,602	15.60	6.51	9.74	10.92 <40.05>	81.7
1995	242.80	1,623	15.78	6.68	9.73	11.04 <40.47>	83.4
1996	243.43	1,690	16.21	6.94	9.59	11.35 <41.62>	84.6
1997	250.98	1,820	17.08	7.25	9.38	11.95 <43.81>	85.0
1998	243.40	1,790	16.73	7.35	9.34	11.79 <43.22>	85.8
1999	241.10	1,850	16.77	7.67	9.06	11.94 <43.78>	86.3
2000	242.78	1,869	16.61	7.70	8.89	11.90 <43.64>	87.1
2001	235.21	1,865	16.58	7.93	8.89	11.95 <43.83>	88.3
2002	235.36	1,854	16.50	7.88	8.90	11.84 <43.40>	87.9
2003	237.53	1,888	16.75	7.95	8.87	11.96 <43.85>	87.5
2004	236.33	1,898	16.69	8.03	8.80	11.87 <43.54>	88.8

(4) = (2)/[(1) + (3)]

Reclaiming the Desert: Towards a Sustainable Environment in Arid Lands – Mohamed (ed.)
© 2006 Taylor & Francis Group, London, ISBN 0 415 41128 9

Modeling of methane air combustion

N. Boukhalfa
Department of Industrial Chemistry, University of Biskra, Algeria

M. Afrid
Department of Physics, University of Constantine, Algeria

ABSTRACT: The aim of the present work is the study of the time evolution of the concentrations of species between the initial and the final state of methane air combustion with the consideration of a closed system and stoechiometric conditions. We have used in this modeling work a detailed mechanism involving 213 elementary reactions and 48 species. The mathematical model consists of a system of coupled ordinary differential equations resulting from a set of rate equations. The solution is performed with a computer code in FORTRAN language based on the numerical 4th order Runge-Kutta method with a step of time equal to 10^{-9} seconds. The results illustrate the consumption of the main reactants (CH_4 and O_2) and the stabilization of the concentrations of the products (CO_2 and H_2O) after 1.89 10^{-3} seconds. This permanent mode time is in good agreement with the experimental one which is about 2 milliseconds.

1 INTRODUCTION

The elementary reactions have a great importance in chemical kinetics. A total chemical reaction can represent a great number of elementary reactions. The interaction between these elementary reactions produces a total reaction.

Knowledge of the detailed chemistry of hydrocarbon combustion processes has greatly increased during the last two decades (Kee et al. 1987, Warnatz 1981, 1984, Warnatz et al. 1996, Westbrook et al. 1984, Revel et al. 1994, Glarborg et al. 1986).

At the present time, methane combustion chemistry has been extensively studied, and is now rather well known. Several methane combustion kinetic detailed mechanisms have been proposed such as the detailed mechanism used by Westbrook et al. (1984) involving 150 elementary reactions and 39 species and the mechanism proposed by Glarborg et al. (1986) which contain 213 elementary reactions and 48 species.

In this modeling work, we have used the detailed mechanism proposed by Glarborg et al. (1986). This mechanism appears more complete. Our work is based on the study of the time evolution of the concentrations of species between the initial and the final state of the combustion with the consideration of a closed system and stoechiometric conditions.

2 MATHEMATICAL MODEL

2.1 *Rate laws*

If the equation of an elementary reaction, r, is given by:

$$\sum_{s=1}^{S} \vartheta_{rs}^{(e)} A_s \rightarrow \sum_{s=1}^{S} \vartheta_{rs}^{(p)} B_s \tag{1}$$

where $\vartheta_{rs}^{(e)}$ = stoechiometric coefficients of reactants; $\vartheta_{rs}^{(p)}$ = stoechiometric coefficients of products; s = chemical species; S = number of reactants in an elementary reaction; r = elementary reaction.

The rate law of a species i in the elementary reaction r, is given by the expression:

$$\left(\frac{\partial C_i}{\partial t}\right)_r = k_r\left(\vartheta_{ri}^{(p)} - \vartheta_{ri}^{(e)}\right)\prod_{s=1}^{S} C_s^{\vartheta_{rs}^{(e)}} \tag{2}$$

where C_i = concentration of a species i, C_s = concentration of reactant s.

If we write the mechanism of methane-air combustion in the form:

$$\sum_{s=1}^{S} \vartheta_{rs}^{(e)} A_s \rightarrow \sum_{s=1}^{S} \vartheta_{rs}^{(p)} B_s \quad ; r = 1,.....,213, \tag{3}$$

The rate law for a species i will be:

$$\left(\frac{\partial C_i}{\partial t}\right) = \sum_{r=1}^{213} k_r\left(\vartheta_{ri}^{(p)} - \vartheta_{ri}^{(e)}\right)\prod_{s=1}^{S} C_s^{\vartheta_{rs}^{(e)}}; \; i = 1,.....,48 \tag{4}$$

where k_r = rate constant of an elementary reaction r; t = time of combustion

Using the modified Arrhenius equation, we have:

$$k_r = AT^{\beta}\exp\left(-\frac{E_0}{RT}\right) \tag{5}$$

where A = frequency factor; β = pre-exponential temperature exponent; E_0 = activation energy; R = constant of ideal gases; T = Temperature.

The rate laws equations with the initial conditions can be put in the following form:

$$\begin{cases} C_i = C_i^0 \quad at\ t = 0 \\ \dfrac{dC_i}{dt} = F_i(C_1,.....,C_S,k_1,.....,k_R);\ i = 1,....,48 \quad for\ t \succ 0 \end{cases} \tag{6}$$

The numerical solution is necessary to solve the system of modeling differential equations. The solution is performed with a computer program using the fourth order Runge-Kutta method (Press et al. 1986). We have elaborated a computer code in FORTRAN language for this purpose. The computation procedure is shown in Figure 1.

2.2 *Initial conditions*

We consider in this work the case of combustion of a stoechiometric mixture at atmospheric pressure. The overall reaction of methane-air combustion is written as:

$$CH_4 + 2O_2 \rightarrow 2H_2O + CO_2$$

We consider that the air contains 78% of Nitrogen (N_2), 20% of oxygen (O_2) and 2% of Argon (Ar), the molar fraction of methane is given by:

$$x_{CH_4} = \frac{1}{1+5\vartheta_{O_2}} \tag{7}$$

where ν_{O_2} = stoechiometric coefficient of oxygen (ν_{O_2} = 2).

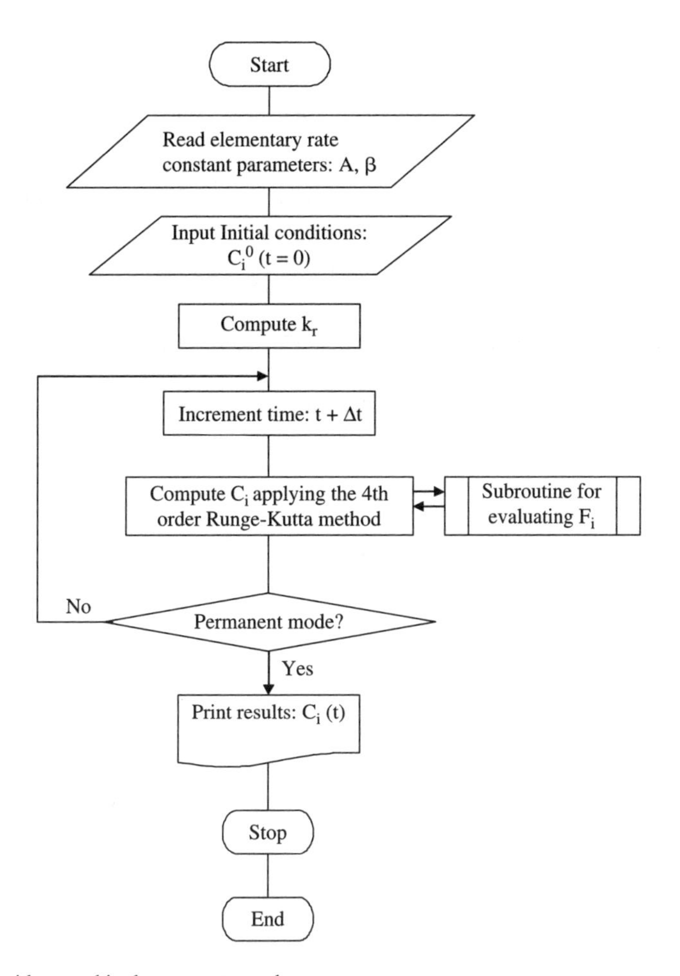

Figure 1. Algorithm used in the computer code.

Thus, the initial mixture will contain the following stoechiometric proportions:

$$x_{CH_4} = \frac{1}{11}, \quad x_{O_2} = \frac{2}{11}, \quad x_{N_2} = \frac{7.8}{11}, \quad x_{Ar} = \frac{0.2}{11}$$

The initial concentrations are calculated using the following equation:

$$C_i^0 = x_i \frac{P}{RT} \tag{8}$$

where C_i^0 = Initial concentration of species i; P = Pressure; R = constant of ideal gases; T = Temperature; and x_i = mole fraction of species i;

Knowing that the temperature of the methane flame is of 2222°K (Warnatz et al. 1996). Thus the initial concentrations are:

$C^0_{CH_4} = 0.498909\,mol/m^3$; $C^0_{O_2} = 0.997818\,mol/m^3$; $C^0_{N_2} = 3.89174\,mol/m^3$;
$C^0_{Ar} = 0.09978\,mol/m^3$

3 RESULTS

The results illustrate the consumption of the main reactants (CH_4 and O_2) (Figs 2 & 3) and the stabilization of the concentrations of the products (CO_2 and H_2O) (Figs 4 & 5).

We note also the appearance of carbon monoxide CO (Fig 6), this is surely due to a partially incomplete reaction. In fact from experience, we know that the stoechiometric mixture of the reactants does not involve necessarily a complete reaction, an excess of air is always necessary and desirable to complete the combustion.

Figure 7 illustrates a negligible decrease on the concentration of nitrogen N_2 characteristic of the chemical inertia of nitrogen.

We note also the appearance of the nitrogen monoxide NO (atmospheric pollutant) with a very small concentration (Fig 8).

The permanent mode for all species is reached after a time of 1.89×10^{-3} seconds. This result is in good agreement with the experiment time which is about 2×10^{-3} seconds.

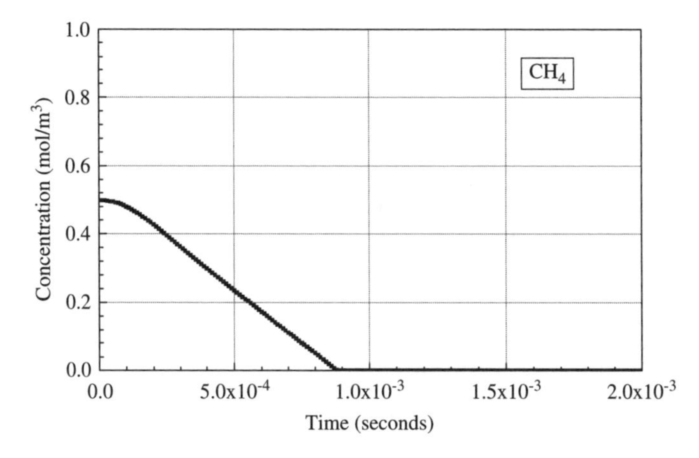

Figure 2. Time evolution of the concentration of methane (CH_4).

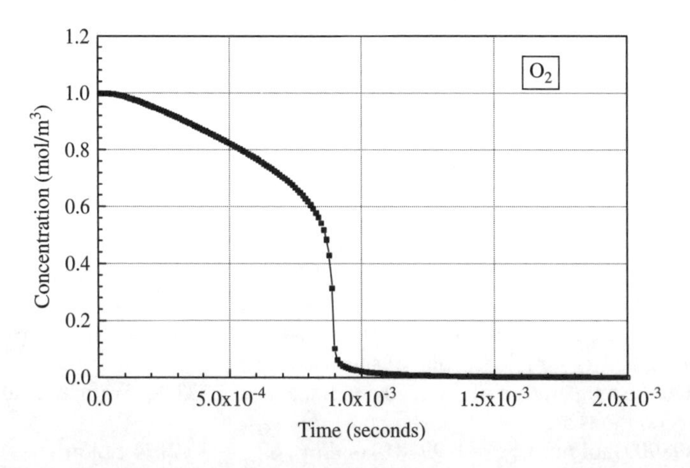

Figure 3. Time evolution of the concentration of oxygen (O_2).

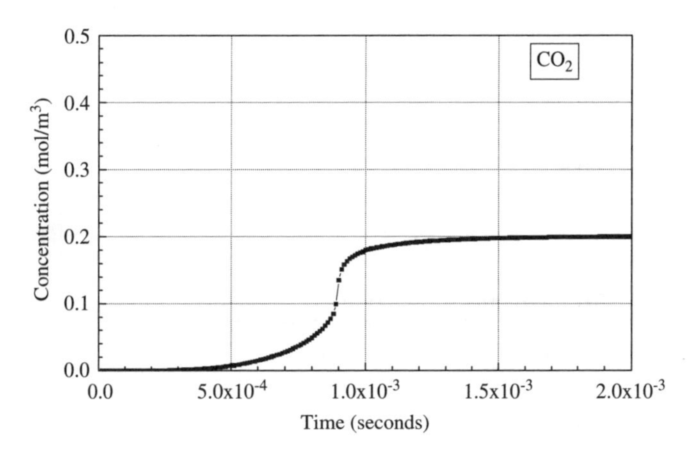

Figure 4. Time evolution of the concentration Carbon dioxide (CO_2).

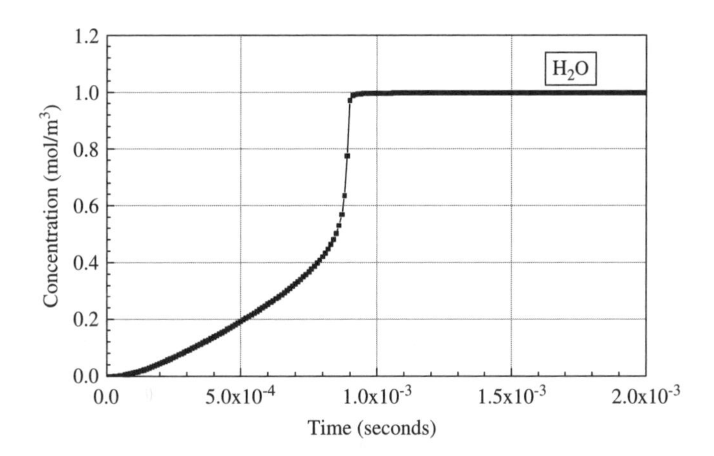

Figure 5. Time evolution of the concentration of H_2O.

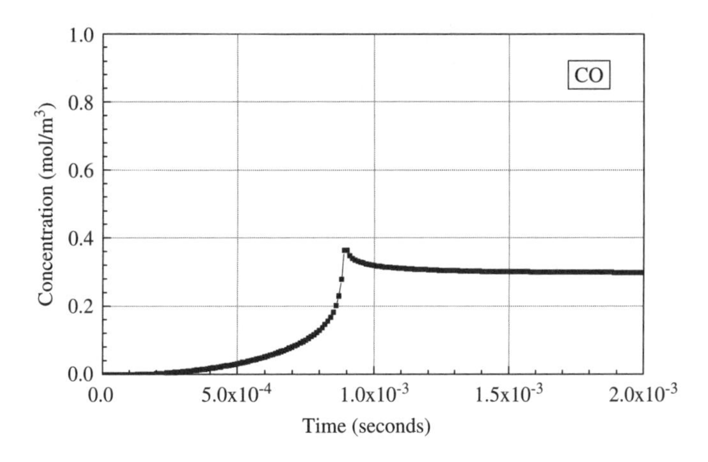

Figure 6. Time evolution of the concentration of CO.

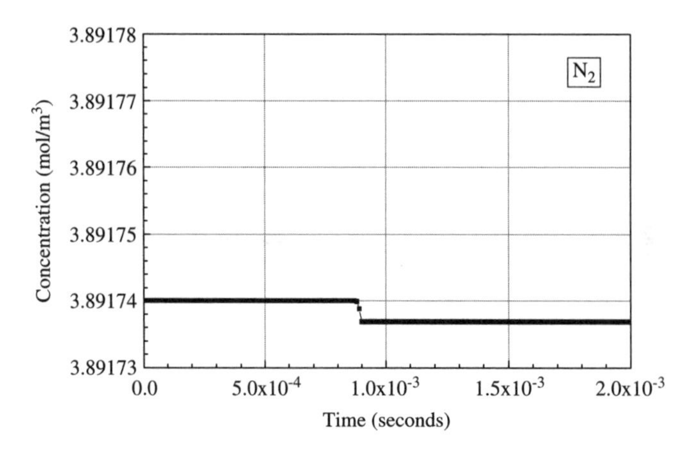

Figure 7. Time evolution of the concentration of N_2.

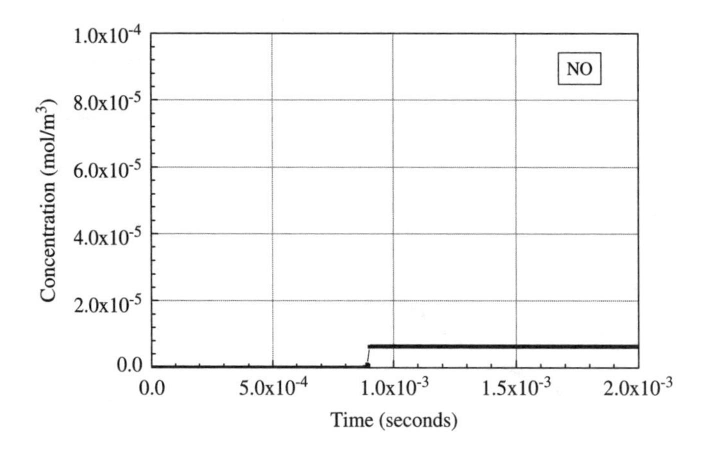

Figure 8. Time evolution of the concentration of NO.

4 CONCLUSION

In this modeling work, we have simulated successfully the problem of methane air combustion by using a detailed mechanism including 213 elementary reactions and 48 chemical species. The system of modeling differential equations is solved by the numerical 4th order Runge-Kutta method with a step of time equal to 10^{-9} seconds. We have elaborated a computer code in FORTRAN language for this purpose.

Concerning the main species, the results illustrate the consumption of the main reactants (CH_4 and O_2) and the formation and stabilization of the products (CO_2 and H_2O).

The results indicate also the appearance of the nitrogen monoxide NO (atmospheric pollutant) with a very small concentration.

The permanent mode is reached after a time of 1.89×10^{-3} seconds. This result is in good agreement with the experiment time which is about 2×10^{-3} seconds.

REFERENCES

Kee, R. J. & Miller, J. A. 1987. Complex chemical reaction systems: Mathematical modeling and simulation. J. Warnatz & W. Jager (eds). Heidelberg. FRG: Springer-Verlag.

Glarborg, P., Miller, J. A. & Kee, R. J. 1986. Kinetic modeling and sensitivity analysis of Nitrogen Oxide formation in well-stirred reactors. Combustion and Flame 65: 177–202.

Press. W. H., Flannery, B. P., Teukolsky, S. A. & Vetterling, W. T. 1986. Numerical Recipes. The art of Scientific Computing. New york: Cambridge University Press.

Revel, J., Boettner, J. C., Cathonnet, M. & Bachman, J. S. 1994. Derivation of a global chemical kinetic mechanism for methane ignition and combustion. J. Chim. Phys. 91: 365–382.

Warnatz, J. 1981. Combustion in reactive systems. J. R. Bowen, N. Manson, A. K. Oppenheim & R. I. Soloukhin (eds). New York: AIAA.

Warnatz, J. 1984. Rate coefficients in the C/H/O system. In W. C. Gardiner Jr (ed.). Combustion Chemistry: 197–360. New York: Springer-Verlag.

Warnatz, J., Maas, U. & Dibble, R. W. 1996. Combustion, Physical & Chemical fundamentals, modeling & simulation, experiments, pollutant formation. Berlin: Springer-Verlag.

Westbrook, C. K. & Dryer, F. L. 1984. Chemical kinetic modeling of hydrocarbon Combustion. Progr. Energy combust. Sci. 10: 1–57.

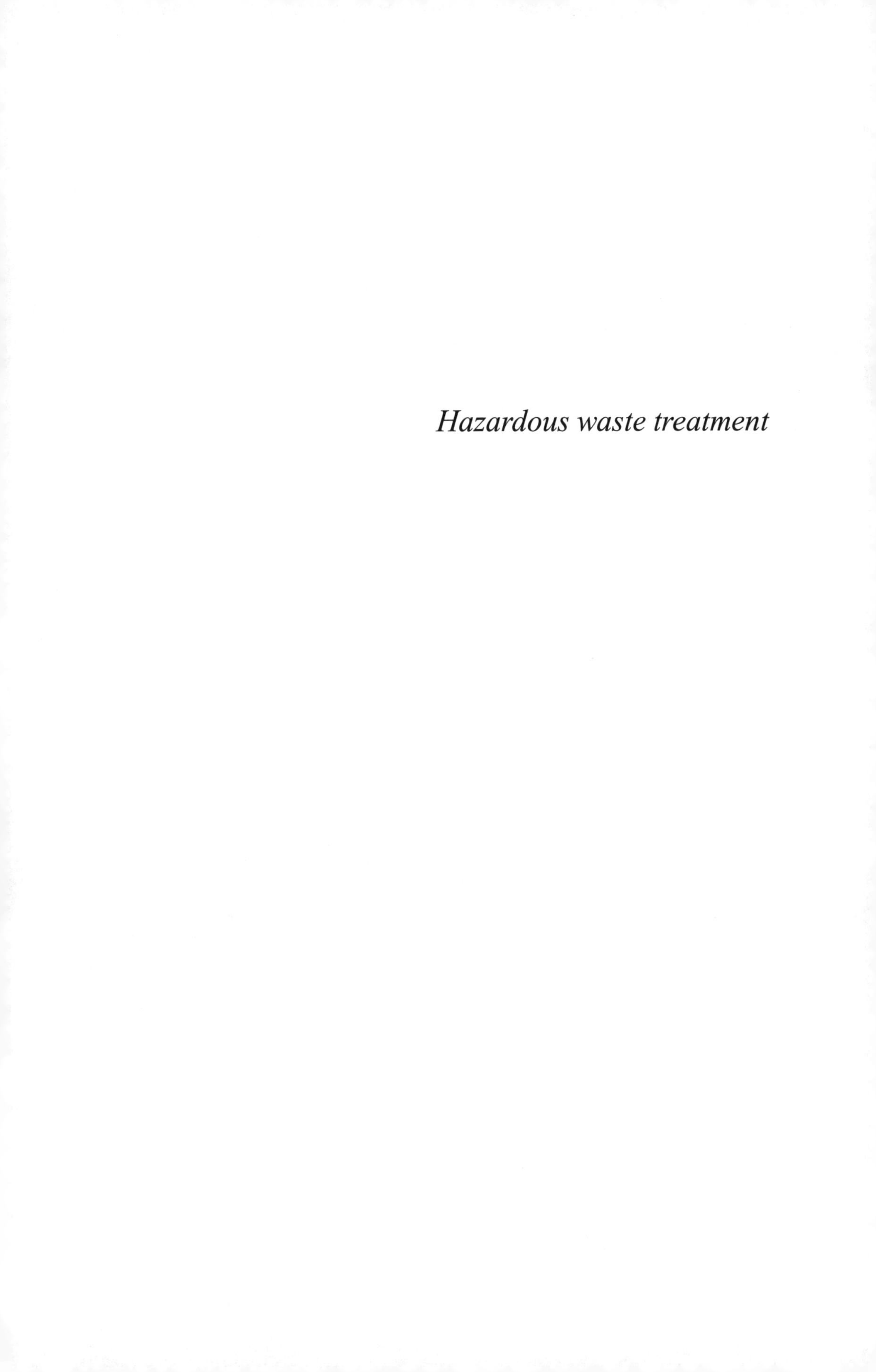

Hazardous waste treatment

Reclaiming the Desert: Towards a Sustainable Environment in Arid Lands – Mohamed (ed.)
© 2006 Taylor & Francis Group, London, ISBN 0 415 41128 9

Bioremediation for fuel oil contaminated soil in Japan

K. Okamura
Institute of Technology, Shimizu Corporation, Tokyo, Japan

ABSTRACT: Bioremediation is a practical technology for treatment of fuel oil contaminated soil, where microbes in soil degrade the oil into harmless compounds such as carbon dioxide and water. In this paper, bioremediation for fuel oil contaminated soil sites in Japan is discussed.

1 INTRODUCTION

There have been a number of plans in recent years to use former factory sites in Japan and these plans have highlighted the problem of contaminated soil which could be described as a negative legacy of the past. The Geo-Environmental Protection Center (GEPC) has already estimated that as many as 44,000 workplaces and former factory sites throughout Japan require a survey on contaminated soil.

Against this background, the Soil Contamination Control Law was enforced in February, 2003, introducing environmental standards for 24 substances. Even though there are no environmental standards for petroleum compounds at present, there is a phenomenon of the fall of the land value of land with contaminated soil. To increase the land value, there has been an increase of the work to remedy fuel oil contaminated soil by those companies planning to sell their land.

Up to the present, physical and chemical methods have mainly been used to remedy the soil quality. These include soil vapor extraction method, soil washing method, and thermal desorption method and among others. Under these circumstances, there has been growing use of the bioremediation method to purify soil as this method is believed to be capable of decomposing and rendering contaminants harmless at a low cost. Bioremediation is defined as a technology to remedy contaminated soil or groundwater using biological function. In Japan, this technology is mainly used to deal with contaminated soil and the maximum scale of such work so far is as large as 100,000 m^3.

As bioremediation is designed to decompose and render contaminants harmless using the biological function, it cannot remedy contamination due to heavy metals, etc. However, it is believed to be capable of dealing with the following organic substances.

- Petroleum hydrocarbons (fuel oils, such as petrol, kerosene, gas oil and Bunker a (Light Marine Fuel))
- BTEX (benzene, toluene, ethylbenzene and xylene)
- Halogen compounds (normal chain halogen compounds, such as trichloroethylene and dichloromethane, etc.)

Further development efforts are earnestly in progress with a view to applying this technology to the decomposition of aromatic halogen compounds, a typical example of which is dioxine.

2 BIOLOGICAL ANALYSIS OF DECOMPOSITION OF FUEL OIL

A test was conducted to prove that oils are decomposed by microbes. In preparation for this test, black soil (fertile soil) was placed in sealed beakers and gas oil, Bunker A and Bunker C (Hevy

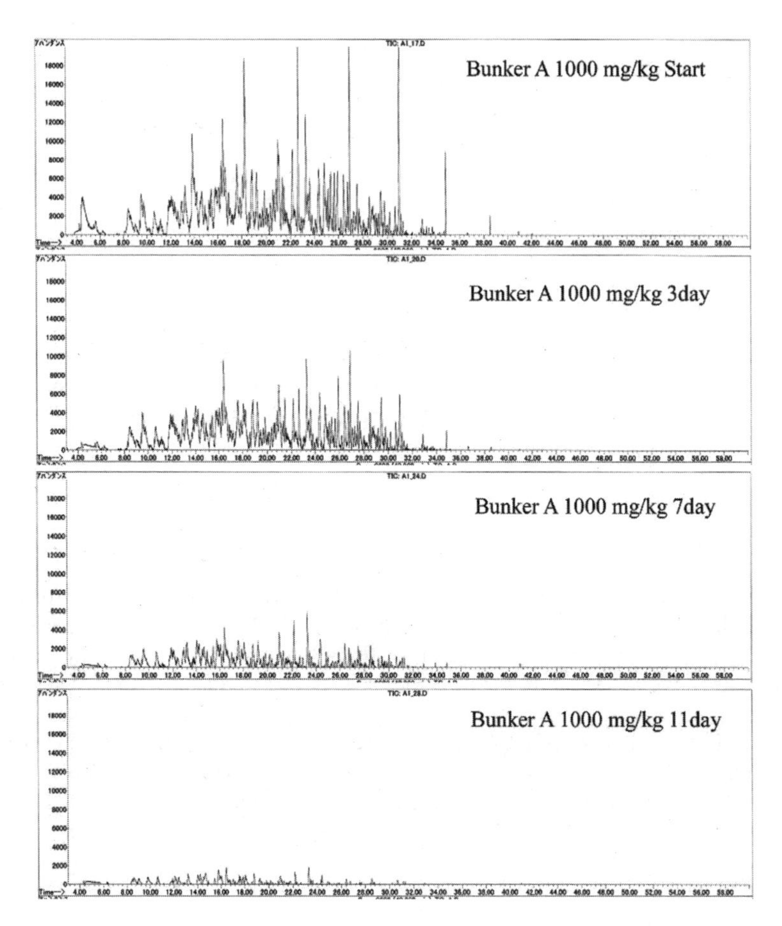

Figure 1. GC-MS analysis results of a heavy oil decomposition test.

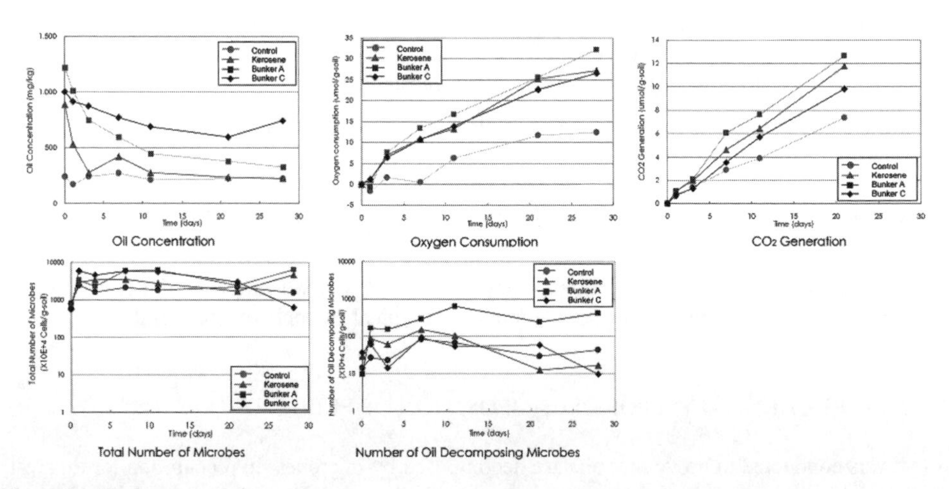

Figure 2. Beaker test results.

Marine Fuel) were added to the soil in separate beakers to achieve a concentration of 1,000 mg/kg. The regular check items were the oil concentration, oxygen consumption, generation of CO_2 and number of aerobic microbes.

Figure 1 shows the GC-MS analysis results of the head space gas for the beaker containing Bunker A. With the passing of time, the peak values steadily decreased to the trace level after 11 days.

The concentrations of gas oil, Bunker A and Bunker C used in this test began to decrease after the start of culture as shown in Fig. 2.

The most noticeable decrease was observed with kerosene, followed by a heavy oil. In contrast, the concentration of Bunker C little changed. With the decline of the concentration, the oxygen consumption increased for each type of fuel oil together with an increase of the generated CO_2. Throughout the test period, the total number of microbes was in the order of 10^7 cells/g-soil and no significant change was observed. Meanwhile, the number of oil decomposing microbes showed the most noticeable increase with the kerosene contaminated soil from 10^5 cells/g-soil at the start to 10^6 cells/g-soil after one day. However, the number dropped to the original level after 20 days. With the Bunker A contaminated soil, the figure increased from 10^5 cells/g-soil at the start as in the case of the kerosene contaminated soil to 10^6 cells/g-soil after one day. This level was maintained up to the 28th day.

The above results indicate that the types of oil used in the test are decomposed by microbes which consume oxygen while generating CO_2. In this process, the number of oil decomposing microbes increases, suggesting the involvement of microbes in the decomposition of oil.

3 OUTLINE OF REMEDIATION METHOD

Many methods of bioremediation have been conceived as shown in Fig. 3. Land farming which involves the excavation of soil and the addition of fertilizer, etc. to purify the soil is often used to deal with soil contamination near the ground surface. Others include Biomixing to purify the soil under the ground for the purification of soil in cold districts, Biosparging to inject air into the ground to deal with oil contaminated soil deep under the ground surface and in-situ Circulation to circulate and treat contaminated groundwater to remove trichloroethylene and other substances which can easily dissolve into the groundwater. The appropriate technique is selected to conduct the required remediation depending on the type of the soil as well as the type of the contaminant. While guidelines on how to conduct surveys on contaminated soil have been issued by the Ministry of Environment, they do not include a survey method for oil contaminated soil. Because of this, the processes shown in Fig. 4 are generally employed to conduct a survey and remediation.

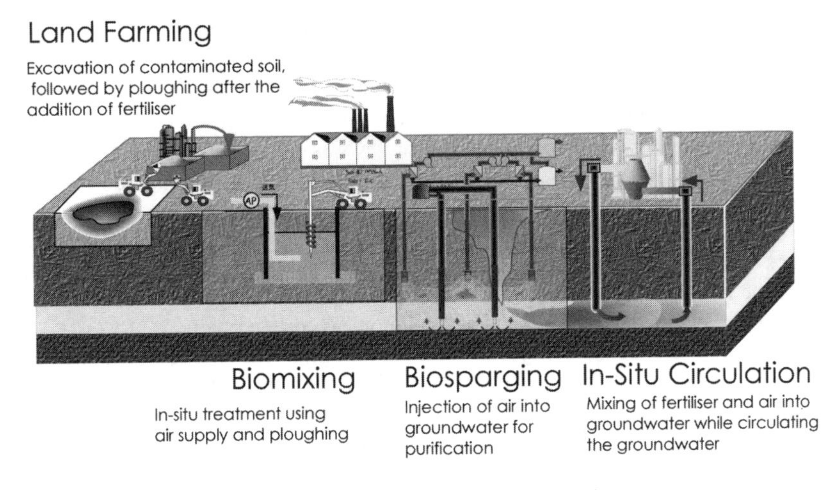

Figure 3. Soil/groundwater purification technologies using bioremediation.

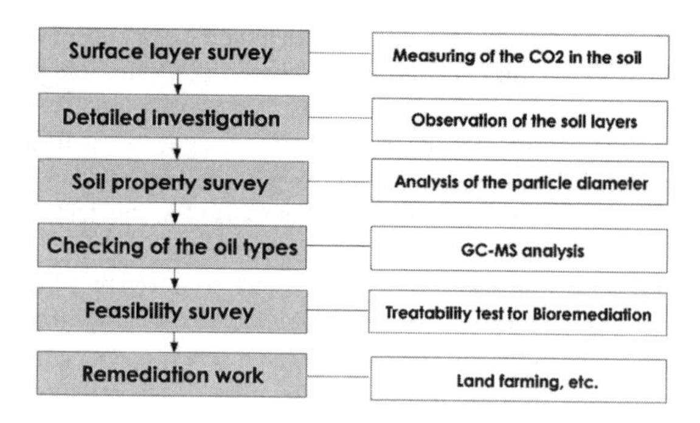

Figure 4. Survey flow.

3.1 *Survey using CO_2 concentration of soil as an indicator*

As fuel pipes and electrical cables, etc. are criss-crossed in the ground of any oil refinery or oil storage facility, it is not easy to conduct a soil survey using a boring machine without careful planning to avoid a disaster, making it difficult to safely, quickly and effectively establish the state of soil contamination at such places. It is, therefore, highly desirable to examine the state of soil contamination without boring. One method to determine the state of soil contamination without involving an underground survey is a surface layer survey. However, this method is currently capable of analysing only simple compounds, mostly such organic chlorides as trichloroethylene, apart from benzene, and no reliable surface layer survey method dealing with such petroleum compounds with many constituents as fuel oil has been firmly established so far. The present study team conducted a surface layer survey using a soil contamination gas monitor to establish the state of soil contamination at a former oil storage and refinery site. In this survey, survey points were set up at intervals of some 10–30 m and a boring bar was used to create holes of 10 mm in diameter and 50–100 cm in depth to measure the soil gases in each hole. Equipment called Ecoproof 5 was used for the measurement of the soil gases to simultaneously measure the concentrations of TP, PID, CH_4, CO_2 and O_2 contained in the soil gases. The concentration of each gas was then shown on a contamination contour map. Fig. 5 shows the oil analysis results and some examples of the analysis results of the said monitor.

The contours of the PID and oil concentrations in the soil gases at GL −0.2 m were similar but did not suggest the presence of oil below GL −1.0 m. As CO_2 in soil is generated by the decomposition of organic matters by microbes, a higher concentration of organic matters generally results in the detection of a higher concentration of CO_2. At the site in question, a CO_2 concentration of as high as 40,000 ppm was detected at one point. In general, the contour of the CO_2 concentration coincides well with the contour of the oil concentration at each depth. It is inferred that the contour of the CO_2 concentration may imply the presence of oil contamination in the deep ground rather than roughly indicating the oil concentration of the surface layer.

To further investigate this possibility, a detailed survey was conducted at a different contaminated site. Fig. 6 shows the outline of this surface layer survey at an actual contamination site. At the site, 24 gas monitoring points were set up at intervals of 30 m and the survey depths to sample the soil gases were 0.5 m, 1.0 m, 1.5 m and groundwater level (2.0 m; hereinafter referred to as "WL").

A hammer drill was used to open a hole in the ground, an aluminium gas sampling tube was inserted into the hole and the sampled gases were collected and analysed by the Ecoproof 5. The target gases for measurement were TC (total hydrocarbon gas concentration), PID, CO_2, O_2 and CH_4. Following the completion of the soil gas survey, the ground at each survey point was excavated using a backhoe to check the state of contamination. Soil samples by depth were collected and the WL was checked at each point.

The collected soil samples underwent oil contents analysis by the FTIR together with a breathing test as well as measurement of the ion concentration, total number of microbes and number of oil decomposing microbes.

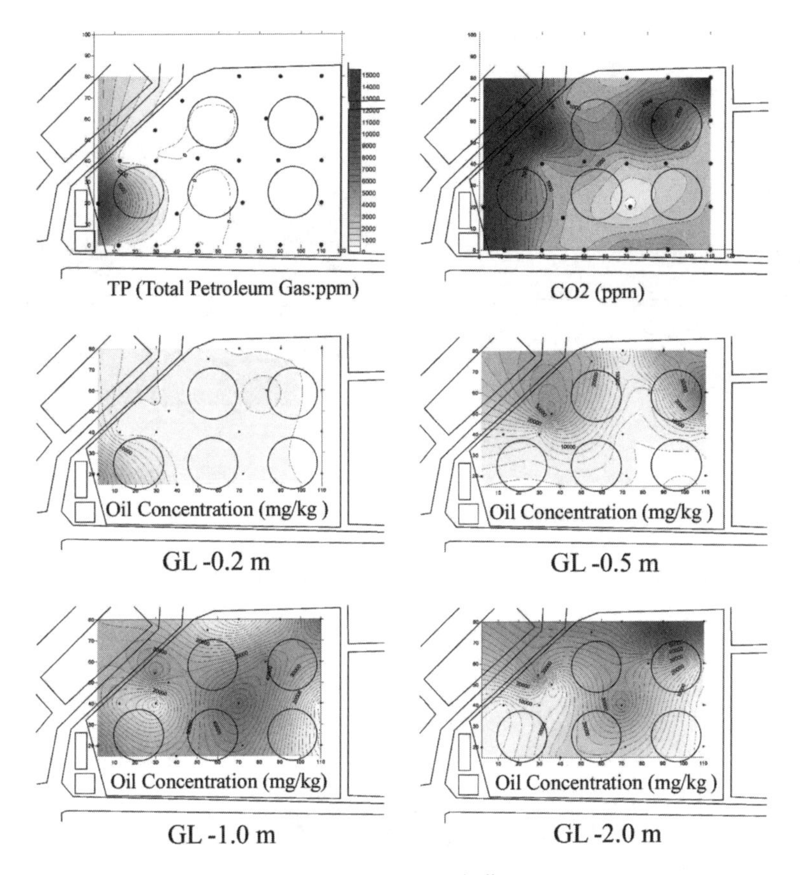

Figure 5. Example of surface layer survey using CO_2 as an indicator.

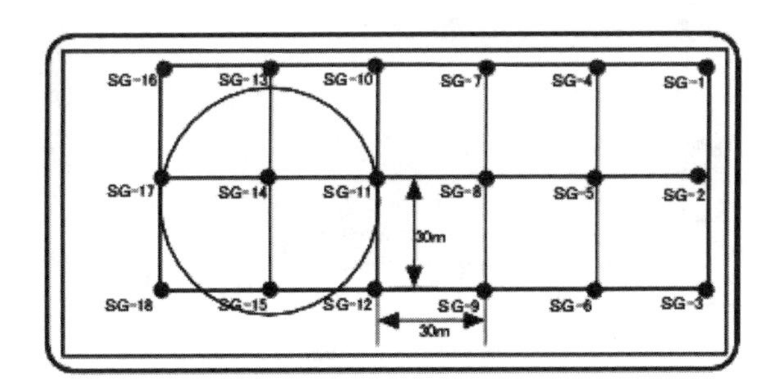

Figure 6. Outline of survey layer survey at actual contamination site.

3.2 *Surface layer survey results*

Figure 7 shows the general state of contamination at the survey points. The denser red in Fig. 7 indicates a higher CO_2 concentration or a lower O_2 concentration depending on the chart. The oil contents analysis results are shown in terms of the respective ground depths of 0.5 m, 1.0 m, 1.5 m and WL. A high oil concentration is found at a depth of 1.5 m and the WL. Apart from a high oil contamination point at the left-hand bottom corner of the chart, contamination appears to progress from the centre bottom point upwards. The CO_2 concentration at each depth shows a similar tendency to that of the oil concentration. The CO_2 contour at a depth of 0.5 m shows a low CO_2 concentration,

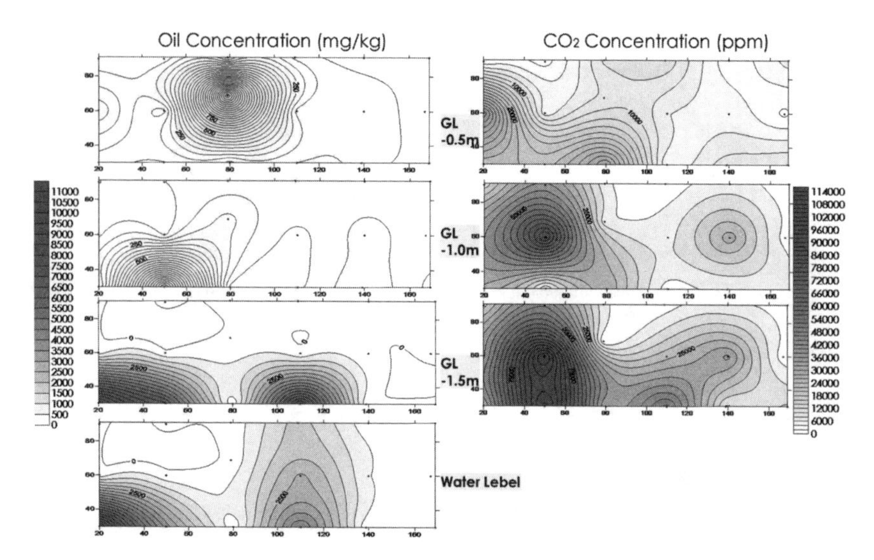

Figure 7. Comparison of oil concentration with surface layer survey results.

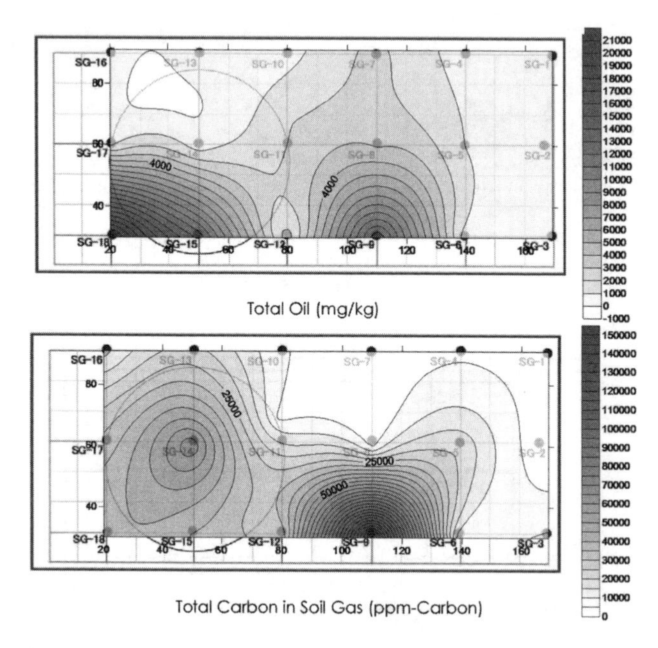

Figure 8. Comparison of total oil concentration and total carbon.

presumably because the ground at this depth is affected by fresh air because of its proximity to the ground surface. Accordingly, it is desirable to conduct a soil gas survey at a depth of 1 m or lower to eliminate the influence of fresh air. The O_2 concentration tends to show a low concentration at points where the oil concentration is high.

Given the fact that the CO_2 concentration of the atmosphere is around 300 ppm compared to 21% for O_2, measurement of the CO_2 concentration can better assess the state of soil contamination provided that the influence of boring on the soil is minimal and that the boring depth is not deep.

Figure 8 compares the total oil contents of the soil at each survey point of the contaminated site in question with the total carbon (converted value of the total TP and CO_2 in the soil gases at each

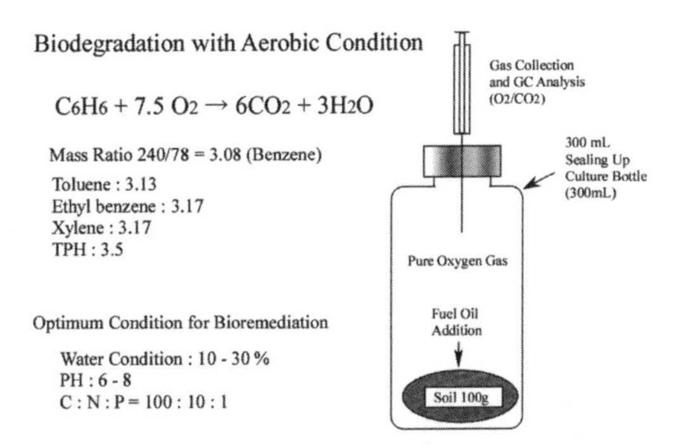

Figure 9. Feasibility test method.

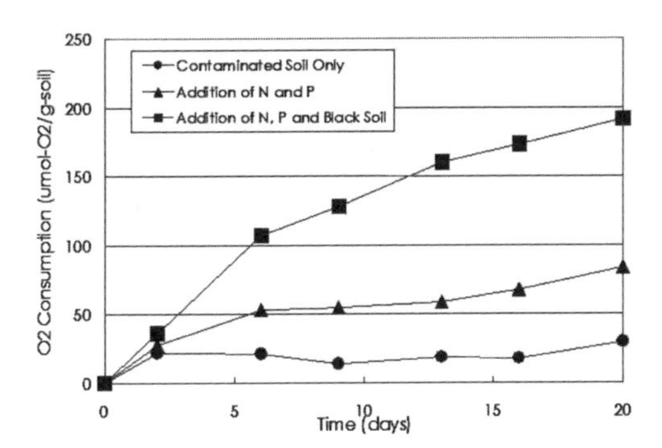

Figure 10. Feasibility test results.

depth to the carbon concentration). The results show a similarity between the oil concentration of the soil and the total carbon in the soil gases.

3.3 *Feasibility survey*

The fuel oil decomposition test using the sealed beaker shown in Fig. 9 was conducted to examine the applicability of bioremediation. 100 g of a contaminated soil sample was placed in a 300 ml beaker with a butyl rubber stopper which was then cultured for 20 days at a constant temperature of 30C. During this period, the O_2 concentration was regularly measured to calculate the speed of O_2 consumption. After 20 days, the oil concentration of the sample was also measured. When the O2 concentration in the sealed beaker became low, the gas inside the beaker was replaced by O_2. Four types of samples were used: conventional soil, contaminated soil with the addition of urea and K_2HPO_4 as nutrients and contaminated soil with the addition of two different types of black soil as a source of microbes in addition to nutrients.

Figure 10 shows the results of the feasibility survey. The O_2 consumption of the contaminated soil and the contaminated soil to which nutrients were added was low, indicating that the addition of nutrients alone is insufficient to facilitate the process of decomposition. Meanwhile, the O_2 consumption in the case of the two types of samples to which black soil was added as a source of microbes was considerable, suggesting a low level of oil decomposing microbes in the contaminated soil in question. It was, therefore, decided to add black soil as a source of microbes in the actual remediation work. As shown in Table 1, the decomposition rate of the oil after 20 days was 85.3% with the addition of black soil. Given the initial O_2 consumption per 1 kg of soil with the addition of black soil, replacement of the air every three and a half days was found to be necessary based on an assumed air void ratio of 30% in the contaminated soil. A backhoe was used for ploughing. In view of the use of fresh air, the O_2 utilisation efficiency was believed to be lower than in the case

Table 1. Changes of oil content. (Unit: mg/kg)

	Initial content	After 20 days	Removal Rate (%)
Contaminated soil only	5,500	2,430	55.9
Addition of N and P	5,500	1,660	69.7
Addition of N, P and black soil	5,500	810	85.3

Figure 11. Scene of the remediation work.

of the feasibility test. As such, it was judged that ploughing to resupply O_2 should be conducted every day, at least at the initial stage of the remediation work.

3.4 *Remediation plan*

Based on the feasibility survey results, a remediation plan was formulated for some 800 m^3 of soil contaminated by gas oil. The feasibility survey results suggested the necessity for nitrogen (N) and phosphorous (P) as nutrients. Sources of N and P were added to make the ratios of N and P to the oil content of the contaminated soil 10% and 1% respectively. Black soil was also added as a source of microbes as it was clear that such a source was required. For the remediation of the soil contaminated by gas oil, the land farming technique was employed. A scene of this work is shown in Fig.11. In the first month, ploughing was conducted every day using a backhoe. For the purpose of properly managing the remediation work, the oil concentration as well as the temperature, O_2 concentration, CO_2 concentration, water content and pH value of the soil were measured at the ground surface and at depths of 40 cm and 60 cm from the ground surface immediately before ploughing at four sampling points.

3.5 *Remediation results*

(1) Changes of Oil Concentration

Figure 12 shows changes of the average oil concentration of the remediated soil. At the beginning, the oil concentration was 2,100 mg/kg but rapidly fell in approximately two weeks. Approximately one month after the start of the remediation work, the oil concentration was 450 mg/kg, recording a removal rate of 78.5%. Soil remediation in such a relatively short time was possible because the oil concentration of the subject soil was relatively low due to light contamination by gas oil. After the fall of the oil concentration as well as the O_2 consumption, the remediation work continued for two months with a ploughing frequency of once a week or so to completely eliminate the oil film and odour.

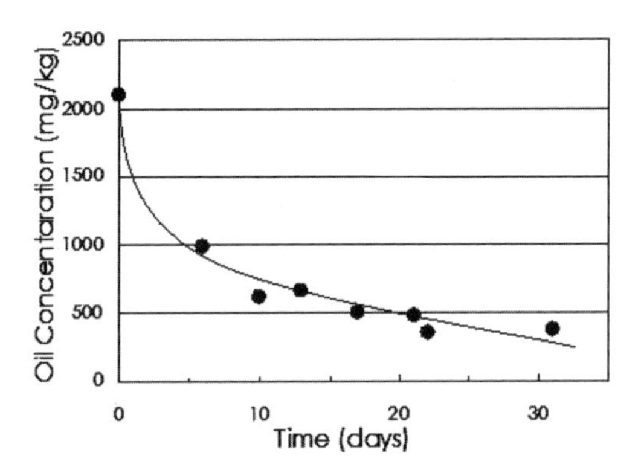

Figure 12. Changes of oil concentration.

(2) Changes of O_2 and CO_2 Concentrations of Soil

Figure 13 shows changes of the O_2 concentration of the soil. Despite ploughing every day, the O_2 concentration of the soil in the early days was as low as some 5%, suggesting a vigorous biological decomposition process. Around two weeks after the commencement of the work, the O_2 concentration of the soil began to increase to reach a level of more than 15% after the third week. Meanwhile, the CO_2 concentration of the soil was as high as around 20% at the beginning but showed a declining trend with the progress of the work. This value was much higher than could be explained by the theoretical O_2 consumption rate and the reason for this is unclear. It is inferred that CO_2 tends to remain in ploughing work using a backhoe, etc.

4 FUTURE PROSPECTS

4.1 *Method to use soil microbes (biostimulation)*

The biostimulation method activates the soil microbes which exist in a contaminated environment by means of adding O_2 and fertiliser to facilitate the decomposition process. When microbes can directly decompose such contaminants as petroleum hydrocarbons, the contaminated soil is supplied with nitrogen and phosphorous as well as fresh air. Ploughing using a backhoe is sufficient to supply fresh air and special equipment is unnecessary. In the case of such chlorinated organic compounds as trichloroethylene, etc., as these cannot be directly decomposed by microbes, a process called cometabolism is used where methane, nitrogen, phosphorous and fresh air are supplied to the soil to propagate those microbes which produce the enzymes which decompose trichloroethylene, etc. It is currently believed that tetrachloroethylene cannot be decomposed by aerobic microbes and, therefore, dechlorination treatment under an oxygen-less anaerobic condition is used by adding organic matters to the groundwater.

4.2 *Method to use microbes (bioaugmentation)*

When the number of decomposing microbes is small or few in a contaminated environment, the bioaugmentation method is used to facilitate the decomposition of contaminants by means of culturing and introducing decomposing microbes in a large quantity. Two demonstration tests on bioaugmentation have so far been conducted in Japan, both of which aimed at the decomposition of trichloroethylene. In these tests, microbes which were believed to be effective to decompose trichloroethylene were added with toluene or phenol and those producing an enzyme to decompose trichloroethylene were fed to the soil.

4.3 *Trends of R & D and practical application*

R & D activities on the purification of oil contaminated soil commenced after a pioneering study by the JPEC (Japan Petroleum Energy Center) on the purification of oil contaminated soil as a result of

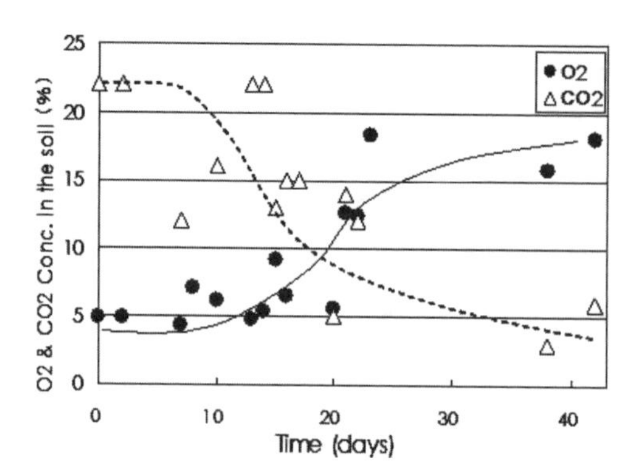

Figure 13. Changes of O_2 and CO_2 concentrations.

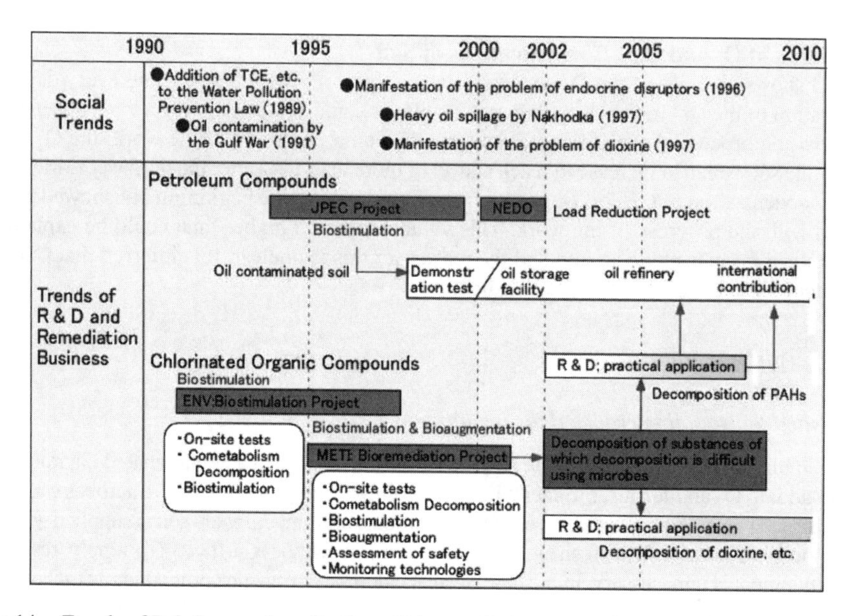

Figure 14. Trends of R & D on and application of bioremediation.

the Gulf War. After a series of demonstration tests, soil remediation work is in progress for soil purification at former oil storage sites and petrol station sites with the application of the techniques developed through these tests. In regard to the decomposition of chlorinated organic compounds, R & D activities have been conducted on biostimulation and bioaugmentation systems, evaluation of the safety of these systems and technologies to monitor microbes, etc. through the activities of the Biostimulation Project with a grant from the Ministry of Environment (ENV) and the Bioremediation Project with a grant from the Ministry of Economy, Trade and Industry (METI), etc.

It is believed that the purification of petroleum compounds which can be purified by soil microbes will be well underway in the coming years. However, in the case of such substances as dioxine and polynuclear aromatic compounds, their decomposition by soil microbes will be difficult, making reliance on the bioaugmentation method for their decomposition highly likely.

When releasing mass-cultures decomposing microbes in the case of the bioaugmentation method, etc., it is essential to assess their safety. At present, ENV and METI have their own guidelines for the safety of releasing microbes into the environment. These two ministries have begun talks this year with a view to unifying their guidelines. The surface layer survey referred to in this report was conducted as part of a technological development project which is in progress by the JPEC with a grant from METI.

Reclaiming the Desert: Towards a Sustainable Environment in Arid Lands – Mohamed (ed.)

Air-sparging, soil vapor extraction and pump and treat remediation system for groundwater and soil contaminated by diesel – a case study

R.S. Al-Maamari & A.S. Al-Bemani
Sultan Qaboos University, College of Engineering, Department of Petroleum and Chemical Engineering, Sultanate of Oman

A. Hirayama & M. Sueyoshi
Shimizu Corporation, Energy Solutions Department, Tokyo, Japan

ABSTRACT: This paper addresses the investigation and remediation work carried out at a contaminated site in the Sultanate of Oman. Groundwater and soil are contaminated by diesel.

Tasks conducted in this work include, drilling/core-sampling of boreholes, permeability and pumping tests, collection and analysis of water and soil samples for hydrocarbon contamination, groundwater level monitoring, air-sparging (AS), soil vapor extraction (SVE), free product recovery and pump and treat (P&T).

After six months of SVE, benzene in the soil gas was reduced from a range of 15–60 ppm to a range of 0–1 ppm. No benzene has been detected during the last 12 months. Also during the past 12 months, AS with P&T treatment has reduced the total petroleum hydrocarbons (TPH) in the groundwater from as high as 25–50 ppm to less than 0.5 ppm. Treatment processes are being shutdown in stages and groundwater and soil gas quality are being observed for any rebound in contamination.

1 INTRODUCTION

Groundwater is a precious resource for the Sultanate of Oman, where desalination accounts for less than 4% of all consumed water, and the remaining 96% comes almost exclusively from groundwater. Rainfall is both scarce and irregular, and surface water is transitory or insignificant. Moreover, the demand for groundwater has increased in line with rapid economic development and its side effects, i.e. higher living standards, population growth, and increased demand for agricultural products. Unfortunately, the primary driving force behind such development, i.e. the oil industry, is also one of the main threats, in terms of hydrocarbon contamination, to increasingly limited groundwater supplies. Protection of groundwater resources is a top priority issue for the sustainable development of Oman.

Hydrocarbon contamination of groundwater has been discovered at more than ten locations in Oman (MRMEWR, 2001). Such contamination can be mainly attributed to leakage of and spillage from filling stations, power plants, and oil pipelines. Oil industry activities are the other major source in all oilfield concession areas and along pipeline routes, i.e., produced water, sludge, leakage, drilling activities, other hazardous wastes (Al-Baloushi, 2000).

Oman recognizes the principle of "the polluter pays," i.e., persons/bodies responsible for contamination are responsible for restoring the contamination site to original conditions. Unfortunately, in actual practice, few remedial measures have been implemented following the discovery of contamination, resulting in the unfortunate further contamination of groundwater.

This study was started with the goal to evaluate the conditions of hydrocarbon contamination of groundwater, select and remediate one of the contaminated sites in Oman.

2 STUDY SITE

The study site occupies approximately 20 hectares, roughly 150 km southwest of the capital Muscat. The site is situated on hills of a few meters in height surrounded by wadis, and slopes downwards from south to north. The wadis flow from south to north on the rare occasions when surface water is available. The groundwater at the study site is also flows in the same direction. The annual rainfall observed at the study area is around 80 mm (Fig. 1). The ground formation in the area consists of wadi alluvium at the top and fractured bedrock below (Fig. 2).

In November 1998, approximately 10,000 liters of diesel were accidentally released from an above-ground storage tank of 25,000 liter capacity. Following the accident, the tank was removed, and a block of soil measuring 6 m × 6 m × 6 m around the Contamination Source was excavated and removed. The site was backfilled with clean soil and six polyvinyl chloride (PVC) pipes were installed for venting gas.

Site investigation revealed the following: a free oil film (~2 mm) was observed at one borehole (20 m from the source) at the beginning of the study and was not seen later; groundwater level

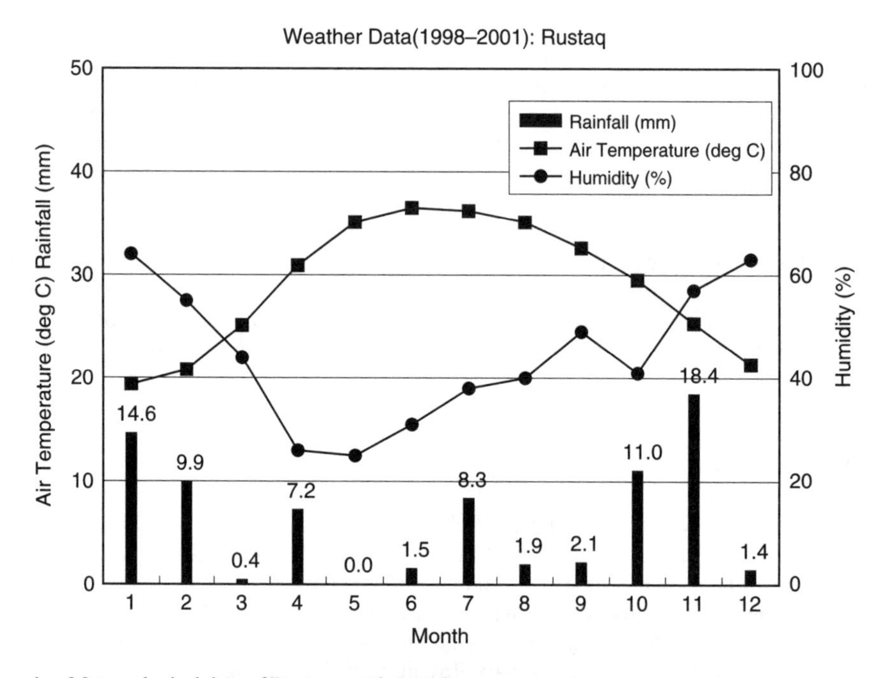

Figure 1. Meteorological data of Rustaq weather station.

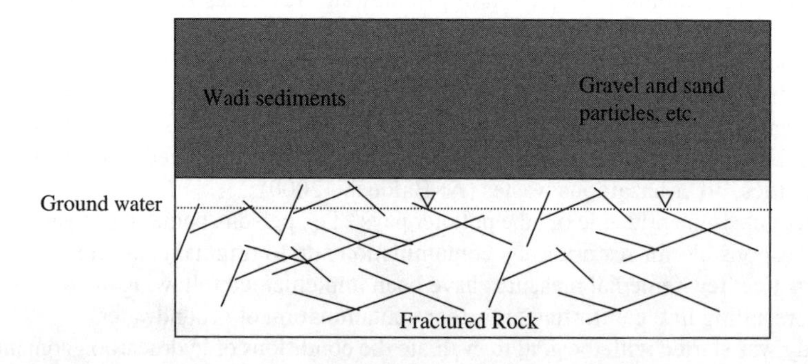

Figure 2. Schematic sketch of geological conditions of Rustaq site.

fluctuates between −13 m and −15 m; the permeability coefficient is in the range of 10^{-4} to 10^{-3} cm/s and is not significantly affected by depth, suggesting that the formation forms a single large aquifer with many fractures; benzene was detected at four out of the six PVC pipes (maximum concentration 5 ppm); oil contamination in the water was detected at several boreholes at the contamination source and at some of the upstream wells (0.6–9 mg/l); no BTEX and poly-cyclic aromatic hydrocarbons (PAH) were detected in the water samples; oil was detected in the soil samples in some boreholes at different depths (490–610 mg/kg).

A sample of the observed free oil was analyzed for oil components (thin-layer chromatography–flame ionization detection (TLC-FID), BTEX (JIS, 1995), and poly-cyclic aromatic hydrocarbons (PAH) (EPA-8270)), and compared with analysis of retail diesel from Oman. In terms of oil components, results indicated that the ratios of different components in the free oil and Oman retail diesel are basically the same. In terms of BTEX, results indicated that lighter fractions of free oil were probably volatilized or degraded. In terms of PAHs, results indicated that while lighter PAHs were probably volatilized or degraded in the free oil, heavier PAHs (from C_{18} onwards) in the free oil remained at basically the same levels as for Oman retail diesel. The analytical results from these analyses are shown in Tables 1, 2 & 3.

3 SITE CLEANUP METHODS

Different cleanup methods are appropriate for different conditions (EPA, 2005). The effectiveness of some of these technologies for certain conditions are well-established, while the effectiveness of the same technologies for other conditions are uncertain and require some field condition verification and laboratory testing prior to application.

Based on site conditions, four different treatment methods were considered for application at the Study Site; pump & treat, bioremediation, soil vapor extraction, and air-sparging. Pump & treat is a widely adopted groundwater cleanup method that is effective for treating both free oil and dissolved oil, including heavier compounds contained in diesel.

Bioremediation is another possible method, whereby natural biodegradation capacities are accelerated. However, prior to application of this method, site conditions should be checked to assess its suitability. As such, field measurements of dissolved oxygen levels in groundwater were taken, and a laboratory test was conducted with soil from the Study Site.

Table 1. Oil component analysis results.

Component	Oman retail diesel (%)	Free oil (%)	Solvent
Straight chain	74.9	73.6	Hexane
Polynuclear aromatics	24.1	24.4	Toluene
Resin	1.0	1.6	Dichloromethane 95vol% + methanol 5vol%
Asphaltenes	0.0	0.4	Residue
TOTAL	100.0	100.0	

Table 2. BTEX analysis results.

	Concentration (mg/l)	
	Oman retail diesel	Free oil
Benzene	54.3	<1
Toluene	214	<1
Ethyl-benzene	89.5	<1
m,p-Xylene	384	<2
o-Xylene	188	<1

Table 3. PAH analysis results.

		Concentration (mg/l)	
		Oman retail diesel	Free oil
Naphthalene	$C_{10}H_8$	455	<1
Acenaphthylene	$C_{12}H_8$	283	<1
Acenaphthene	$C_{12}H_{10}$	326	<1
Fluorene	$C_{13}H_{10}$	744	<1
Phenanthrene	$C_{14}H_{10}$	488	<1
Anthracene	$C_{14}H_{10}$	<1	<1
Fluoranthene	$C_{16}H_{10}$	14.1	<1
Pyrene	$C_{16}H_{10}$	<1	<1
Benz(a)anthracene	$C_{18}H_{12}$	<1	<1
Chrysene	$C_{18}H_{11}$	32.1	31.9
Benzo(b)fluoranthene	$C_{20}H_{12}$	<10	<10
Benzo(k)fluoranthene	$C_{20}H_{12}$	<10	<10
Benzo(a)pyrene	$C_{20}H_{12}$	18.3	18.2
Indeno(1,2,3-cd)pyrene	$C_{22}H_{12}$	<40	<40
Dibenz(a,h)anthracene	$C_{22}H_{14}$	<40	<40
Benzo(g,h,I)perylene	$C_{22}H_{12}$	<40	<40

Soil vapor extraction (SVE) is another method, whereby volatile contaminants are removed by vacuum. Air sparging (AS) is a method used in conjunction with SVE where air is forced into the saturated zone from where volatilized oil is flushed to the unsaturated zone and sucked to the surface by SVE method. However, while both of these methods are most appropriate for treatment of volatiles in diesel, the effectiveness of these methods for the heavier compounds in diesel is uncertain. As such, a laboratory air-sparging test was conducted to check the method's effectiveness on diesel contamination in soil.

3.1 *Dissolved oxygen measurements*

As lower oxygen levels in water can indicate greater bioactivity due to the degradation of oil contamination, a Horiba multi-sensor probe was used to take dissolved oxygen (DO) levels in five monitoring wells at the site in June 2002. Lower DO levels observed at the groundwater surface in three wells most likely to have highest levels of oil contamination, indicate probable microbial degradation of such contamination in those wells (Fig. 3).

3.2 *Biodegradability test*

Laboratory tests were conducted to test the suitability of site conditions for the biodegradation of oil contamination in soil. Oman retail diesel was mixed with three types of soil (Site soil, Site soil with Nitrogen (N) and Phosphorus (P), and Japanese black soil with N and P at concentration of 2000 mg of diesel per 1 kg of soil. Source of N was urea, and source of P was K_2HPO_4. Diesel:N:P ratio was 100:10:1. 100 g of each of the three soil mixtures were placed in three separate 300 ml incubation bottles, after which the bottles were filled with oxygen and sealed. Samples were kept in the bottles for 30 days at 30°C, and headspace gas was analyzed for oxygen consumption associated with oil degradation, as seen in Figure 4. As oxygen reduction translates into oil degradation, biodegradation of diesel in the site soil appeared to be slow when compared to that in rich Japanese soil. Therefore, bioremediation could not be recommended as a primary treatment technology.

3.3 *Air-sparging test*

Laboratory tests were conducted to gauge the effectiveness of air-sparging for diesel contamination of soil. Oman retail diesel was mixed with Toyoura Japanese standard sand at a concentration

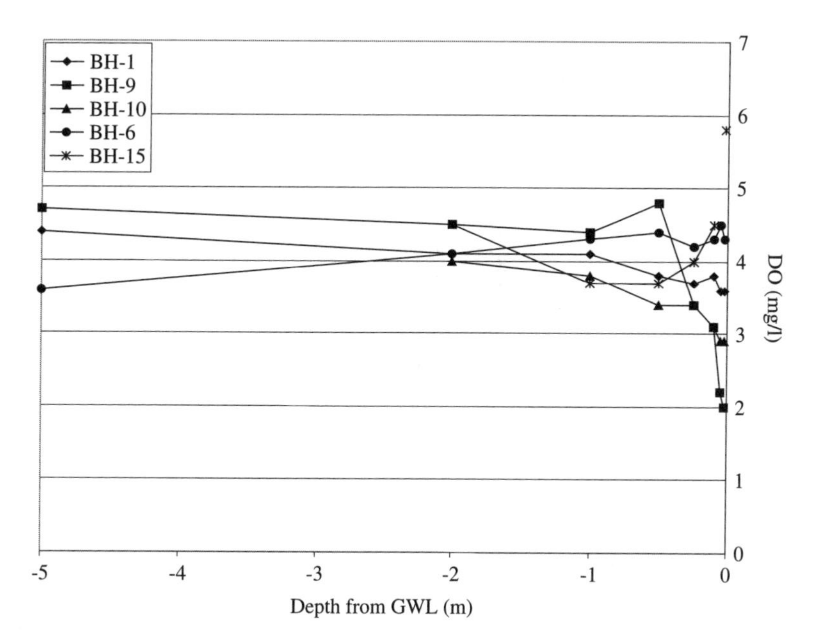

Figure 3. Dissolved oxygen measurements in some boreholes.

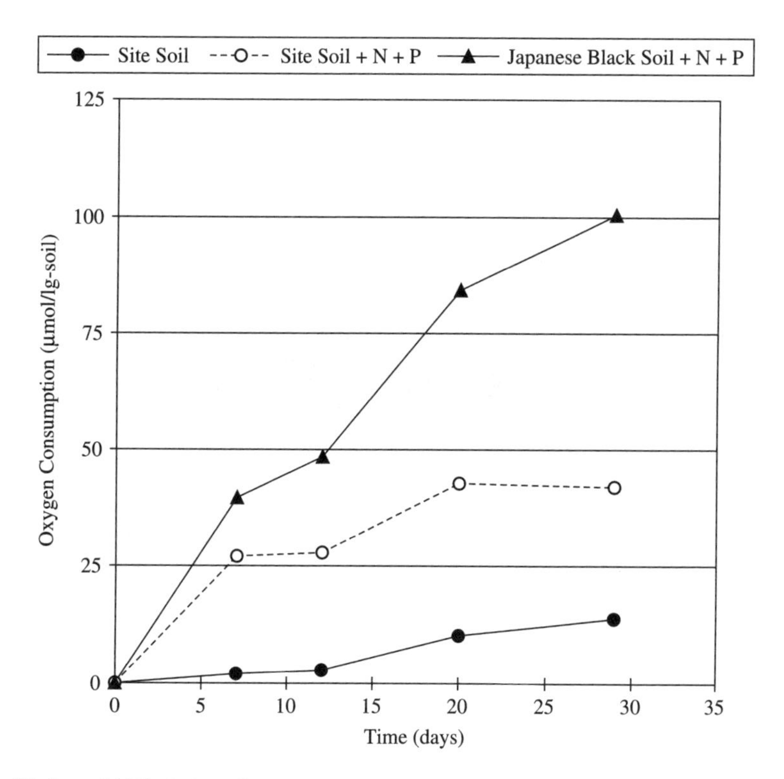

Figure 4. Biodegradability test results.

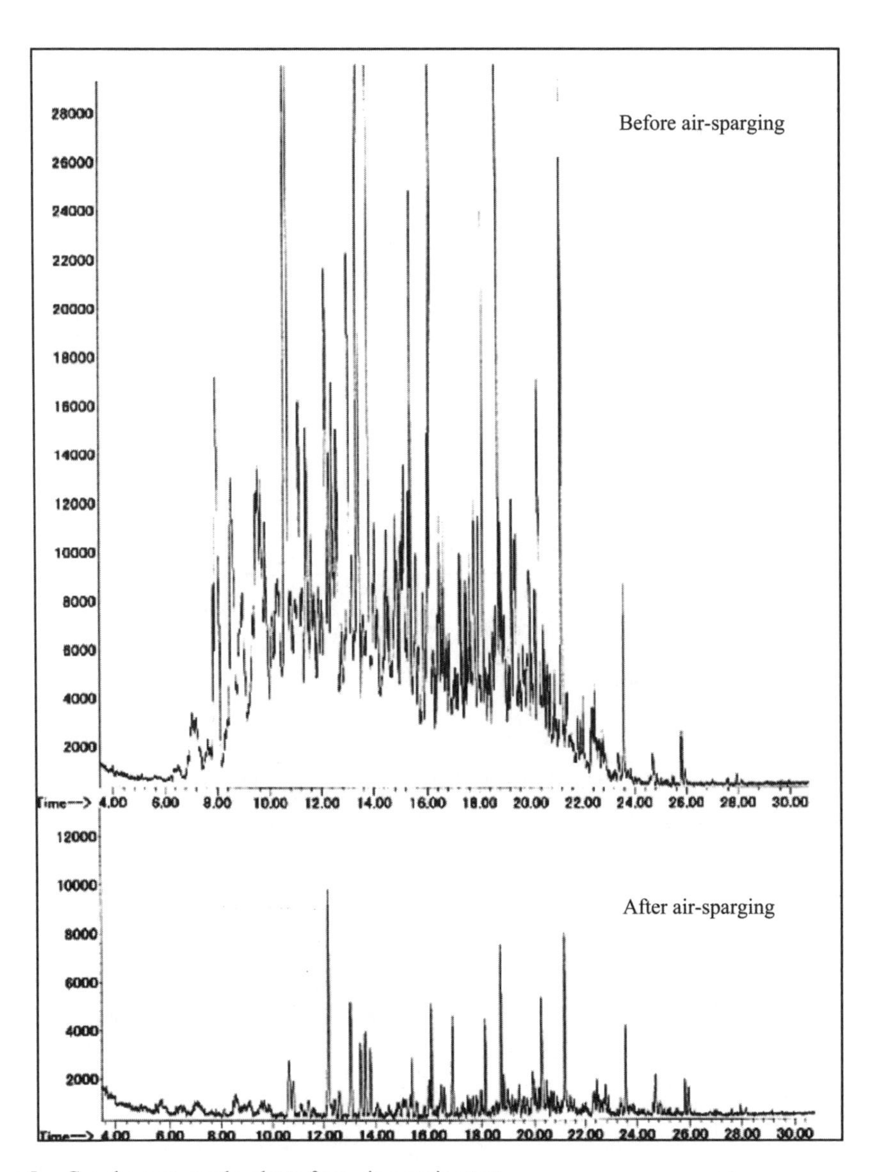

Figure 5. Gas chromatography charts from air-sparging test.

of 2000 mg of diesel per 1 kg of sand. The sample was aerated for three weeks at 1.2 L of air per 1 g of sand per day. The headspace gas was sampled and analyzed by GCMS, as shown in Figure 5. The oil removal rate was 40% of TPH. Results indicated that while heaviest fractions of hydrocarbons remained, lighter fractions were mostly removed. Therefore, air-sparging is expected to be effective for treating the BTEX detected in soil at the contamination source area.

Based on the results from the above tests, an integrated air-sparging (AS) system was selected as the remediation method for the site (Fig. 6). AS removes volatile components of hydrocarbons through the sparging of pressurized air. While AS is conventionally applied to homogeneous soil, not fractured conglomerate; the application of this method together with soil vapor extraction (SVE) and pump and treat (P&T), was expected to be effective based on various selection criteria, i.e. necessity of in-situ treatment, relatively deep groundwater table, good permeability, and the positive results of laboratory tests. One AS well and four surrounding extraction wells, were installed. P&T was also applied from the two down gradient extraction wells.

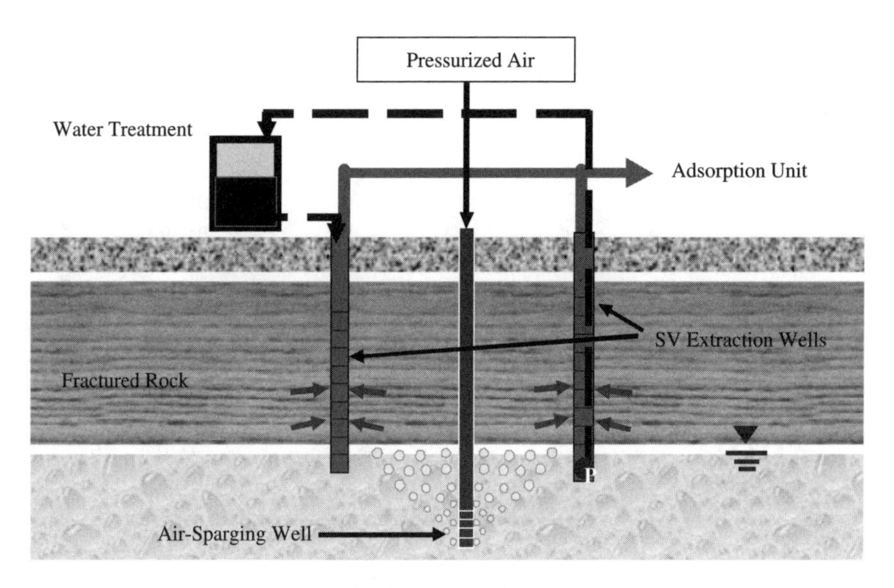

Figure 6. Integrated air-sparging systems implemented at the study site.

To quantitatively establish the effects of this system, the test was conducted in accordance with the following: (1) determination of appropriate extraction volume by the SVE only, (2) identification of the areal coverage of impact of P&T operation, and (c) identification of the areal coverage of stepped air-sparging.

4 FIELD SCALE RESULTS AND DISCUSSION

The results indicate that stable operation covering the contaminated area can be maintained with an air-sparging volume of 160 liters/min, an extraction volume of 150 liters/min at each extraction well and a pumping up volume of 15 liters/min at each well.

4.1 *Surface gas measurements*

Figure 7 shows the change in the concentration of the benzene in the first three months of treatment. After six weeks of SVE, benzene in the soil gas was reduced from a range of 15–60 ppm to a range of 1–6 ppm, representing a roughly 90% reduction A total of an additional 18 weeks of SVE, over three different periods, further reduced benzene in the soil gas from a range of 1–6 ppm to a range of 0–1 ppm. No benzene has been detected during the last 12 months. The SVE unit was shut down for monitoring any rebound in the benzene level in the extracted air. Several measurements were taken after that and the benzene concentration was zero.

4.2 *Groundwater analysis*

The total hydrocarbon concentration in the groundwater continued to decline in line with the treatment time as shown in Figure 8. During the past 12 months, air-sparging and P&T treatment has reduced TPH in the groundwater averaging between 5–8 ppm, and as high as 25–50 ppm during the first three months, to less than 0.5 ppm during the past six weeks. Similar to the SVE, the P&T unit was shut down for monitoring any rebound in contamination. The results show that after stopping P&T, small fluctuation in the level of contamination was observed. However, the level started to stabilize after sometime.

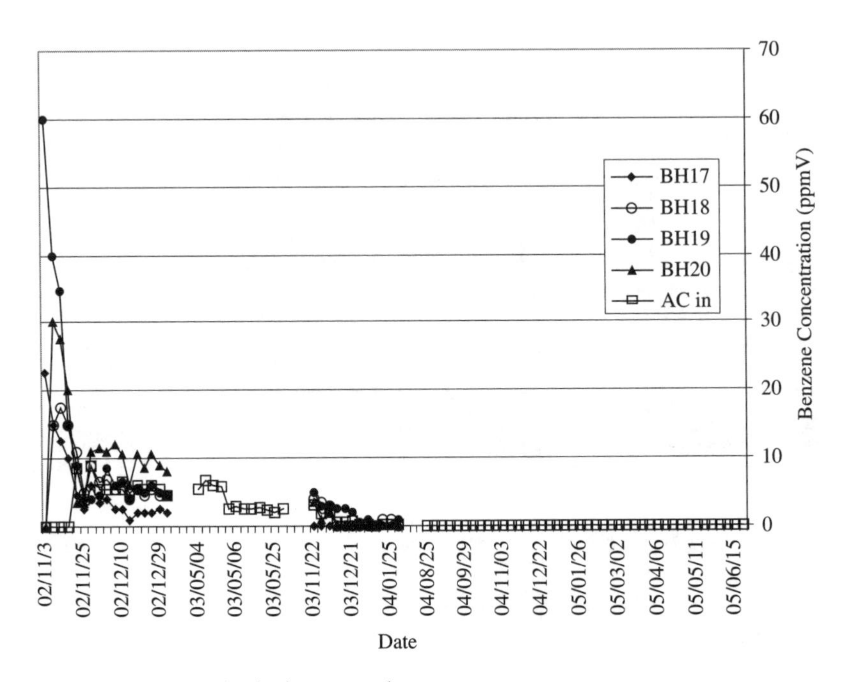

Figure 7. Benzene concentration in the extracted vapor.

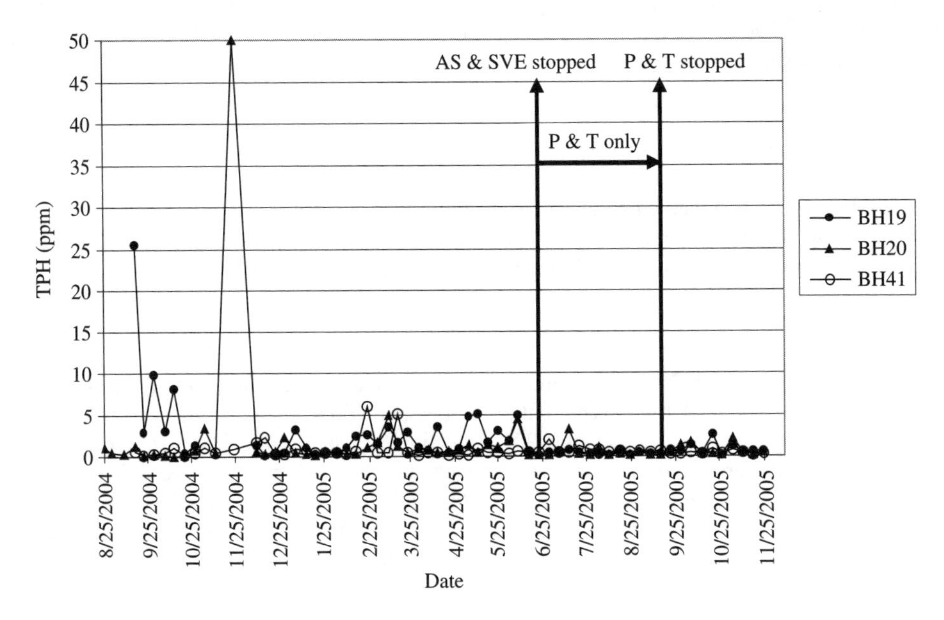

Figure 8. Water analysis results.

5 SUMMARY

In this study, investigation and remediation of a hydrocarbon contaminated site have been carried out. Tasks conducted in this work included, drilling/core-sampling of boreholes, permeability and pumping tests, collection and analysis of water and soil samples for hydrocarbon contamination, groundwater level monitoring, air-sparging, soil vapor extraction, and pump and treat.

Benzene in the soil gas was reduced from a range of 15–60 ppm to zero. The TPH in the groundwater was reduced from a range as high as 25–50 ppm to less than 0.5 ppm.

Results clearly indicate that with appropriate control, application of integrated AS system can remediate hydrocarbon-contaminated groundwater in fractured formation.

ACKNOWLEDGEMENTS

The authors would like to thank Japan Cooperation Center, Petroleum (JCCP), Japan, for their kind sponsorship of this project, and Ministry of Regional Municipalities, Environment, and Water Resources (MRMEWR), Oman for their contribution to the project.

REFERENCES

Al-Balushi, J.A. 2000. Source of Groundwater Pollution in the Sultanate of Oman. MRMEWR. Sultanate of Oman http://www.epa.gov/tio/pubitech.htm. 2005.

Japan Industrial Standard (JIS). 1995. K 0125. Testing Methods for Volatile Organic Compounds in Industrial Water and Waste Water.

Ministry of Regional Municipality, Environment, and water Resources (MRMEWR). 2001. Directorate of Pollution Prevention Database. Sultanate of Oman.

Reclaiming the Desert: Towards a Sustainable Environment in Arid Lands – Mohamed (ed.)
© 2006 Taylor & Francis Group, London, ISBN 0 415 41128 9

Waste management aspect analysis for Cylingas company L.L.C., Dubai, UAE

A.A. Busamra & W.S. Ghanem
Emirates National Oil Company L.T.D. (ENOC), Dubai, UAE

ABSTRACT: This study illustrates how a sequential review of data collection, analysis and prognosis can be applied in the specific context of waste management. The outcome of this review has been to develop a strategical framework known as the Waste Management Plan (WMP) for the waste handling logistics starting from generation through to final treatment and disposal.

A principle feature of the review has been to develop the plan for the waste management system based upon providing technical and financial benefits as well as reducing the risks and liabilities of any inefficiencies or man practices in the generation, handling and management of the waste.

Through the case study shown in this paper of CYLINGAS, the authors have shown below how the generic systematic approach to waste management reviews can be applied to other companies in general.

1 INTRODUCTION

As part of increasingly tighter controls from environmental legislation and the increasing awareness of the public and occupational hazards of waste, it is necessary that companies must take proactive stance towards developing sustainable waste management strategies.

In order to develop frame works within which effective waste management strategies can be planned, it is therefore essential to know the amount of waste generated, their sources, the materials in which waste streams, their properties, potential toxicity and hazards to human health and the environment.

1.1 *Background*

CYLINGAS is one of the Affiliates of the Emirates National Oil Company Ltd (ENOC), which operates within the Environmental regulations of the UAE National Laws and Regulations. It was established in 1974 and become fully operational in 1976. It is located in Al Quoz Industrial Area in Dubai and occupies an area of more than 300,000 square feet. CYLINGAS manufactures pressure vessels, storage tanks, LPG cylinders and oil and gas pipelines. In addition, the company is equipped with in house radiography facilities for examination of weld joints. CYLINGAS has two main divisions:

- Cylinder Division: where welded LPG, Acetylene and Refrigerant gas cylinders, refillable and non-refillable are designed and all of which are made to international standards.
- Tank Division: where storage tanks for petroleum products and chemicals, pressure vessels, gas and oil pipelines are designed and fabricated in accordance with international standards and the technical requirements of the clients.

Cylinders are manufactured under strict quality control to meet the specific requirements of individual customers. Production of quality cylinders starts with the selection of steel produced to rigid strength, ductility and thickness tolerances. Cylinder manufacturing process consists of four different operations including center portion manufacturing, top portion manufacturing, collar formation and foot ring formation.

CYLINGAS undertake bulk LPG installation for factories, hospitals, filling plants and hotels, fabrication of custom made tanks (with valves and fittings) and connecting them with pipelines to deliver LPG gas up to the point of use. The tank division at CYLINGAS consists of two main operations including shell manufacturing and saddle formation.

As per ENOC EHS Principles fully adopted by CYLINGAS, under Element 4, Expectation 4.1: "All applicable EHS-related Local, National, and International regulatory requirements are complied with, or shall have been granted permission to operate outside these requirements by the regulating Authorities or Senior Management, after due consideration of the impacts of non-compliance. Waiver of compliance to a regulatory requirement is documented."[GEHS/GL/001, 2002]

It is therefore, part of the management commitment of CYLINGAS to conduct a waste management aspect analysis study for all types of waste generated through various operational activities within CYLINGAS main office located at Al-Quoz, EPPCO Terminal site in Jebel Ali and Dubai International Airport site in Deira where CYLINGAS is currently involved in operational activities.

1.2 *Waste management system process*

Proper waste management starts with pollution prevention, which refers to the elimination, change or reduction of operating practices, which result in discharges to land, air or water. If elimination of waste is not possible, then minimizing the generated amount of waste should be examined. Responsible waste management may be accomplished through hierarchical application of the practices of sources reduction, reuse, recycling, recovery and responsible disposal. The waste management hierarchy sets the order of preference for various options for managing prescribed industrial waste. The principles of waste management include the incorporation of a hierarchy of waste management practices in the development of waste management plans. The below flow diagram shows how waste management practices could be applied.

The following UAE legislative drivers pertain to handling of waste:

1. Federal Environmental Protection Law (Law 24: 1999) and the subsequent Ministerial Order No. 37 for the Year 2001, Regulation for Handling Hazardous Materials, Hazardous and Medical Waste.
2. Federal Law No. (1) Of 2002 Concerning Regulation and Control of the Use of Sources of Radiation and Protection against its Hazards.

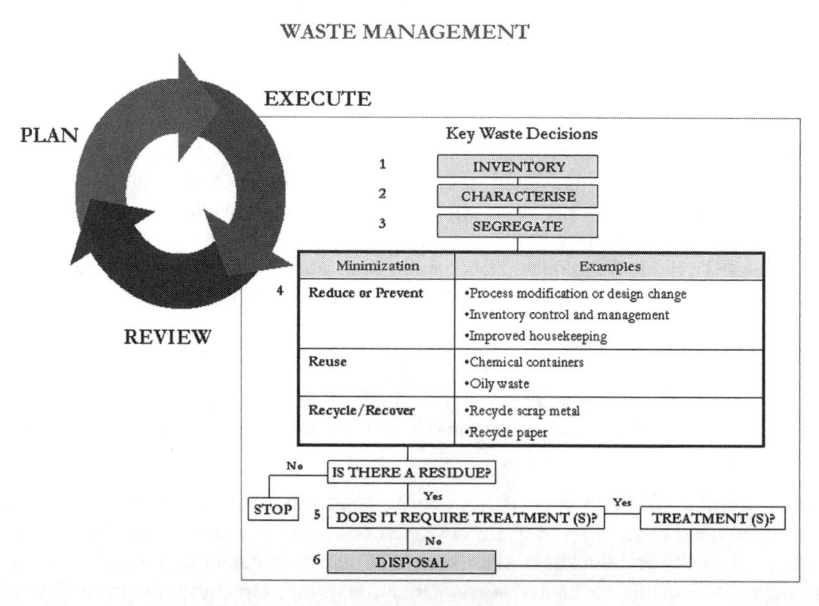

Figure 1. Generic flow diagram of waste management practices.

3. Dubai Municipality Local Order 61/1991 – The Environmental Protection Regulations for the Emirate of Dubai.
4. Dubai Municipality Technical Guideline of Waste Management Legislation.
5. Dubai Municipality Local Order 7/2002 – On Management of Waste Disposal Sites in the Emirate of Dubai.
6. Dubai Municipality Environmental Information Bulletin of Waste Management Legislation.
7. Dubai Municipality Environment Codes of Practice.
8. Dubai Municipality Code of Practice for the Management of Dangerous Goods in the Emirate of Dubai issued in 1997.

1.3 *Study objectives and methodology*

The purpose of this study was to assess CYLINGAS current WMP in terms of the applicable UAE current legislation and provide them with a full waste aspect analysis.

The recommendations of this study give a vital look at areas where performance improvement may be possible with respect to managing the generated quantities of waste both at the management and operators level in terms of minimization, documentation, and training programmes to achieve the pollution prevention goals.

The following simple steps show the methodology that has been applied.

1. Preliminary Preparation
2. Data Collection and Review
3. Legislative and Policy Review
4. Data Analysis and Assessment
5. Conclusions and Recommendations
6. Report Preparation

2 ACTIVITIES GENERATING WASTE

Table 1 summarizes types of waste generated at CYLINGAS.

3 CYLINGAS WASTE CLASSIFICATION AND DISPOSAL METHOD

The categories of waste generated at CYLINGAS can be categorized as follows:

- Trade/Industrial wastewater from painting booth: This wastewater is mainly generated from the plant particularly from the painting booth area. It is stored in a concrete base holding tank nearby CYLINGAS entrance gate as a trade wastewater. This classification is based on Dubai Municipality (DM) wastewater discharge limits to sewerage system and the laboratory analysis of the sample. The collected amount of wastewater is disposed off to DM sewerage treatment plant located in Al-Aweer to use it for irrigation through normal tankers.
- Hazardous wastewater from the plant: This wastewater comes from the cylinder division specifically the degreasing area (Alkadet, SONA, castor oil) and deep drawing process (deep drawing compound). This wastewater is deemed to be hazardous as the quality of the generated wastewater is exceeding the DM wastewater discharge limits to sewerage system as per the sampling analysis of the approved DM laboratory. Alkadet is a blend of chemicals having surfactant, detergency power with specific ionic characters and other speciality chemicals such as sequestering agent, precipitator, foam control agent with aesthetic and preservative in small quantities. It has low toxicity and completely soluble in water and may produce foam, as well as it is biodegradable. SONA is the detergent that is used for the cleaning.

CYLINGAS has one underground concrete base holding tank for the degreasing area. It is located nearby the boundary and two aboveground holding tanks located nearby the compressor room. The

Table 1. Summary of CYLINGAS activities generating waste.

No.	Operation/Activity	Generated Waste
		a. Liquid Waste
1	Decoiling and shearing	Hydraulic oil
2	Deep drawing	Wastewater including drawing compound (castor oil) and hydraulic oil
3	Clean cutting	Hydraulic oil
4	Joggling	Hydraulic and lubricating oil
5	Degreasing process	Wastewater including degreasing compound, detergent (SONA), lubricant/castor oil and water
6	Forming cylinder foot ring	Hydraulic oil
7	Painting cylinders and tanks	Wastewater including paint, sand and metal particles
8	Cylinders pneumatic test with soap solution	Wastewater including paint, sand and metal particles
9	Cylinders and tanks hydro-testing	Normal water
10	Domestic usage (offices, staff accommodation and canteen)	Domestic wastewater and solid particles
		b. Solid Waste
1	Raw Materials (mild steel)	Packing sheets (wooden pallets and plastic sheets), packing steel sheets and strips
2	Cylinder Manufacturing (center portion, top portion, and collar and foot ring formation)	Scrap steel plates and grinding disk, MS electrodes, flux and welding slag, metal scrap and metal grit and dust, contaminated cottons and gloves, blasting grit, paint sludge and paint accessories, incinerated paints and empty paint drums, labels, teflon balls and plastic waste, packing sheets and wooden pallets, empty containers and different types of drums, old rubber carpets, general waste (contaminated cloths/rags, sand dust, etc.)
3	Tank Manufacturing (shell manufacturing and saddle formation)	Molten and rust metals and metallic cans, MS and carbon electrodes, metal powder and grinding disk, welding slag and flux, paint sludge and paint accessories, empty drums and cans, blasting grit and general waste, radioactive decayed source
4	Cylinders, tanks and plant maintenance	Metal scrap and molten metals, empty drums and scrap cylinders, general waste (gloves and cleaning cloths), electrical and mechanical fittings
5	Repairing Tank 114 at EPPCO Terminal site in Jebel Ali Free Zone Area	MS electrodes, metal slag and powder, steel metal scrap and general waste
6	Constructing four tanks (E, F, G and H) for Dubai International Airport	Cleaning muffing brush, scrap steel and grinding disk, MS electrodes and metal slag, wooden pallets and packing sheets, general waste
7	Offices, Canteen and Staff Accommodation	Papers, plastic bags and packaging sheets, canteen equipment and food waste, accommodation and toilets outlets

collected amount of hazardous wastewater generated from the degreasing area is transported through DM approved transporter and disposed off to DM disposal sites located in Jebel Ali.

The deep drawing compound wastewater is sent to an approved DM waste management contractor to treat it and reuse it in different purposes.

- Domestic wastewater: This type of wastewater is generated from the offices, canteen and from the staff accommodation. It is collected in a two sewage collection pits each one is having a capacity of 5000 gallons. One is located nearby CYLINGAS offices and other in the staff accommodation. The generated amount of domestic wastewater is sent to DM Sewerage Treatment Plant in Al-Aweer via normal tankers.

- Waste oil: is generated mainly from the maintenance activities, gear boxes, hydraulic machines, compressors, generators, etc. There are different types of oil used by CYLINGAS i.e. castor oil (both commercial castor oil and refined castor oil), diesel, etc. CYLINGAS is generating more than 300 gallons/year of waste oil and it is accumulated in small trays located at the bottom of each machine and then the waste oil is transferred into drums in order to sell it to an approved DM waste management contractor.

Solid waste is classified by CYLINGAS into two different types including hazardous and non-hazardous. The following are some of these types:

- Blasting grit powder waste: this type of waste is generated from cylinders shot blasting activities and CYLINGAS consider it to be hazardous. Blasting grit is a grey-black steel abrasive chemical appear in a solid state and it contains less than 1 (0.85) silicon as well as it is insoluble in water. It is carried out at CYLINGAS in an automatic booth. CYLINGAS collects the blasting grit powder in small containers and then transfer it into polyethylene bags to store it in the scrap yard by the use of the forklift. The amount collected is analysed and then disposed off by DM in their Jebel Ali dumping site. In contrast, tank blasting is carried out by a contractor within CYLINGAS premises with full disposal responsibility.
- Paint sludge waste: paint sludge waste is solid or semi-solid waste generated from CYLINGAS painting activities at the paint booth through using of both paint and thinner. This type of waste is deemed to be hazardous as per DM, so it is collected manually in empty paint containers and then shifted to a designated marked place for storage and then analysed and disposed off by DM in their Jebel Ali dumping site.
- Empty paint containers: these containers are generated from the painting section. They are deemed to be hazardous. They are collected and stored in a designated area in order to sale them to an approved DM waste management contractor.
- Packing steel sheets and strips: the steel as a raw material comes to CYLINGAS in a packing sheets and strips. These packing sheets and strips are considered as a general waste. They are collected and stored in a scrap yard in order to sell them to a waste management contractor via pickup.
- Steel scrap and molten metals: they are generated from steel fabrication at cylinder and tank division. They are considered as a general waste. They are collected in skip and pallets and then segregated and kept in a designated waste area with a help of forklift in order to sell them.
- Welding slag: it is generated from SAW welding machine. Welding slag is a mixture of flux and steel. The flux is removed after filtration. Then the flux is stored in the waste bin in order to reuse it again in welding. Some traces of flux will remain with the welding slag which can not be removed. The welding slag then collected in bins and when it's full, the content is offloaded into the main skip with the help of a forklift in the waste area.
- Welding electrodes and grinding disks/wheels: these types of waste are considered as solid inorganic material. They are collected into small bins and then disposed into the general waste skip located in the waste area. CYLINGAS uses SMAW welding electrode (7016/7018/Baso 100) for their welding process. This electrode is made of manganese and/or manganese alloys and compounds (as Mn) and no harmful effects of this product on the environment are known.
- Radioactive decayed source: CYLINGAS uses Iridium192 as a radioactive substance or source in testing tanks to check if there are any defects in tank welding portions. CYLINGAS owns three projectors that are used for industrial radiography. These projectors are stored inside a small underground room in a concrete base store located in CYLINGAS yard clearly marked as radioactive substances store. In addition, the transporting containers are stored as well in the same concrete base store, but aboveground. The half-life of radioactive substance is over 74 days and its activity is 50 curies. When the radioactive substance activity drop into 5 curies or less, the radioactive substance is no more used and it is considered as a radioactive decayed source. CYLINGAS deems the decayed source as a radiographic dangerous waste. For that reason,

CYLINGAS stores the radioactive decayed source in the transporting container and clearly mark it as a decayed source as per the DM regulations in order to return it back to the supplier through Dubai Civil Defense approved transporters.

- Wooden pallets waste: this type of waste is considered as a general waste and they are generated from cylinder sheet metal bundles, packaging materials from purchased goods or equipment. Wooden scraps are segregated, useful scraps are kept for reuse in a separate area in the shop and unwanted scraps are kept in a designated waste area to through them later into the general skip.
- General domestic solid waste: all domestic solid waste including cotton waste and rags, grass, gloves, damaged cloths, damaged PPE, normal sand and dust, etc are collected and put into municipality skip which is free of charge.

4 QUANTIFICATION OF WASTE

In this study, CYLINGAS waste quantities are quantified in terms of mass and volume and cost of off site transportation and treatment/disposal.

Domestic wastewater represents the largest quantity of waste generation. It can be seen from Table 2 shown below that over the last few years the generation has been around 470,000 gallon. The total of off site transportation and disposal was around AED 180,000. In the years 2002 and 2003 there was no discharge of industrial wastewater from both holding tanks of paint-water mixture located in the painting booth area to the main holding tank as there were no instructions to the operators to drain the wastewater from the painting booth to the main holding tank. Some amount of wastewater was used to be evaporated naturally and the remaining amount was sent to the main holding tank whenever the operator feels to drain. The instructions came to be known to the operator in 2004.

By comparing between CYLINGAS solid waste quantities and the cost for the past four years, it can be perceived that general domestic waste was generated with the highest amount for around 1,540 ton as well as has the highest cost. CYLINGAS should aim to reduce the amount of the generated waste from their premises by setting up strict instructions based on the local and international regulations as well as the best practice. However, CYLINGAS has stepped towards one of the cleaner production principles by selling some types of its waste to recycle companies such as waste oil, steel scrap, empty paint containers, etc. and generate a good outcome from them.

Table 2. Summary of CYLINGAS waste quantities and costs*.

	Quantity of waste				
Waste	2002 gallon	2003 gallon	2004 gallon	2005 gallon	Total cost/revenue AED
Industrial wastewater	–	–	3,500	3,500	1,100
Hazardous wastewater	90,000	100,000	70,000	90,000	180,000
Domestic wastewater	110,000	110,000	110,000	140,000	7,100
Waste oil	111	111	111	111	1,200
	2002 ton	2003 ton	2004 ton	2005 ton	
Blasting grit	5	20	21	15	3,500
Paint sludge	17	14	20	17	13,000
Empty paint containers	16	16	16	16	1,600
Steel scrap	500	450	470	480	1,000,000
Wooden pallets	16	16	16	16	2,400
General domestic waste	420	420	260	440	24,000

* Costs given above are approximations and for reasons of commercial sensitivity do not represent actual figures

5 ASSESSMENT OF CYLINGAS WASTE MANAGEMENT

CYLINGAS waste management was assessed based on its applicable local regulations, waste management hierarchy and ENOC Guideline no. 12 on Establishing a Waste Management System. The following points are the detailed assessment.

5.1 *Assessment of CYLINGAS waste management hierarchy*

5.1.1 *Waste generation "point"*

The main waste generation sources (solid, liquid, emission) are well recognized in practice by CYLINGAS. However, the main waste generation sources are not identified and documented properly. It is clear that CYLINGAS seeks for waste minimization in its different activities in order to optimise utilization whenever possible. Listed are some of CYLINGAS initiatives towards waste minimisation.

1. Recycling the water used for hydro-testing by returning back the used water to the water tank after testing.
2. Used flux in welding is collected and reused.
3. The remaining LPG in the cylinders that require repairing is stored to use it as fuel for the furnace.
4. Waste scrap and empty paint containers are sold to some dealers.
5. Paper waste is recycled.

But still there are some points that need to be further investigated by CYLINGAS to reach the cleaner production i.e. waste source reduction, selling the used rubber carpets and wooden pallets to some dealers (DM could provide details), etc.

5.1.2 *Waste handling*

CYLINGAS have the basics of a waste segregation scheme at certain locations where specific types of waste are generated, For example skips are located near to the process areas for collection of specific waste types i.e. oils, paints , fluxes etc.

In line with best industrial practices skips could be identified with color codes and labeled in order to ensure the waste segregation and collection is carried out through to final offsite collection and recycling/reuse or disposal.

5.1.3 *Waste collection and disposal*

CYLINGAS has a general collection point for all the waste where the waste is stored onsite rending collection and offsite transportation by a DM approved waste management contractors to the appropriate locations for the waste described in section 3.

6 CONCLUSION

The paper describes a Waste Management Aspect Analysis study that was conducted for different operational activities associated with CYLINGAS Company based upon their commitment for the protection of the environment and public health.

The recommended approach to preparing a waste management plan is to carry out the following steps:

Step 1: Management Support, Top management must provide visible commitment and support for the plan. Overall waste minimization strategy goals and objectives must be defined.

Step 2: Waste Identification and Categorization, CYLINGAS must identify all waste. The streams generated in the area from all operational sources. Each waste stream must be defined in terms of its type and classification, volume produced and source. This will become the waste inventory. The physical, chemical and toxicological properties of the waste must also be recorded. The sources of all waste streams must be identified on a plot-plan of the facility. Table 3 below shows the DM

Table 3. Recommended waste classification for CYLINGAS as per DM.

No.	Type of Waste	Waste Definition	Generated Waste
1. Solid			
a	General Waste	is the solid non-hazardous waste generated due to domestic, trade, horticulture and industrial activities, and the inert solid waste (such as construction and demolition waste) that can be disposed of in the general waste landfills, and any other waste classified so by the ED.	• Steel scrap and molten metals • Packing sheets, strips-steel • Scrap cylinders • Domestic waste • Wooden pallets • Old rubber carpets • Food waste • Grinding wheels • MS electrodes • Plastic waste
b	Hazardous Waste	is the waste unsuitable for direct disposal into the environment or sewer system or by traditional landfill, and any waste deemed by the Municipality to pose risk on the environment or public health due to the production operation, the existence of hazardous components or chemical or physical properties.	• Contaminated cottons and gloves • Paint empty containers. • Paint sludge. • Waste oil. • Flux • Burned paint from paint residue attached by hooks
c	Difficult Waste	is the non-hazardous waste which requires special handling to avoid any unacceptable annoyance or environmental impact.	• Welding slag • Blasting grit
d	Unwanted material	is any non-hazardous material/goods declared by its owner to be unwanted and requires disposal, or of which a decision for its destruction/disposal is issued by a competent authority, or that its proper disposal requires a special care.	• Electrical and mechanical fittings
2. Wastewater			
a	Trade/Industrial wastewater	Any wastewater generated and discharged from industrial operations or commercial activities.	Wastewater generated from the painting booth.
b	Hazardous wastewater	Any wastewater deemed by the Municipality to pose risk on the environment or public health due to the production operation, the existence of hazardous components or chemical or physical properties.	Wastewater generated from the decreasing area and the deep drawing press.
c	Domestic wastewater	All water-borne human wastes, also called sewage, arising from residential premises as well as from, industrial, commercial and institutional buildings.	Wastewater generated from domestic usage.
3. Emissions			
a	Allowable emission	Any gaseous, smoke, fumes, mist, heat, noise, particulate or airborne dust being released into the air environment and within DM standard allowable limits.	CYLINGAS shall measure the emissions to see whether they are within DM limits
b	Non-allowable emission	Any gaseous, smoke, fumes, mist, heat, noise, particulate or airborne dust being released into the air environment and exceeding DM standard allowable limits.	CYLINGAS shall measure the emissions to see whether they are exceeding DM limits

Note: Radioactive decayed source is considered as a dangerous goods as per DM.

generic classification of waste. Through the training of staff and implementation of appropriate waste management systems and procedures, these generic classifications should be followed in CYLINGAS's overall strategy for the handling of the waste.

Step 3: Waste Minimization, Opportunities for waste elimination or minimization must be examined for each waste stream in the following hierarchy of options:

1. Reuse – After all opportunities for source reduction have been considered, the next stage is to evaluate options for reuse of the waste material. Examples could be through the reuse of rubber carpets to rubber dealers rather than disposal; similarly instead of disposal of wooden pallets they can be collected and sold back to wood dealers.
2. Recycling/Recovery i.e. All opportunities for recycling/recovery of waste materials must be actively pursued.
3. Treatment – After source reduction, reuse, recycling and recovery options have been fully explored pre-treatment of the waste in house to reduce volumes and/or toxicity must be considered.
4. Disposal – Disposal must be seen as a last resort, when all the above options have been fully explored. Within Dubai, the Municipality provides state of the art facilities for landfilling of general waste and for the industrial waste a Hazardous Waste Treatment Facility.

Step 4: Selection of Preferred Options, Once the list of management options has been compiled; the preferred option for each waste stream can be selected. This must be based on a life cycle approach, which takes into account the HSE risks associated with storage, transport, treatment and disposal.

Step 5: Plan Implementation, The information collected in the above steps must be used to prepare the Management Plan. This plan must be a concise document, suitable for use by operations personnel. It must contain all relevant information on the preferred disposal option(s). In addition as part of the implementation of the plan it would be necessary to provide the appropriate training to the CYLINGAS employees in order to raise the awareness on the basis of waste management and pollution prevention.

Step 6: Plan, Review and Update, Effective waste management is an ongoing process. The plan must be reviewed periodically to ensure that it is still appropriate for the types of waste being handled. In particular:

1. CYLINGAS management should establish EHS targets and key performance indicators and specifically for waste reduction on a yearly basis and to monitor these KPI's to achieve continual improvement.
2. CYLINGAS management should review progress and performance targets.
3. The waste management plan should be reviewed and updated to reflect changing technologies, needs or legislation.
4. CYLINGAS management to arrange a waste management audit on a yearly basis as the views of findings of a waste audit can be value for improving performance and provide independent view of the management system in place.

REFERENCES

CYLINGAS, 2003, *EHS Policy*, Dubai: CYLINGAS.

CYLINGAS, 2005, *Waste Audit Report*, Dubai: CYLINGAS.

DM, 1997, *Dubai Municipality Code of Practice for the Management of Dangerous Goods in the Emirate of Dubai*, Dubai: DM.

DM, 1996, *Dubai Municipality Environmental Bulletin, Air Monitoring Summary*, Dubai: DM.

DM, 2003, *Dubai Municipality Information Bulletin, Environmental Standards and Allowable Limits of Pollutants on Land, Water, and Air Environment*, Dubai: DM.

DM, 2002, *Dubai Municipality Information Bulletin, Explanatory Notes on Local Order No. (7) For year 2002 on Management of Waste Disposal Sites in the Emirate of Dubai*, Dubai: DM.

DM, 2005, *Dubai Municipality Information Bulletin on Waste Oil Collection and Reuse*, Dubai: DM.

DM, 1991, *Local Order No. 61 of 1991 on the Environment Protection Regulations in the Emirate of Dubai, Dubai*: DM.
DM, *Dubai Municipality Technical Guidelines applicable to planning, construction and waste management in Emirate of Dubai*, Dubai: DM.
ENOC, 2003, *ENOC EHS Guideline Manual on Establishing a Waste Management System*, GEHS/GL/012, Dubai: ENOC.
ENOC, 2002, *ENOC EHS Guideline on Environment, Health and Safety Principles*, GEHS/GL/001, Dubai: ENOC.
FEA, 1999, *Federal Environmental Protection Law (24), Protection of the Environment*, Abu Dhabi: FEA.
FEA, 2002, *Federal Law No. (1) Of 2002 Concerning Regulation and Control of the Use of Sources of Radiation and Protection against its Hazards*, Abu Dhabi: FEA.
FEA, 1999, *Regulation for Handling Hazardous Materials, Hazardous Wastes and Medical Waste*, Abu Dhabi: FEA.
http://www.wef.org/Whoweare/WWIndustry/industrial.jhtml, 27th March 2005.
John J. Munro & Francis E. Roy, 1986, *Gamma Radiography-Radiation Safety Handbook*, Burlington: Amersham Corporation, 40 North Avenue.
Singh B K, 2003, *CYLINGAS EHS Management System*, Dubai: CYLINGAS.
Singh B K, 2004, *CYLINGAS EHS Objectives, Targets and Performance Review*, Dubai: CYLINGAS.
Singh B K, 2003, *CYLINGAS EHS Regulatory Compliance*, Dubai: CYLINGAS.
Singh B K, 2004, *CYLINGAS Employee Training Procedure*, Dubai: CYLINGAS.
Singh B K, 2003, *CYLINGAS Hazard Identification, Assessment and Management*, Dubai: CYLINGAS.
Singh B K, 2004, *CYLINGAS Internal Audit and Inspection Procedure*, Dubai: CYLINGAS.
Singh B K, 2004, *CYLINGAS Waste Disposal Procedure*, Dubai: CYLINGAS.

Reclaiming the Desert: Towards a Sustainable Environment in Arid Lands – Mohamed (ed.)
© 2006 Taylor & Francis Group, London, ISBN 0 415 41128 9

Investigation of mercury levels in sediments and fish from coastal areas of Abu Dhabi city

S.R. Al-Hassani, A.E. Elsaiid & S. Bashir
Main Chemical Laboratory, Armed Force, Abu Dhabi, United Arab Emirates (UAE)

J.H. Dennis
Department of Environmental Science, University of Bradford, Bradford, UK

ABSTRACT: Mercury levels in sediments and fishes of the coastal marine environment of Abu Dhabi city, the capital of United Arab Emirates, were preliminary investigated during February of 2005. Superficial coastal sediment samples were collected from 12 sites located between longitude 54°16′55″ to 54°37′50″ E and latitude 24°15′35″ to 24°32′05″ N. Nine types of traditional edible fish consumed by the local population, obtained from the same region above: shaam, gabot, hamra, feskar, shaary, badah, anfoos, farsh, and saafi were inspected and its tissues were examined for its mercury content. Analysis of mercury was conducted using cold vapor atomic absorption spectrometry. The determined mercury concentration values in the 12 sediment samples, expressed as mg/kg dry sediment weight, were ranged from 0.015 to 0.107 with a mean value of 0.049. For the fish species, the determined mercury concentrations (mg/kg wet weight) in the fish tissues were ranged from 0.008 (for shaam) to 0.064 (for shaary). The concentration of mercury was below the maximum allowed limit by the Emeriti and international legislations for fish human consumption permissible limit. The precision of analyses was checked by analyzing certified reference materials.

1 INTRODUCTION

Mercury is a metal that is commonly found in the environment in several forms, all of which are toxic. Mercury contamination has been an environmental issue for several decades. It was established that emissions of mercury to the environment could have serious effects on human health (Niencheski et al. 2001, Lindqvist 1994). Mercury in the environment is derived from both natural sources and human activities. The metal is naturally occurring element and exists in soils, sediments, rocks, oil, and coal. Mercury deposits are generally found as cinnabar and this is the most important mercury ore (Covelli 2001). Mercury also occurs naturally in very small quantities in barite, a major component of drilling fluids used by the offshore oil and gas industry (Lindqvist 1991, Porecella 1994).

The unique properties of mercury have resulted in a long history of use by the enterprising human race. Mercury is used in a number of consumer and commercial products. Some of these products are more commonly recognized as containing mercury than others. Mercury is found in varying amounts in batteries, fluorescent and high intensity light bulbs, thermometers, thermostats, and light switches. Mercury is also used to make chlorine and caustic soda and certain types of dental fillings. Some paints and pesticides contain mercury (as a preservative and fungicide). Thus, citizens, hospitals, dental offices, farmers, builders, and certain types of manufacturing operations all use and eventually discard products containing mercury into the municipal solid waste stream. Following disposal the mercury in these items may ultimately be released into a landfill or the atmosphere following combustion in a waste combustor (Lindquvist 1991).

In addition to mercury emissions associated with disposal and incineration of municipal wastes, mercury is also released into the atmosphere by the burning of fossil fuels such as coal and oil, medical wastes, and wood. Releases also occur when products containing mercury, such as fluorescent lights, are broken, from volatilization during laboratory and industrial uses, during cremation of human bodies, due to mercury use in amalgam fillings, and, in the purification, or roasting, of ores.

In addition to industrial activities, worldwide agriculture and mining have also contributed major amounts of mercury to soils, water and air.

Mercury released into the environment can either stay close to its source for long periods, or be widely dispersed on a regional or even world-wide basis. The metal can then undergo chemical transformations including oxidation, reduction, methylation, and demethylation. Biological processes play an important part in these transformations; depending on local conditions, bacteria may ultimately convert some of the deposited mercury to methyl mercury, which is taken up by organisms through ingestion and absorption (e.g. Compeau & Bartha 1984, Baldia et al 1989).

Mass balance studies show that atmospheric transport and deposition of mercury is the dominant pathway delivering mercury to many of the world's lakes and oceans. The atmospheric deposition of mercury is sufficient to account for almost all the mercury in many aquatic ecosystems. Mercury enrichment in superficial lake sediments suggests that atmospheric mercury deposition has increased by a factor of 3–5 globally since the industrial revolution (Fitzgerald 1989, Rada et al 1989, Swain et al 1992 & Niencheski et al 2001).

While it circulates in the environment and changes its form, mercury is persistent and is not biodegradable. It tends to accumulate in sediments – in rivers, streams, lakes and the ocean. Global and regional atmospheric circulation and the relative importance of regional and local sources of mercury must be understood for assessing the local risks with this contaminant (Niencheski et al 2001).

Sediments are being used to monitor coastal environments because sediments faithfully record and time-integrate the environmental status of an aquatic system. Contaminant concentrations are high in sediments, and thus they are easily, cheaply, and accurately analyzed. Sediments can be an important secondary source of pollutants, and because they integrate contaminants over time, sediments provide useful spatial and temporal information. Sediments quality influences the nature of overlying and interstitial waters through physical, chemical and biological processes. Because sediments play a major role in the transport and storage of contaminants, they are important in identification of contaminant sources and determining dispersion pathways. Sediments also provide an important habitat for animals and are a food source for many species. Sediments quality thus determines, to a large degree, biodiversity and ecological health in aquatics systems, and they are economically attractive in environmental assessment of coastal environments (e.g. Horowitz 1991, D Daskalakis & O'Connor 1995, von Gunten et al 1997 & Lorry & Driscoll 1999).

To manage the risks associated with mercury contamination, the first step is to understand the exposure of target populations to mercury contamination. In case of human populations, the greatest risk is associated with exposure to mercury – contaminated fish. Thus an assessment of the levels of mercury in local edible fish is the first step in assessing the risks to human populations due to environmental contamination by this element.

The present work is a part of a larger study concerned with the contamination of coastal marine sediments of Abu Dhabi city and adjacent zones with various contaminants especially heavy metals. The present paper presents the results of a preliminary investigation of mercury levels in local fishes and coastal marine sediments of Abu Dhabi city coastal areas. The aim of this pilot study was to assess whether mercury levels are related to environmental contamination, and to assess the potential risk to human populations consuming these fishes. The study can serve as an indicator for the extent of pollution by mercury metal in the Arabian Gulf.

2 MATERIALS AND METHODS

2.1 *Sampling*

The coastal areas of Abu Dhabi Island were divided into 12 zones located between longitude 54°16′55″ to 54°37′50″ E and latitude 24°15′35″ to 24°32′05″ N (Al-Hassani et al 2005).

Table 1. Location of sampled sites and concentration of mercury (mg kg^{-1} dry sediment weight) in sediment samples collected from them.

Zone and Site number	Longitude	Latitude	Mercury concentration (mg kg^{-1})
A1	54° 23′ 40″	24° 28′ 50″	0.023
B11	54° 21′ 20″	24° 29′ 0″	.039
C20	54° 22′ 25″	24° 25′ 30″	0.015
D34	54° 24′ 40″	24° 26′ 40″	.034
E42	54° 25′ 50″	24° 29′ 30″	0.073
F51	54° 20′ 30″	24° 30′ 0″	.046
G59	54° 25′ 25″	24° 17′ 10″	0.046
H66	54° 27′ 35″	24° 19′ 45″	.035
I76	54° 29′ 40″	24° 26′ 20″	0.107
J80	54° 27′ 30″	24° 30′ 0″	.062
K90	54° 2′ 30″	24° 28′ 50″	0.056
L98	54° 37′ 50″	24° 30′ 0″	0.038

Superficial coastal sediment samples were collected from 12 sites near-shore line of Abu Dhabi Island, one site from each zone. The designations and the exact locations for these sites are given in Table 1 and their positions on Abu Dhabi map area were given elsewhere (Al-Hassani et al 2004 & Al-Hassani et al 2005). The sediment samples were collected during February 2005. Sampling was manually conducted from the water course sediments using PVC tubes. The depth of the sampled sediment layer under water surface was around one meter. A sampling tube was pushed in the sediment layer and a sediment core (10–15 cm in length and 5.5 cm in diameter was collected. The upper 2-centimeter slice was cut off, dried, powdered and employed in the analysis. Nine of fish species: shaam, gabot, hamra, feskar, shaary, badah, anfoos, farsh, and saafi were fished from the same area of the sediment sites as given above. The weights of these fishes were chosen to be around one kg, except for anfoos which was 0.5 kg. Fish tissues were used in the analysis.

2.2 *Analysis*

Mercury concentration in various samples was determined by cold vapor atomic absorption spectrometry, CVAAS, (Chou & Naleway 1984). In this technique, a sample is digested with acids in a water bath and strong oxidizers to convert all mercury to Hg^{+2}. Tin chloride is then added to reduce the mercury ion to Hg^0 where it is stripped and carried to a mercury lamp where atomic absorption occurs by mercury vapor at wavelength 253.7 nm and is measured. The applied used procedure was as follows. To exactly about 2 gram of sample material, 10 ml of aqua regia ($HCl:HNO_3$ in 3:1 ratio, v/v) was added. The mixture was digested in a microwave (CEM Corporation, US) using the microwave program (CEMMDS. 81D), and finally diluted to 25 ml with distilled deionized water. Mercury concentration in extracted acid solution was determined by atomic absorption spectrometer (Spectra AA. 800 Varian, with VGA77 vapor generation accessory).

2.3 *Quality control*

The reliability of the analytical method was evaluated by analysis of standard reference solutions and blanks as well as certified reference materials (Lobster hepatopancreas marine reference material for fish and HISS-1 and MESS-3 for sediments from National Research Council of Canada). Relative standard deviation was generally less than 5%. The detection limit was determined by spiking reagent water and digested food samples with several diluted concentration of mercury and found to be 0.1 μg/L.

3 RESULTS AND DISCUSSION

3.1 *Mercury in sediments*

Results of analysis of surficial marine sediments from the sampled coastal areas for total mercury content are given in Table 1 and compared in Figure 1. The range of mercury concentrations in the sediment samples is 0.015–0.107 mg kg^{-1} dry sediment weight, the average is 0.049 mg kg^{-1}, the median is 0.043 mg kg^{-1} and the mode is 0.046 mg kg^{-1}.

Mercury in sediments is derived from both natural sources and human activities. For assessment of sediment quality (the ability of sediment to support a healthy benthic population) a scheme was developed by the US National Oceanic and Atmospheric Administration (NOAA) (Long et al 1995, Long & MacDonald 1998). The scheme provides two values, ERL and ERM, which delineate three concentration ranges for a particular chemical. The concentrations below ERL values represent a minimal – effects range, which is intended to estimate conditions where biological effects would be rarely observed. Concentrations equal to, or greater than ERL, but below ERM, represent a range within which biological effects occur occasionally. Concentrations at, or above ERM values represent a probable – effects range within which adverse biological effects frequently occur.

From this scheme, the ERL value may be used as a threshold level that triggers the requirement for additional investigative work concerning contamination of sediments with a given chemical. The ERL and the ERM limit values in the NOAA scheme for mercury are 0.15 and 0.71 mg kg^{-1} dry sediment, respectively.

All the determined mercury concentrations in the examined sediment samples were less than the ERL value for mercury, perhaps due to absence or negligible anthropogenic inputs. For this reason, probably these mercury concentrations may provide a good regional representation of the natural levels of the element in the sediments of the region.

3.2 *Mercury in fish tissues*

Table 2 shows the total mercury content in the tissues of the nine inspected types of fish. Highest determined mercury level, 0.064 mg kg^{-1}, was found in shaary and lowest determined level, 0.008 mg kg^{-1} wet weight for whole tissue was found in shaam.

Nearly all of the mercury in fish muscle (greater than 95%) occurs as methylmercury (Huckabee et al 1979, Grieb et al 1990, Bloom 1992) and most of it is transferred to fish through their food. Methylation of mercury, the addition of ($-CH_3$) to form methylmercury (the organic form of mercury which is the most toxic to animals) may occur in water, sediments or soil. The extent of this process may be attributed to the following. Mercury is a metal that is commonly found in the environment in several forms, as mercury metal and as inorganic and organic mercury compounds.

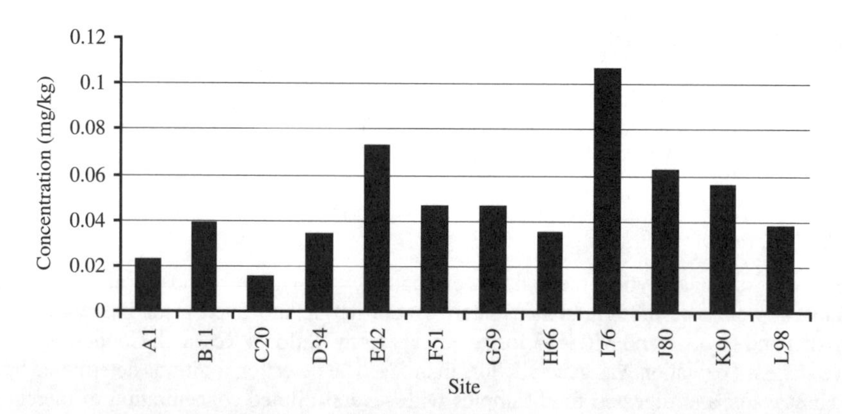

Figure 1. Mercury concentrations in tested sediment samples.

The metal is mobile and can change its forms. It can form many stable complexes with organic (carbon-containing) compounds. In addition, inorganic mercury can be methylated by microorganisms indigenous to soils, sediments, fresh water, and salt water, to form organic mercury (Press & Siever 1978). In water and sediments the amount of methylation is affected by several factors such as the amount of dissolved oxygen present, the amount of sulfur present, the pH of the water or sediment, and the presence of particles of clay or organic material (Grieb, et al 1990). Where the amount of oxygen is limited, as in deeper layers of the surface water or sediments, more methylmercury is formed. The presence of sulfur may be important because it is thought that sulfate-dependent bacteria are involved in the methylation process (King et al 2002). Low pH is associated with an increase in methylation. (This means that methylation may occur more readily in water affected by acid rain.) (Grieb et al 1990, Gilmour & Henry 1991). If clay particles are present in the water, the mercury may attach to the particles, and may not be as available for methylation.

Small organisms take up methylmercury as they feed. Fish accumulate methylmercury directly from the water in which they live (the compound is fairly soluble in water) or from prey. As this process, (known as bioaccumulation), continues, levels of methylmercury increase up the food chain. Fish species that prey on other fish have higher concentrations of mercury and the mercury concentration increases in these predators as they age (Grieb et al 1990). The main pathway for human exposure to methylmercury is through consumption of freshwater and marine fish.

Many countries (e.g. Health & Welfare Canada 1984) have set a threshold safety value of 0.5 mg kg^{-1} wet weight for mercury concentration in fish sold commercially. Recently human toxicology data suggest that allowable mercury concentrations in fish be lowered to 0.30 mg/kg wet weight (NRC 2000, USAEPA 2001). Fish inspected in the present study were found to be below levels currently considered to be safe for regular consumption. The weight of a fish of the studied types was around one kg, except anfoos which was about 0.5 kg. Bearing in mind that aged and larger fishes presumably will have higher levels of mercury content, this study needs to be further extended to comprise other fish types of different ages and sizes.

3.3 *Reduction of mercury exposure*

The above results for mercury analyses were within normal levels for sediments and fishes and well below regulatory levels. Since mercury is listed as a priority pollutant by the international agencies in charge of marine environmental protection. While the metal circulates in the environment and changes its form, it is persistent and is not biodegradable. Furthermore, once mercury present in a biological system, it can be passed up the food chain, "bioaccumulating" (increasing its concentrations) accordingly (Svobodova et al 1999). For these reasons, routine mercury analysis should be included in all the environmental monitoring programs. Release of mercury to the environment should be prevented or at least minimized to reduce the health risks associated with its release. It is virtually impossible to minimize mercury from natural sources, such as leachate from mercury-containing ores. However, control through legislation should be imposed on sources that are releasing mercury to the environment such as factories, power plants, incinerators and hospitals. Reduction of mercury content in various products and promoting collection and recycling of mercury-containing items will help prevent mercury release to the environment.

Table 2. Determined mercury concentrations in tissues of the studied types of fished.

Fish type	Hg concen. (mg kg^{-1})	Fish type	Hg concen. (mg kg^{-1})	Fish type	Hg concen. (mg kg^{-1})
Shaam	0.008	Feskar	0.043	Anfoos	0.012
Gabot	0.018	Shaary	0.064	Farsh	0.049
Hamra	0.032	Badah	0.057	Saafi	0.024

4 CONCLUSIONS

From the results of this study, the following conclusions can be drawn.

- Since levels of mercury in sediments of the studied areas are below the ERL concentration of US NOAA (Long et al. 1995, Long & MacDonald 1998) for sediment quality, therefore, these levels are within minimal - effects range.
- The concentration of mercury was below the maximum allowed limit by the Emeriti and international legislations for fish human consumption permissible limit. In other words, there is little risk of mercury exposure to human populations from the consumption of fish from the indicated coastal areas assuming mercury concentrations in inspected species observed in this preliminary study are representative of the local fish.
- Because of all the results for mercury analyses were within normal levels for sediments and fishes and well below regulatory levels, thus the mercury in the region is interpreted to be natural in origin rather than the result of contamination.
- Mercury concentration in fish from the studied coastal areas may provide a good regional representation of natural levels of the element in the region.
- Since mercury is released into the environment from a wide variety of natural and human activities and because the mercury's combined qualities of potential toxicity, environmental persistence, and potential for bioaccumulation, this metal should be continuously monitored in the environment and legislation to be issued to minimize its potential health risks associated with it.

ACKNOWLEDGEMENT

Authors would like to acknowledge gratefully the help received from Municipalities and Agriculture Department, Food & Environmental Control Centre, Emirate of Abu Dhabi.

REFERENCES

Al-Hassani, S.R., Elsaiid, A.E., Bashir, S. & Dennis, J.H. 2004. Seasonal variations of heavy metals in marine surface sediments of Abu Dhabi city coastal areas. *Chemical and Environmental Engineering; Proceeding of the 2nd International Conference, 23–25 November 2004.* Cairo: Military Technical College.

Al-Hassani, S.R., Elsaiid, A.E., Bashir, S. & Dennis, J.H. 2005. Heavy metals distribution in coastal marine surface sediments of Abu Dhabi city. *Environmental Protection is a must, Proceeding of the 15th International Conference, 3–5 May 2005.* Alexandria :Middle East Research center-Ain-Shams Univeristy.

Baldi, F., Filippelli, M. & Olson, G.J. 1989. Biotransformation of mercury by bacteria isolated from a river collecting cinnabar mine waters. *Microb. Ecol* 17: 263–274.

Bloom, N.S.1992. On the chemical form of mercury in edible fish and marine invertebrate tissue. *Canadian Journal of Fisheries and Aquatic Science* 49: 1010–1017.

Chou, H.N. & Naleway, C.A. 1984. Determination of mercury by cold vapor atomic absorption spectrometry. *Analytical Chemistry 56 (9)*: 1737–1738.

Compeau, G. & Bartha, R. 1984. Methylation and demethylation of mercury under controlled redex, pH, and salinity conditions. *Appl. Environ. Microbiol.* 48: 1203–1207.

Covelli, S., Faganeli, J., Horvat, M. & Brambati, A. 2001. Mercury contamination of coastal sediments as the result of long-term cinnabar mining activity(Gulf of Trieste, northern Adriatic sea). *Applied Geochemistry* 16: 541–558.

Daskalakis, K.D. & OÇonnor, T.P. 1995. Normalization and elemental sediment contamination in coastal United States. *Environmental Science and Technology* 29: 470–477.

Fitzgerald, W. 1989. Atmospheric and oceanic cycling of mercury. In Riley, J. & Chester, R. (eds), *Chemical Oceanography*: 151–186. New York: Academic Press.

Gilmour, C.C. & Henery, E.A. 1991. Mercury methylation in aquatic systems affected by acid deposition. *Environmental Pollution* 71: 131–169.

Grieb, T.M., Driscoll, C.T., Gloss, S.P., Shofield, C.L., Bowie, G.L. & Porcella, D.B. 1990. Factors affecting mercury accumulation in fish in the Upper Michigan Peninsula. *Environmental Toxicology an Chemistry* 9: 919–930.

Health Welfare Canada. 1984. Methylmercury in Canada. Exposure of Indian and Inuit Residents to methylmercury in the Canadian Environment. 164 pg Health and Welfare Canada, Medical Services Branch.

Horowitz, A.G. 1991. *A primer on sediment trace element chemistry,* second ad. Chelesea, MI: Lewis Publishers Inc.

Huckabee, J.W., Elwood, J.W, & Hilebrand, S.G. 1979. Accumulation of mercury in freshwater biota. In J.O. Nrigu (ed). *Biogeochemistry of mercury in the Environment*: 277–302. New York: Elsevier.

King, J.K., Harmon, S.M., Fy. T.T. & Gladden, J.B. 2002. Mercury removal, methylmercury formation, and sulfate-reducing bacteria profiles in wetland mesocosms. *Chemosphere* 46: 859–870.

Lindquvist, O. 1991. Mercury in the Swedish environment – recent research on causes, consequences and corrective measures. *WaterAir Soil Pollution* 55: 1–261.

Lindquvist, O. 1994. Atmospheric cycling of mercury: an overview. In C.J. Watras, J.W. Huckabee (eds), *Mercury Pollution-Integration and Synthesis*: 181–185. Chelsea, MI: Lewis Publishers.

Long, E.R., MacDonald, D.D., Smith, L. & Calder, F.D. 1955. Incidence of adverse biological effects within ranges of chemical concentrations in marine and estuarine sediments. *Environmental Management* 19: 81–97.

Long , E.R. & MacDonald, D.D. 1998. Recommended uses of empirically derived sediment quality guideline for marine and estuarine ecosystems. *Human and Ecological Risk Assessment* 4/5: 1019–1039.

Lorry, P. & Driscoll, C.T. 1999. Historical trends of mercury deposition in Adirondack lakes. *Environmental Science and Technology* 33: 718–722.

Niencheski, H.L., Windom, H.L., Baraj, B., Wells, D. & Smith, R. 2001. Mercury in fish from Ptos and Mirim Lagoons, southern Brazil. *Marine Pollution Bulletin* 42: 1403–1406.

NRC. 2000. Risk characterization and public health implications. In *Toxicological Effects of Methylmercury*: 304–332. Washington, DC: Academic Press.

Porcella, D.B. 1994. Mercury in the environment: biogeochemistry. In C.J. Watras, J.W. Huckabee (eds), *Mercury Pollution-Integration and Synthesis*: 3–19. Chelsea, MI: Lewis Publishers.

Rada, R.G., Wiener, J.G., Winfrey, M.R. & Powell, D.E. 1989. Recent increases in atmospheric deposition of mercury to north central Wisconsin lakes inferred from sediment analyses. *Archives of Environmental Contamination and Toxicology* 18: 175–181.

Svobodova, Z., Dusek, L., Hejtmanek, M., Vykusova, B. & Smid, R. 1999. Bioaccumulation of mercury in various fish species from Orlik and Kamyk water reservoirs in the Czech Republic. *Ecotoxicol. Environ. Safety* 43:231–240.

Swain, E.B., Engstrom, D.R., Brigham, M.E., Henning, T.A. & Brezonik, P.L. 1992. Increasing rates of atmospheric mercury deposition in midcontinental North America. Science 257: 784–787.

USEPA. 2001. EPAA-823-R-01-001. Methylmercury. In *Water Quality Criterion for the Protection of Human Health:* Washington, DC: United States Environmental Protection Agency.

von Gunten, H.R., Sturm, M. & Moser, R.N. 1997. 200-year record of metals in lake sediments and natural background concentrations. *Environmental Science and Technology* 31: 2193–2197.

Reclaiming the Desert: Towards a Sustainable Environment in Arid Lands – Mohamed (ed.)
© 2006 Taylor & Francis Group, London, ISBN 0 415 41128 9

Quantification of exhaust emissions from fleet vehicles at Dubai international airport

A.A. Busamra & W.S. Ghanem
Emirates National Oil Company L.T.D., (ENOC), Dubai, UAE

ABSTRACT: The objective of this study was to develop a model for quantifying the amount of emissions generated from the various categories of both diesel and gasoline road vehicles utilized in the daily operation of the Dubai International Airport (DIA).

The development of the model was based upon emission factors which considered the mileage of the vehicles and consideration of engine conditions such as idling, acceleration, cruising and deceleration.

The main considered pollutants in this study were Carbon Monoxide (CO), Hydrocarbons (HC), Nitrogen Oxides (NO_x), Aldehydes, Sulphur Dioxides (SO_2), and Particulates Matter (Soot/Carbon Black).

The model which is based upon standard EXCEL Windows based software has been created to be 'user friendly' and adaptable for the future predictive vehicle emission modelling in DIA.

1 INTRODUCTION

In today's modernized society, transportation, whether by car, bus, or rail, serves as a catalyst for economic growth and trade and for improving the quality of life globally. In general, also the demands for transport services increment as social growth rise.

Unfortunately on the negative side, transportation has also been traditionally been blamed as the cause for many multifarious types of health and environmental problems. Although many people believe that the cause of air pollution is industry, several statistics show otherwise with the highest contributor to air pollution being the transportation sector, especially from cars and trucks. In addition, it has been indicated that the transportation sector occupies around 49% of world total consumption magnitude of oil–with the distribution as follows: Land Transportation 78%, Aerial Transportation 13% and Marine Transportation 9%. [Shareb, 2005]

The main factors that influence the amount of emissions attributable to the transport sector could be attributable to a number of contributory factors such as excessive vehicle use, age of fleet and technology used, poor maintenance of vehicles and unavailability or improper use of appropriate fuels. In general, it should be noted that the last issue is not a primary cause of concern in the UAE due to the government regulatory compliance checks.

It has been found from the literature that an automobile and its internal combustion engine emit three major types of gaseous pollutants. These are carbon monoxide (CO), oxides of nitrogen (NO_x), and hydrocarbons (HC). Additionally, an engine produces some solid emissions of lead, sulphur, carbon, and other particulates, as well as some sulphur oxides. In general also, diesel engines emit less HC and CO but more particulates and sulphur oxides than gasoline engines. NO_x emissions are slightly less from a diesel engine but are still enough to require controls on late-model automobiles. [Layne, 1993]. Within todays urbanized society it is reported that automobile exhausts can be responsible for more than 75% of total air pollution with automobiles, cars, scooters, motors, taxis, trucks, buses etc. releasing large hydrocarbons (about 14%); sulphur monoxide (about 77%); nitrogen oxides (about 8%). [Sharma, 1999] Listed below are a few major transportation air pollutants which pose risk to the human health and the environment.

- Carbon Monoxide (CO): In general, transportation accounts from 70 to 90% of total carbon monoxide emissions in today's modernised society. CO as the pollutant which is most strongly associated with transportation is produced as a result of the incomplete combustion of hydrocarbons.
- Nitrogen Oxides (NO_x): In general, transportation accounts from 45 to 50% of total emissions of nitrogen oxides. NO_x results when combustion temperature exceeds around 1370°C and the oxygen and nitrogen molecules combine in large quantities to form NO_x.
- Hydrocarbons and Volatile Organic Compounds (HC/VOC): Transportation accounts from 40 to 50% of total emissions of HC/VOC. Hydrocarbons (HC) are a group of chemical compounds composed of carbon and hydrogen. When in a gaseous form, HC are called Volatile Organic Compounds (VOC). They are mostly the result of the incomplete combustion of gasoline and include methane (CH_4), gasoline (C_8H_{18}) and diesel vapours, benzene (C_6H_6), formaldehyde (CH_2O), butadiene (C_4H_6) and acetaldehyde (CH_3CHO).
- Particulates Matter (PM): Transportation accounts for around 25% of total emissions of particulates with diesel engines being the main emitters. Particulates include various solids in suspension in the atmosphere such as smoke, soot, and dust and results of the incomplete combustion of fossil fuels, notably coal. They may also carry traces of other toxic substances like HC/VOC. [http://www.people.hofstra.edu/geotrans/eng/ch8en/app18en/ch8a1en.html]
- Lead (Pb): Lead is a toxic metal mainly used as an anti-knock agent in gasoline in the form of Lead tetraethyl-Pb $(C_2H_5)_4$ and in batteries in the form of lead dioxide as the anode and lead as the cathode. Until recently, lead tetraethyl was a main source of atmospheric lead emissions in developing countries. This contribution has dropped in absolute numbers but still accounts for 30–40% of total emissions.

  On January 1st 2003 a new milestone was made in UAE's history through the adoption of legislation to phase out leaded gasoline and introduce two new grades of unleaded gasoline. This major event was in line with the directives of the Gulf Cooperation Council on the phase out of leaded gasoline and marked the Federal Government's commitment to provide the people of the UAE with a pollution-free environment.
- Sulphur Dioxide (SO_2): Transportation activities in general accounts for around 5% of total sulphur dioxide emissions to the environment and is produced as a result of the combustion of fossil fuels like coal (particularly bituminous coal) and hydrocarbons.

  Investigated general comparison between diesel and petrol engines Table 1, below, shows the level of pollutants from both types of engines under similar operating conditions. [Watkins, 1981]. It should be noted that, although the exhaust of diesel engines contains significantly lower concentrations of pollutants that the diesel engines are very liable to emit smoke and to make a

Table 1. Representative composition of exhaust gases*.

Engine Type	Pollutant	Idling (ppm)	Acceleration (ppm)	Cruising (ppm)	Deceleration (ppm)
Petrol engines	Carbon monoxide	69000	29000	27000	39000
	Hydrocarbons	5300	1600	1000	10000
	Nitrogen oxides	30	1020	650	20
	Aldehydes	30	20	10	290
Diesel engines	Carbon monoxide	Trace	1000	Trace	Trace
	Hydrocarbons	400	200	100	300
	Nitrogen oxides	60	350	240	30
	Aldehydes	10	20	10	30

* Watkins, 1981.

nauseating smell if they are inefficiently operated and/or maintained. Under such 'abnormal' conditions it should be mentioned that amount of carbon monoxide and hydrocarbons produced can increase to levels of carbon monoxide reaching as high as 2000 ppm.

Following extensive research, particular reported amongst some Western countries where significant attention has been paid to long term epidemiological studies, is the opinion that air pollution, which can disperse over thousands and hundreds of kilometers may have harmful effects even at typical ambient concentrations. Health organizations which set the exposure limits for different pollutants for the protection of both occupational and public health include the Occupational Health and Safety Administration (OSHA), Work Health Organization (WHO), Occupational Health and Safety (OHS) standards in Germany, The National Institute for Occupational Safety and Health (NIOSH).

Examples can be seen in legislative empowerment directives of governments around the globe to establish the powers linking the public, occupational health and environment issues of air quality control.

Examples include the Environmental Protection Agency (EPA) in USA, the European Union (EU) as well as UAE's own Federal Environmental Agency (FEA) and Dubai Municipality (DM) who have all taken stringent legislative measures to establish and enforce limits on discharges of pollutants into the environment and in particular the necessary controls to monitor and enforce the exhaust emissions which come from vehicles.

In this study the related emission regulations have been divided into the following three classes including: ambient air quality standards, tailpipe emission regulations, and workplace exposure limits. In addition, the study covered only the US EPA, and EU standards in the international level. Besides, the FEA, DM and the United Arab Emirates Directorate of Standardization standards that are considered as a Gulf Standards approved by the Metrology Standardization and Metrology Organization for G.C.C (GSMO) in the local level, which set the standards with respect to ambient air quality and the tailpipe emissions respectively. Moreover, the study covered OSHA and NIOSH limits with respect to exposure limits for different exhaust pollutants. (Please refer to Table 4.1–4.11 in the full study to see the standards)

In recent years, several estimates have been published of health damage due to air pollution from cars, and different types of land vehicles by the Environmental Protection Agency (EPA) to capture the extent of pollutant dispersion, and find out areas for improvement in reducing the amount of generated emissions to protect public health environment.

As part of Dubai Civil Aviation commitment towards maintaining a sustainable development and environmental protection, this paper describes a simple model that has been developed by ENOC Group Environment, Health, Safety and Quality Compliance Department to calculate the amounts of generated emissions from the different types of vehicles operating within the airport by DIA/DNATA. This quantification model will help DNATA in identifying the different types of generated emissions and their quantities over the years in terms of vehicles operating with diesel or petrol as well as finding the gaps in order to eliminate or reduce it in the future.

The recommendations of this study give an in-depth and critical look at areas where performance improvement may be possible with respect to exhaust emissions – both at the management and operators levels.

2 QUANTIFICATION MODELING

2.1 *Background on the types of vehicles operating in DIA*

Within this study all the various types of vehicles operating within Dubai International Airport (DIA) have been classified into the following 30 categories:

1. Forklift truck (FKL): which is used to hoist and transport materials by means of steel forks inserted under the load.

2. Tow bar-less Tractor (TBL): which is used for towing almost every kind of commercial aircraft in every kind of situation.
3. Toilet Service Unit (TSU): which is mounted on a commercial vehicle chassis for supplying water to the aircraft for lavatory and collects the refuse from there.
4. Mobile Step Unit (MSU): which is also known as a Passenger Step Unit (STP)-this is a portable unit mounted on a vehicle that is attached to the aircraft for passengers entering or leaving the aircraft.
5. Mobile Conveyor Belt (MCB): which provides for fast and efficient loading of trucks and trailers by delivering a steady flow of product to loading personnel.
6. Water Service Unit (WSU): which is mounted on a commercial vehicle chassis for supplying water to the aircraft and the pump is driven by the vehicle.
7. Lower Deck Loader (LDL): which is used for lifting containers-including luggage and any cargos to the small kinds of aircrafts.
8. Rapid Delivery Vehicle (RDV): which offers a smooth and speedy link with the aircraft on ground and other terminals at the airport.
9. Pushback Tractor (PBT): which consists of a drawbar pull to provide strong handling for narrow-body aircraft. The unit is fitted with four-wheel drive, disc brakes, power steering and automatic transmission which allows it to go forward or reverse with the same drawbar pull.
10. Mobile Diesel Browser (MDB): which is used for supplying all diesel operating vehicles by a portable diesel fuel browser
11. Ground Power Unit (GPU): This portable device was used to generate electrical power for ground operation and for the starting of certain types of aircrafts.
12. Air Startup Unit (ASU): which is a trailer-mounted power unit capable of providing AC and DC electrical power as well as high volume air for starting certain types of aircraft engines.
13. 30-Seater Bus: which is used for carrying up to a seating capacity of 30 passengers for public transport within the Dubai International Airport premises.
14. Air Conditioning Unit (ACU): which is a mobile air conditioning system mounted on a trailer for use on the apron or unit inside the hangar to cool the cabin during maintenance.
15. Main Decker Loader (MDL): which is used for lifting containers including luggage and any cargos to the large kinds of aircrafts.
16. Transporter (TPTR-10'): which is a truck that moves all the containers, materials and equipment around the location of DIA and for dispatching container to the lower or main deck loader.
17. 15-Seater Panel Van (VAN): which is a small van with two rear doors – It is distinguishable in having a large window with seats in the rear.
18. Medical Lift (MLIFT): which is used for passengers of special needs entering or leaving the aircraft.
19. Ambulance (AMB): which is a medic unit used to transport emergency medical services patients within the Airport to local hospitals.
20. Tow Trucker (TOW TR): which is used as a heavy-duty cargo/baggage towing tractor as well as a pushback for smaller short-haul aircraft.
21. Transporter (TPTR-20'): which has the same function of TPTR-10' Transporter but it is used for moving large containers, materials and equipment around the location of DIA and for dispatching container to the lower or main deck loader.
22. Pick-up Truck (PKP): which is a small truck with an open-top cargo area designed to carry goods and an opening rear gate.
23. Container Loader Transporter (CLT): which is used for the same purpose of Transporter, but is faster, much modern and is just used for moving single containers to a lower/main deck loader.
24. Petrol Car (PT): which is a four wheeled motor vehicle, propelled by an internal combustion engine, used for carrying passengers from one place to another.
25. Aircraft Washer (ACW): which is used for washing the aircrafts by mean of a washer person who can stand on for distributing the water on the aircraft body and washing it.

26. Brake Cooling Unit (BCU): which is used for cooling the aircrafts brake by means of cold air.
27. Carpet Cleaning Unit (CCU): which is used for cleaning the aircraft carpets.
28. Mobile Work Platform (MWP): which is used for lifting the mechanical engineers, workers and other aircraft maintenance personnel.
29. Fuel Tanker (F/TANKER): which is used for supplying the vehicles operating within the DIA premises with fuel.
30. Ramp Bus (BUS): which is used for carrying Airport passengers within the DIA areas.

2.2 *Objectives of the model*

This simple point source model, which has been developed on the Excel spreadsheet programme, and which we have called the *'Dnata Emission Quantification Model'* has been developed to estimate the major emissions of light and heavy-duty gasoline and diesel vehicles in the 'mg/year' unit and is based upon emission factors expressed per mile (km traveled).

The major emissions, as typically applied for models of this nature, include Carbon Monoxide (CO), Hydrocarbons (HC), Nitrogen Oxides (NO_x), Aldehydes, Sulphur Dioxides (SO_2), and Particulates Matter (Soot/Carbon Black).

The literature survey has shown that several models have been developed by certain researchers for modeling light-duty vehicle emissions as a function of unit distance of travel. In contrast it is noted a distinguishing feature of these earlier models is to model heavy-duty diesel truck emissions as a function of brake work output by the engine. Since heavy-duty trucks encompass a wide range of diesel engine sizes and gross vehicle weights it must be noted that emission factors normalized to work output vary less than they would on a distance traveled basis.

Furthermore, performance maps for heavy-duty diesel engines indicate that brake specific fuel consumption (bsfc) varies only slightly as engine operating conditions change. For example, Heywood presents the performance map for a 6.5 liter diesel engine. Over a wide range of operating conditions, bsfc varied from 220 to 260 g/kW.hr for this engine. Therefore, work output by the engine can be directly related to fuel input; and heavy-duty diesel engines are effectively regulated and designed to meet emission targets on a per unit of fuel burned basis.

As mentioned above the estimated amount of heavy-duty diesel vehicle emissions in this model is conducted per mile or km traveled. Although the disadvantage of this choice is that it is not as accurate as considering the model design to meet emission targets on a per unit of fuel burned basis, but the advantage is that it will give the DIA/Dnata an indication of the generated amount of heavy-duty diesel emissions within the DIA premises as the 'baseline' data for the continual strive in performance of the overall airports environmental management systems. The advantages of the model can be summarized as follows:

- As the initial 'baseline review' it will provide the initial step towards continual improvement in terms of environmental protection and the ongoing sustainable and safe environmental management system for the airport.
- It will give a rapid comparison between the generated amount of emissions in terms of gasoline and diesel that will lead to aiding DIA/Dnata management to review various transportation logistic scenarios with the purpose of choosing the type of fuel with respect to its environmental performance – i.e. the BPEO (Best Practicable Environmental Option).
- It will provide a clear and permanent record for the DIA/Dnata with the generated amount of exhaust emissions in the previous years, and the expected amount of emissions in the new coming years in mg/year with taking into consideration the engine percentage performance drop.
- It will help the airport in developing vehicles replacement programme and it will save the operator's time and energy resources. In addition, the operator will be more familiar with the environmental international and local regulation.
- The model can present some useful charts indicating the amount of generated emissions in terms of vehicle type, which can facilitate to find the area where the gap is occurred in order to help the DIA/Dnata to proceed in their continual improvement in the future.

- The model which is based upon a standard Windows programme can be run on any normal PC and has been based upon the need to be 'user friendly' able to be flexible and adaptable for future development and improvements.

2.3 *The methodology for developing the model*

The basis of this model was the following information which was collected from DNATA from the years 2001 up to 2004 for the all types of vehicles operating in DIA:

1. Type of fuel used for the vehicles.
2. Engine size.
3. Average operational number of hours per day.
4. Typical consumption of fuel and fractions including (idling, acceleration, cruising and deceleration).

The methodology used for the model is based on the emissions concentrations including, (CO, HC, NO_x, Aldehydes, SO_2, and Soot/Carbon Black) in ppm during the four main phases of an engines cycle – i.e. idling, acceleration, cruising, and deceleration.

The representative composition of exhaust gases in (ppm) in terms of petrol and diesel engines during idling, acceleration, cruising and deceleration is based on table 1.

The development of the model can be described in the sequential development of 5 main equations which are described as follows:

The initial equation which is used to find out the total concentration of an emitted pollutant per vehicle in (ppm) is as follows:

$$C_T = [(C_{a1} \times x) + (C_{a2} \times y) + (C_{a3} \times z)] \times N \qquad (1)$$

where: C_T is the total pollutant concentration in (ppm), C is the pollutant concentration in (ppm), a_1, a_2, and a_3 are any pollutants including CO, HC, NO_x or Aldehydes in idling, acceleration and deceleration. The parameters (x, y, z) are the fractions ranging between $(0 \leqslant (x, y, z) \leqslant 1)$, representing the percentages of vehicle operational time during idling, acceleration and deceleration, and cruising respectively, and N is the vehicle quantity.

From the above, the total amount of one emitted pollutant from vehicle in terms of mileage in (mg/year) was calculated by using the following equation which was based on equation no. (1):

$$Q_T = C_T \times R \times 1/K_1 \times S \qquad (2)$$

where: Q_T is the total amount of one emitted pollutant per vehicle in terms of mileage in (mg/year), C_T is the total pollutant concentration in (ppm), R is the volume ratio between the exhaust produced from a vehicle in (L) and the burned fuel in (L), $1/K_1$ is the typical fuel consumption rate of a vehicle in (L_{fuel}/km), and S is the mileage rate driven in (km/year).

In terms of estimating the expected number of vehicles for the years 2004, 2005, and 2006, the curve fitting method was used. Curve fitting is the process of computing the coefficients of a function to approximate the values of a given data set within that function. The approximation is called the 'best fit'. The selection of this method was based on data supplied by DNATA for the number of vehicles operating in DIA during the years 2001, 2002, and 2003.

For the model the expected amount of emissions was projected for 2006 based upon the statistical trend seen during the years of 2003, 2004, and 2005 and taking into considerations the percentage of engine performance drop (10%) per year.

The equation used to find out the total amount of one emitted pollutant from vehicle in terms of mileage for the next year in (mg/year) was:

$$Q_{T(x+1)} = (N \times Q_{T(x)} \times P) + (N_{new\,(x+1)} \times Q_{T(x)}) \qquad (3)$$

where: $Q_{T(x+1)}$ is the total amount of one emitted pollutant per vehicle for the next year in (mg/year), N is the vehicle quantity, $Q_{T(x)}$ is the total amount of one emitted pollutant per vehicle of the previous year in (mg/year), P is the estimated factor of performance drop per year of age, and $N_{new(x+1)}$ is the new vehicle quantity for the next year.

The total amount of all emitted pollutants from vehicle in terms of mileage in (mg/year) was calculated by using the following equation:

$$Q_{TT} = \sum Q_T(x_1) + Q_T(x_2) + Q_T(x_3) + Q_T(x_4) \tag{4}$$

where: Q_{TT} is the total amount of all emitted pollutants per vehicle in terms of mileage in (mg/year), Q_T is the total amount of one emitted pollutant per vehicle in terms of mileage in (mg/year), and x_1, x_2, x_3, x_4 are the different emitted pollutants including CO, HC, NO_x or Aldehydes.

The total volume of exhaust emissions was calculated by using the following equation:

$$V_{Te} = (m_{a1} \times 1/\rho_{a1}) + (m_{a2} \times 1/\rho_{a2}) + (m_{a3} \times 1/\rho_{a3}) + (m_{a4} \times 1/\rho_{a4}) + \ldots \tag{5}$$

where: V_{Te} is the total volume of exhaust emissions in (cm^3), $m_{a1}, m_{a2}, m_{a3}, m_{a4}$ are the gaseous masses in (kg), and $\rho_{a1}, \rho_{a2}, \rho_{a3}, \rho_{a4}$ are the gaseous densities in (kg/cm^3).

The volume ratios between the produced exhaust gases from vehicles in (L) and the burned fuel in (L) in terms of gasoline and diesel were calculated based upon the toxic compositions of burned fuel of carburetor-gasoline and diesel engines as mentioned in the below table 2.

The densities of different exhaust gases in kg/cm^3 are based in the following table 3.

Table 2. Toxic compositions of burned fuel of carburetor and diesel engines.

Exhaust gases*	Carburettor-Gasoline (kg/1000 L)	Diesel (kg/1000 L)
Carbon Monoxide (CO)	200	25
Hydrocarbons $C_m H_n$ (unburned hydrocarbons)	25	8
Nitrogen Oxides (NO_x)	20	36
Soot	1	3
Sulphur Dioxide (SO_2)	1	30
Total	247	102

*http://www. rec.vsu.ru/vestnik/pdf/chembio/2003/02/djuvelikyan.pdf.

Table 3. Toxic compositions densities of burned fuel of carburetor and diesel engines.

Exhaust gases	Density (kg/cm^3)
Carbon Monoxide (CO)	1.25*
Nitric Oxide/Nitrogen Oxide/Nitrogen Monoxide (NO)	1.27 at 1.013 bar and 15°C (59°F)**
Nitrogen Dioxide (NO_2)	1.95 at 1.013 bar and 21°C (70°F)**
Sulphur Dioxide (SO_2)	2.77 at 1.013 bar and 15°C (59°F)**
O-Xylene/O-Dimethylbenzene/Ethyl-Benzene (C_8H_{10})	0.000867 at 20°C***
Decane ($C_{10}H_{22}$)	0.0000049 at 1.013 bar and 15°C****
Soot (Carbon Black)	0.00186*****

*http://www.simetric.co.uk/si_materials.htm.
**http://www.airliquide.com/en/business/products/gases/gasdata/index.asp?formula=&GasID=27.
***http://funnel.sfsu.edu/courses/laforce/geo476/258,3,Examples Found In Groundwater.
****http://www.vadilalgases.co.in/decane.htm.
*****http://www.a-m.de/englisch/lexikon/russ.htm.

2.4 *Basis of the model*

The following assumptions were made based upon the information provided by DNATA/DIA:

1. The estimated percentage drop in the vehicle's engine performance per year of age is around 10%.
2. Since the data provided by Dnata, showed clearly the equivalency of number of vehicles during years 2001, 2002 and 2003, and no significant change or difference. The number of certain vehicles was assumed to be constant.
3. The trace generated amount of carbon monoxide from diesel engines in terms of PPM, as mentioned in the previous table 1 is negligible during idling, cursing and deceleration.
4. Ramp Bus (BUS) was assumed to be operated by petrol, since the operated number of ramp buses by petrol was much higher than the amount operated by diesel.
5. The distributions of 'NO_x' gases are assumed to be Nitrogen Oxide (NO) and Nitrogen Dioxide (NO_2) in the proportion 50%:50% as specified in Perry's Chemical Engineering Handbook, seventh edition.
6. Since the density of gasoline is very light the unburned hydrocarbons in terms of gasoline are assumed to be aromatics – in predominance being O-Xylene/O-Dimethylbenzene/Ethyle-Benzene (C_8H_{10}) (as also confirmed by ENOC Processing Company (EPCL) Laboratory.) Conversely, since diesel is considered as a heavy fuel consisting of complex mixture of normal, branched, and cyclic alkanes (paraffin), especially the cyclic alkanes that occupies the highest range for around (60% to >90% by volume), usually between (C_9 and C_{30}), the alkanes compound assumed to be Decane ($C_{10}H_{22}$). [http://www.inchem.org/documents/ehc/ehc/ehc171.htm].
7. Based on table 2, the ratio between Hydrocarbons (HC), Soot and Sulphur Dioxide (SO_2) in terms of petrol during idling, acceleration, cruising and deceleration is 1:0.04:0.04, and in terms of diesel the ratio is 1:0.375:3.75. The following table 4 provides the calculated concentrations of soot and sulphur dioxide in PPM based on hydrocarbons concentration. This assumption was made since soot and sulphur dioxide are generated mainly from burning fuel during the engine internal combustion. The burned fuel consists mostly of hydrocarbons – the compounds of which are hydrogen and carbon, and non-hydrocarbon fractions the compounds which might include nitrogen, sulphur, oxygen, or traces of metals such as vanadium or nickel, such elements often constitute less than 1% of the whole. [http://www.nationmaster.com/encyclopedia/Petroleum].

2.5 *How to use the model – developing the operations manual*

The model consists of one table and it is divided into two parts. The first part is the main facet of the table that represent/estimate the major generated amount of exhaust emissions from vehicles operating by gasoline and diesel in mg/year. This estimation is based on the calculated accurate pollutants concentrations. This part consists of several columns including type of equipment and its ID code, quantities of vehicles in 2001, vehicle engine size, type of fuel used by the vehicles enclosed with their typical fuel consumption rate. In addition, the estimated percentage drop in engine performance per year of age is included along with the mileages driven and the four vehicle fractions.

This first part of the table will require from the operator to enter the mileage driven by the vehicle (last and new mileage) and it is specified clearly in blue bolded colour except E-TBL vehicle which

Table 4. Representative calculated composition of soot and sulphur dioxide.

Engine type	Pollutant	Idling (ppm)	Acceleration (ppm)	Cruising (ppm)	Deceleration (ppm)
Petrol engines	Hydrocarbons	5300	1600	1000	10000
	Soot	212	64	40	400
	Sulphur dioxide	212	64	40	400
Diesel engines	Hydrocarbons	400	200	100	300
	Soot	150	75	38	113
	Sulphur dioxide	1500	750	375	1125

is obliterated from the table since it is operating by electricity. The equation used to calculate delta mileage was:

$$\Delta d = d_n - d_o \tag{6}$$

where: Δd is the difference between the new mileage and the old mileage in (km), d_n is the new mileage in (km) and d_o is the old mileage in (km).

The second part of the table is created to calculate the generated amount of exhaust emissions for the new coming years of the year 2001 including 2002, 2003, 2004, 2005, and 2006. The first three columns in this table represent the type of equipment used and its ID code enclosed with the vehicles quantities in 2002, 2003, and the expected quantities in 2004, 2005, and 2006 along with the expected amount of generated exhaust emissions.

The followings are the steps that must be followed by the user to operate the model:

1. Collect the data about the driven mileages (km) for all types of vehicles.
2. Open the Dnata Emission Quantification Model file in Excel Programme by double clicking it. An Excel Spread Sheet will appear in the screen.
3. Enter in the respective column of the spreadsheet 'the new driven mileage' and in another column 'the last driven mileage' for the types of vehicles as mentioned earlier in section 2.1 of this paper.
4. Also entered in columns of the spreadsheet are dates of last and new service respectively.
5. Through a number of macros that are built into the programme using the methodology described earlier in this paper the expected amounts of emissions are then provided automatically.
6. The programme is then saved and the changes are automatically saved until used again.

3 CONCLUSIONS

This paper describes the point source model used for quantifying the major emissions generated from the vehicles (light and heavy-duty gasoline and diesel vehicles) at DIA and was developed based upon the airport's commitment for the protection of the environment and public health.

The model is considered as the initiative towards the strive for continual improvement against the benchmarks of international standards and best practices.

The following recommendations were developed based upon the work which was carried out for the development of this model:

1. The DIA management should establish environmental targets and key performance indicators (KPI's) specifically for emissions reduction on a yearly basis and to monitor and review these KPI's to achieve continual improvement.
2. The DIA management should arrange an environmental audit on a yearly basis as the outcome of emissions audit generally assist in improving performance and providing an independent view of the management system in place.
3. The DIA may look to measure and monitor the generated emissions from its premises especially from vehicles to see whether they are within the legislative requirements or not in order to take a positive action regarding it.
4. The DIA responsible supervisor could keep an electronic record of all information related to their operated vehicles such as type of vehicle, its quantity, operating hours, consumption of fuel, engine size, type of fuel, etc.
5. The DIA to prepare an annual assessment report that summarizes the quantities of generated emission from the vehicles. The report should also compare, evaluate, and analyze the emission reduction rates with the emission reduction targets. This report should be presented to the Department of Civil Aviation to demonstrate the progress made towards achieving the emission reduction targets, and also to ensure continued support from the management.
6. The DIA to provide environmental awareness training to its employees to raise the awareness on basis of emission reduction and pollution prevention.

7. The DIA to use vehicles with a high technology that are capable of burning the un-complete combusted gases.
8. The DIA to put a strict programme for the used vehicles maintenance and follow up the international countries who have been conducting the same since long time ago.
9. The DIA to develop an intensive awareness campaigns that aimed to finger the operators who are functioning the vehicles to get rid of the unnecessary use of vehicles and as a result of that polluting the environment.
10. The DIA to look for forming a committee consisting of expertise that can represent the DIA in various conferences and meetings with the top Environmental Agencies to coordinate and follow up the latest standards related to different environmental aspects and issues.
11. The DIA to follow national, regional and International guidelines for vehicle emissions and monitor developments in the international market place on emissions testing and pollution-control devices-i.e. in particular on more efficient vehicles in terms of less fuel usage and fewer emissions.
12. The DIA could consider conducting a complete study of vehicle emissions where they include other parameters e.g. particulate matter (which will consist of several components), tire wear particulate matter, brake wear particulate matter, ammonia (NH_3), carbon dioxide (CO_2), other minor toxic gases e.g. butadiene, etc. as well as very minor items such as wire wear particulate matter, brake wear particulate matter emissions-of which the latter will probably be more if considering heavy vehicles and plane traffic also within airport.
13. The DIA could consider how to benchmark and produce targets of vehicle emissions as against other airports in terms of vehicle emissions (e.g. British Airports Authority web site: (http://www.baa.co.uk/) that relate to CO_2 emissions and the use of LPG and electric cars).
14. DIA could consider further study including for example the project management of an air monitoring programme that will also consider airplane emission – In particular this could focus on the impact of environmental health from airplane emissions on adjacent areas to the airport. With regard to the later it is noted that a growing trend now in airports in Europe/USA is to add an air quality related element to the landing fees. – i.e. the charge is based on the amount (kilograms) of nitrogen oxides calculated to be emitted by each aircraft during a standard landing and take-off cycle – as aircrafts can be fitted with a range of engines with different emissions performance this charge is calculated individually for each aircraft and engine fit.

REFERENCES

Colls, Jeremy First Edition, 1997, *Air Pollution-An introduction*, London: E and FN Spon, An Imprint of Chapman and Hall.

Dreher David B and Harley, Robert A, *A fuel-Based Inventory for Heavy-Duty Diesel Truck Emissions*, Berkeley: University of California, Department of Civil and Environment Engineering.

Diesel Engines and Emissions, http://www.dieselnet.com/standards/eu/Idhtml, 11/8/2005.

Dr. Shareb, Husain 11/12/2005, *Arab Oil Refineries Capabilities in Producing a Clean Fuel, the GCC Fuel and Diesel Specification Workshop*, Abu Dhabi: FEA.

Dr. Rodrigue, Jean-Paul 11/8/2005, Air *Pollutants Emitted by Transport Systems, Local and Regional Impacts*, http://people.hofstra.edu/geotrans/eng/ch8en/appl8en/ch8a1en.html.

Ellinger, Herbert E and Halderman, James D Second Edition, 1991, *Automotive Engines; Theory and Servicing*, New Jersey: Regents/Prentice Hall, Englewood Cliffs.

Environmental Protection Section, May 2003, *Information Bulletin, Environmental Standards and Allowable Limits of Pollutants on Land, Water, and Air Environment*, Dubai: Dubai Municipality.

Faiz Asif, Weaver, Christopher S and Walsh, Michael P 1996, *Air Pollution from Motor Vehicles; Standards and Technologies for Controlling Emissions*, Washington: The International Bank for Reconstruction and Development/The World Bank.

Federal Environmental Agency, 1999, *Federal Environmental Protection Law (24), Protection and Development of the Environment, Executive Order, Draft Order of Air Pollution Protection System*, Abu Dhabi, FEA.

Haley, J Thomas, Berndt, D William 1987, *Hand Book of Toxicology, Hemphere Publishing Corp. Watkins, L. H., 1981, Environmental Impact of Roads and Traffic*, England: Applied Science Publishers Ltd.

Harrington JM, Gill FS, 4th edition, 1998, *Occupational Health*; Pocket Consultant, T.C. Award K. Cavdina: Blackwell Science.
http://www.airliquide.com/en/business/products/gases/gasdata/index.asp?formula=&GasID=27
http://www.a-m.de/englisch/lexikon/russ.htm
http://www.baa.co.uk/, 18/8/2005.
http://www.inchem.org/documents/ehc/ehc/ehc171.htm, 16/8/2005.
http://www.nationmaster.com/encyclopedia/Petroleum
http://www.people.hofstra.edu/geotrans/eng/ch8en/app18en/ch8a1en.html
http://www.rec.vsu.ru/vestnik/pdf/chembio/2003/02/djuvelikyan.pdf.
http://funnel.sfsu.edu/courses/laforce/geo476/258,3, Examples Found in Groundwater, 16/8/2005.
http://www.simetric.co.uk/si_materials.htm
http://www.vadilalgases.co.in/decane.htm, 16/8/2005.
Layne, Ken Volume I, Second Edition, 1993, *Automotive Engine Performance; Tuneup, Testing, and Service*, New Jersey: Regents/Prentice Hall, Englewood Cliffs.
Perry, Robert H and Green, Don W, Seventh Edition, 1998, *Perry's Chemical Engineering Handbook*, McGRAW-Hill International Editions.
Sharma BK, 1999, *Air Pollution*, Merrent: GOEL Publishing House.
Standardization and Metrology Organization for G.C.C (GSMO), 2003, *Catalogue of Gulf Standards (1679 H/2003 G).*

Reclaiming the Desert: Towards a Sustainable Environment in Arid Lands – Mohamed (ed.)
 ISBN 0 415 41128 9

Emergency response planning of oil spills from offshore platforms

A.L. Hanna
GASOS Bin Hamoodah HSE Training Consultants, Abu Dhabi, United Arab Emirates

ABSTRACT: Oil spills in the marine environment raise a major concern among platform management, government authorities and the public. One of the key elements for an efficient spill control is the preparedness for an adequate and prompt response. The subject of this study (mentioned here as the OOP) is an oil producing platform that lies about 200 kilometers offshore Abu Dhabi. The OOP has some unique characteristics which influence the type of response options. The geological and oceanographic and metrological characteristics of the field were investigated. The 3 main wind directions dominant in the field were identified and used to establish the oil spill scenarios. The Oil Map 97 modeling software was used and the results obtained were essential in the preplanning process and the selection the combating methods and equipments of oil spills in the OOP field. This study suggests a framework for formulating an oil spill emergency response plan. This framework can also be modified and adopted by any other offshore platform.

1 INTRODUCTION

Offshore oil exploration and production are exposed to several types of hazards such as fires, explosions, accidental gas release and oil spills. The risks involved can have devastating effects on people's lives, the oil installations, neighboring facilities as well as the marine and coastal environment. Major accidents such as Piper Alpha 1987 (167 fatalities) and EXXON Valdez (more than $4 billion dollars worth of compensations and unprecedented damage to the environment of Alaska) have resulted in changes in the offshore industry's safety and environmental regulations both on the local and international levels. Learning from these accidents, applying new technologies and implementing preventive measures are all necessary tools to prevent reoccurrence. However accidents will continue to occur and spills can affect areas far from their place of origin. Successful oil spill emergency response planning is not a simple task and can only be achieved through teamwork and requires total management commitment and it incorporates the use of all available resources.

1.1 *Scope and aim of the study*

The study is concerned with the environmental parameters (meteorological, oceanographic, and geographical) which govern the effect, fate and control of an oil spill in the vicinity of the OOP. The study also necessitates the formulation of a practical and well tested emergency response plan for oil spill control in the OOP field.

1.2 *The objective of the study*

There are four main objectives of this study these are:

- To describe the environmental conditions for the area surrounding the OOP platform;
- To predict the fate and effect of oil spills resulting from the OOP activities;

- To justify the use of oil spill response options for the OOP case;
- To develop the OOP's emergency plan which can be integrated with the platform operational procedures and can be also integrated with local and national plans if necessary.

2 UNDERSTANDING THE PROBLEM

It is imperative for the oil spill emergency planner to understand the different factor affecting the response activities.

2.1 *Offshore platform characteristics*

Offshore oil platforms have some unique characteristics that affect oil spill response operations and make them different from onshore and near-shore operation. The main differences can be listed as follows:

- Remotely located which affect the mobilization of resources during an emergency;
- Some are located on the borders of 2 or more countries (security and logistics);
- Weather and sea conditions affect the fate of spill and the response operations;
- Multicultural crew with great differences in language and background (communication factor) (Hanna, 2003).

2.2 *Oil composition and oil classification*

Oil is a complex mixture of thousands of different compound. However crude oil is primarily composed of five elements: carbon, hydrogen, sulphur, nitrogen and oxygen. There are different ways to classify oil some of them are important for the oil spill emergency planner such as:

- *According to their molecular weight:*
  1. light-weight components (1–10 carbon atoms in each molecule) these are highly volatile, dissolve readily, potentially flammable and are more bio-available to animals (primary entry route is respiratory system);
  2. Medium-weight components (11–22 carbon atoms in each molecule) these may evaporate or dissolve slowly not as bio-available as light-weight group primary entry route is respiratory system and are readily absorbed through skin;
  3. Heavy-weight components (23 or more carbon atoms in each molecule) no or little evaporation, can smother or coat organisms for long time primary entry route through direct contact and can be absorbed through the skin.

As the percentage of residual content increases from light to heavy crude oil, the gasoline content decreases because the heavier oil typically contains reduced quantities of the light weight components. Identifying the type of oil involved in an incident will help the emergency planner and the responder to predict the oil's behaviour and fate in the marine environment.

- *According to their specific gravity/ relative persistence of oil in the marine environment:*

There are 5 groups according to this classification which combine the specific gravity of crude oil with its persistence. A persistence scale was developed by applying a persistence rating to the 5 oil groups (each persistence value was divided by the least persistent oil product gasoline to obtain a scale that ranged from 1 for gasoline as the least persistent to a value of 1600 for residual asphaltene which is a highly persistent compound.

Crude oil with higher proportion of medium and heavy weight components is more persistent than those with light-weight products. This information is very useful in deciding the fate of spilled oil and hence in selecting the most appropriate response options. The way in which an oil slick breaks up and dissipates depends largely on how persistent the oil is. Light products such as kerosene tend to evaporate and dissipate quickly and naturally and rarely need cleaning up.

2.3 *Physical and chemical properties of oil*

The physical properties and chemical composition of oil vary greatly from one type to another and greatly influence the behaviour and fate of spilled oil in water and the efficiency of various clean-up methods.

The toxicity of oil is more closely related to its chemical composition e.g. light fuel oil has a higher proportion of aromatic hydrocarbons than heavy fuel oil, and is generally more toxic to aquatic organisms. On the other hand heavy fuel oil and some crude oil may result in damage to inter-tidal organisms due to smothering or displacement from sure line surfaces.

Following are the main physical characteristic that must be considered:

- *Specific gravity*: most oil and refined products have a specific gravity of less than one which mean they float, knowing the specific gravity of the slick can help the decision maker to determine if the oil is likely to sink or float;
- *API gravity*: The API scale ranges from zero to more than 60, the larger the API gravity value, the greater the amount of light-weight components;
- *Pour point*: If the temperature of the water is as cold as or colder than the oil's pour point, the oil will stiffen and will no longer flow and the oil will drift like semi-submerged strands with the majority of the bulk oil residing just below the water surface. Pour point values range from −60 C for jet fuel to 46 C for waxy No.6 fuel oil;
- *Viscosity*: The natural viscosity of the oil increases with the loss of many of its light products. Viscous oil is difficult to recover using conventional technologies such as disc skimmers;
- *Asphaltene and wax content*: Oil and refined products with high asphaltene content are typic ally heavier and more persistent oil. Waxes are also heavy-weight components of the oil that are in crystal form when the oil is below its pour point, this components of the oil do not undergo any significant weathering;
- *Trace constituents:* Trace chemicals in the oil's make-up such as nickel, vanadium, iron, aluminum … etc. can be important in stabilizing emulsions and affecting the weathering of oil. Large concentrations of these constituent aid oil emulsification.

2.4 *The fate of oil spills*

Environmental conditions together with the specific characteristics of oil will determine the fate of an oil spill in water. Once the oil is spilled in water it will undergo several physical and chemical changes, these changes are collectively called "Weathering".

2.4.1 *Weathering of oil*

There are several weathering processes that are known to affect and influence the fate of an oil spill and consequently any clean-up operation, it is important to note that taking a wrong decision of applying a certain techniques may cost the offshore platforms dearly and this could be the direct result of the responders' failure to understand what happen to the oil during these processes. These weathering processes can be summarized as follows: (1) Spreading; (2) Evaporation; (3) Dissolution; (4) Dispersion; (5) Emulsification; (6) Photo-oxidation; (7) Sedimentation and shoreline stranding; (8) Biodegradation.

It is important to understand how different oil changes over time whilst at sa, the planners and responders need to know how the weathering processes interact.

The weathering processes do not occur separately, but simultaneously as they overlap each other through the course of an oil spill. The processes interact and affect each other, and in turn affect the properties of the spilled oil.

2.4.2 *Predicting the drifting of oil slick on sea surface*

Oil Slicks will not stay in place but will usually drift under the influence of external factors. The most important of these factors are winds, waves, tides and surface currents. The transport of water, particularly surface waters in the sea due to wind, waves and currents is a difficult topic about which much remains unknown.

a. The effect of wind: Friction between wind and water induces a current at the sea surface, which is negligible a few millimeters below the surface. The presence of oil on the water surface alters this vertical current profile. Oil on open water will move more quickly than the water directly beneath it, with the result that oil towards the lee side of a slick will be thicker than that to windward. In addition, the slick will rapidly become elongated and will form windrows. The speed with which an oil slick drifts under the influence of the wind depends on wind strength and oil thickness. Typically it will move at between 2% and 5% of the speed of the wind measured 10 m above the water surface. In open water, 3.4% of the wind speed is normally used to estimate drift rate (IMO, 1988).
b. The influence of currents: If the wind is negligible, which is rarely the case; the oil will move only under the influence of currents or tides. Current regimes may be constant but more commonly vary in strength and direction over time. The strength and direction of tides will influence the movement of the slick in the short term. However, tidal currents rarely cancel each other completely, and this gives rise to a residual current, which will determine the long-term movement of the oil slick. Assessment of the drift of a slick.

Having calculated separately the wind induced effect and the surface water current, the movement of a slick on the surface can be determined by drawing a vector diagram. This can be represented by the following simplified formula:

$$V\ oil = V\ current + V\ wind \times Q$$

In which: V oil is velocity of the oil; V current is velocity of seawater; V wind is velocity of wind at a height of 10 m; and Q is empirically established wind speed factor (usually about 3.4%).

2.5 *The negative effects of coastal and marine oil spills*

Oil spills can have serious economic and environmental impacts on coastal activities and on those who exploit the resources of the sea. In most cases such damage is temporary and is caused primarily by the physical properties of oil creating nuisance and hazardous conditions. The impact on marine life is compounded by toxicity and tainting effects resulting from the chemical composition of oil, as well as by the diversity and variability of biological systems and their sensitivity to oil pollution.

2.5.1 *Impact of oil on coastal activities*

Recreational activities:
The effects of a particular oil spill depend upon many factors, not least the properties of the oil. Contamination of coastal amenity areas is a common feature of many spills leading to public disquiet and interference with recreational activities such as bathing, boating, angling and diving.

Industry:
Industries that rely on a clean supply of seawater for their normal operations can be adversely affected by oil spills. If substantial quantities of floating or sub-surface oil are drawn through intakes, contamination of components may result, requiring a reduction in output or total shutdown whilst cleaning is carried out.

2.5.2 *Biological effects of oil*

The effects of oil on marine life are caused by either the physical nature of the oil (physical contamination and smothering) or by its chemical components (toxic effects and accumulation leading to tainting). Marine life may also be affected by clean-up operations or indirectly through physical damage to the habitats in which plants and animals live.

The main threat posed to living resources by the persistent residues of spilled oils and water-in-oil emulsions ("mousse") is one of physical smothering. The animals and plants most at risk are those that could come into contact with a contaminated sea surface. This include marine mammals and reptiles; birds that feed by diving or form flocks on the sea; marine life on shorelines; and animals and plants in marine culture facilities.

The most toxic components in oil tend to be those lost rapidly through evaporation when oil is spilt. Because of this, lethal concentrations of toxic components leading to large-scale mortalities of marine life are relatively rare.

2.5.3 *Impact of oil on specific marine habitats*

The following summarizes the impact that oil spills can have on selected marine habitats. Within each habitat a wide range of environmental conditions prevail and often there is no clear division between one habitat and another.

Open waters and seabed

Plankton is a term applied to floating plants and animals carried passively by water currents in the upper layers of the sea. Their sensitivity to oil pollution has been demonstrated experimentally. In the open sea, the rapid dilution of naturally dispersed oil and its soluble components, as well as the high natural mortality and patchy, irregular distribution of plankton, make significant effects unlikely. In coastal areas some marine mammals and reptiles, such as turtles, may be particularly vulnerable to adverse effects from oil contamination because of their need to surface to breathe and to leave the water to breed. Adult fish living in near shore waters and juveniles in shallow water nursery grounds may be at greater risk to exposure from dispersed or dissolved oil.

Shoreline

Shorelines, more than any other part of the marine environment, are exposed to the effects of floating oil. The impact may be particularly great where large areas of rock, sand and mud are uncovered at low tide. Whilst inter-tidal animals and plants are able to withstand short-term exposure to adverse conditions, they may be killed by toxic oil components or physically smothered by viscous and weathered oils and emulsions. Decolonization of a shoreline by the dominant plant and animal species can be rapid: on rocks the initial stage is usually the settlement of seaweeds followed by the slower return of grazing animals. However, the complete re-establishment of a normal balance may, in extreme situations, take many years.

Marshes

Marsh vegetation shows greater sensitivity to fresh light crude or light refined products whilst weathered oils cause relatively little damage. Oiling of the lower portion of plants and their root systems can be lethal whereas even a severe coating on leaves may be of little consequence especially if it occurs outside the growing season. In tropical regions, mangrove forests are widely distributed and replace salt marshes on sheltered coasts and in estuaries.

Mangroves

Mangrove trees have complex breathing roots above the surface of the organically rich and oxygen-depleted mud in which they live. Oil may block the openings of the air breathing roots of mangroves or interfere with the trees' salt balance, causing leaves to drop and the trees to die. The root systems can be damaged by fresh oil entering nearby animal burrows and the effect may persist for some time inhibiting re-colonization by mangrove seedlings. Protection of wetlands, by responding to an oil spill at sea, should be a high priority since physical removal of oil from a marsh or from within a mangrove forest is extremely difficult.

Corals

Living coral grows on the calcified remains of dead coral colonies which form overhangs, crevices and other irregularities inhabited by a rich variety of fish and other animals. If the living coral is destroyed the reef itself may be subject to wave erosion. The effects of oil on corals and their associated fauna are largely determined by the proportion of toxic components, the duration of oil exposure as well as the degree of other stresses. The waters over most reefs are shallow and turbulent, and few clean-up techniques can be recommended.

Birds

Birds that congregate in large numbers on the sea or shorelines to breed, feed or moult are particularly vulnerable to oil pollution. Although oil ingested by birds during preening may be lethal, the most common cause of death is from drowning, starvation and loss of body heat following damage to the plumage by oil.

2.5.4 *Impact of oil on fisheries and marine cultures*

An oil spill can directly damage the boats and gear used for catching or cultivating marine species.

Floating equipment and fixed traps extending above the sea surface are more likely to become contaminated by floating oil whereas submerged nets, pots, lines and bottom trawls are usually well protected, provided they are not lifted through an oily sea surface. Cultivated stocks are more at risk from an oil spill: natural avoidance mechanisms may be prevented in the case of captive species, and the oiling of cultivation equipment may provide a source for prolonged input of oil components and contamination of the organisms. The use of dispersants very close to marine culture facilities is not advised since tainting by the chemical or by the dispersed oil droplets may result (ITOPF, 2002).

3 IDENTIFYING THE AVAILABLE OIL SPILL RESPONSE OPTIONS

Oil spill emergency response planners must familiarize themselves with all the possible methods of combating oil spills. When oil is spilled at sea, measures must be taken to minimize physical damage to amenities and pollution of marine resources and the environment. Generally, it is preferable to treat the oil while it is on the sea and before it spreads over a wide area. Measures to deal with oil spills have severe limitations and imperfections. Given unfavourable circumstances, there is a real possibility that no effective response will be available and the oil will have to be left to natural degradation processes. In deciding what action to take, if any, the relative importance of the various resources at risk should be considered, taking account of the likely success of available response measures at sea and the difficulties of the subsequent shore-line clean-up. Conflicting views may be given in an emergency about the importance of different resources and decision-making will be easier if an environmental sensitivity survey has been conducted in advance. The relative order of importance of different resources is likely to vary at different times of the year, with the different fish and bird breeding seasons, and the recreational season.

3.1 *The possible response options (ADNOC, 2002) are:*

1. *No action other than monitoring the oil slick.*

This might be the proper decision if the oil is not moving shoreward, or if no important resources are threatened, or if the oil is breaking up naturally at a considerable rate, or if conditions are such that positive response options are not practicable.

2. *Containment and recovery of the oil at sea*

This is often the preferred option for both environmental and socio-economic reasons. Where practicable it is always better to collect and recover the spilled oil. Physical recovery is not easy; however, various types of booms and skimmers have been developed. Large booms and recovery devices make it possible to recover oil on the open seas as well as in coastal waters, given good weather conditions (ITOPF, 2002). Collection and recovery systems may be either static or dynamic. Static systems rely on the oil drifting into them; dynamic ones actively pursue the oil. Proper attention must be given to the forecasts of weather and sea state. Weather conditions can substantially reduce the success of clean-up operations and may require personnel to work under unsafe conditions (IMO, 1988).

3. *Chemical dispersion at sea*
Oil spilled on the sea surface will float and spread out to form a slick. Wave action and turbulence due to tides and currents will cause some of the oil to break up into small droplets which can be carried down into the water column. This process, which is known as dispersion, can be enhanced by the application of chemical dispersants.

Oil spill dispersants are mixtures of surface-active agents in one or more organic solvents. They are specifically formulated to enhance the dispersion of oil into the seawater column by reducing the interfacial tension between oil and water. Natural or induced movement of water causes a rapid distribution within the water mass of very fine oil droplets formed by the dispersant action, thus enhancing the biodegradation processes. Dispersants also prevent coalescence of oil droplets and re-formation of the oil slick (IMO, 1995). Care must be taken if this method is chosen first because dispersants are toxic so government-approved types are required and because of this toxicity the use of dispersants should not be the first choice unless collection and recovery is not practical. It is also important to note that dispersant will react differently depending on the type of spilled product and the state and condition of the product at the time of application e.g. if the spill takes the form of chocolate mouse use of most dispersants is of little or no value. For criteria considered when selecting a chemical dispersant refer to Emergency Response Planning for Oil Spills from Offshore Platforms (Hanna, 2003).

4. *In-situ burning*
In-situ burning is designed to remove oil from the surface of the sea through combustion. If the operation is successful, only a small fraction (typically 2–4%) of the original spilled oil volume will remain as a residue; the rest of the oil will have been redistributed into the air column in the form of particulates and gases. Care should be taken if the crude oil contains hydrogen sulphide.

5. *Shoreline clean-up*
Given unfavourable circumstances, it is likely that some oil will come ashore and shoreline clean-up will be necessary.

6. *Combination of response options*
A combination of response options will usually be needed in a large spill. Only by having a good understanding of the advantages and limitations of the different options can responders, make the appropriate decisions.

3.2 *The net environmental benefit analysis*

At the emergency planning stage it is necessary to weigh up the advantages and disadvantages of using or not using any of the response options mentioned above under different spill scenarios (that is, in a specific area and during certain climatic conditions). This should be done in order to decide in advance whether the use or non-use of a certain response option is likely to achieve the greatest benefit to the environment.

However the final decision to use a certain response option, for example the use of dispersant in combination with, or instead of, any other method will depend on the specific conditions of each oil spill. In all cases, the aim of the spill response should be to reduce the overall environmental impacts on both natural and economic resources. The predicted trajectories and corresponding fate and effects of untreated versus dispersed oil need to be considered to best identify which spill response/control method will minimize the overall impact on the environment and resources.

3.3 *Factors to be considered when assessing the net environmental benefit analysis*

To conduct a net environmental benefit analysis, it is essential to have the resources listed in view of their protection priorities. Such a list should be drawn up during the emergency planning stage and should consider factors such as possible seasonal variations in the priorities. When drawing up such a list, both natural and economic resources should be considered. In general it can be said that endangered species, highly productive areas, sheltered habitats with poor flushing rates, and habitats that take a long time to recover should receive top protection priority.

When drawing up the emergency plan, the environmental effects of dispersed versus untreated oil for each area should identify areas where dispersant use is not advisable and where it is advisable to a varying degree. It is, however, important that the habitats/resources are not viewed in isolation from each other since applying the decision taken for a particular habitat/resource may affect adjacent ecosystems. For example, in an area where oil occurs above a shallow-water submerged coral reef and is drifting rapidly towards a mangrove swamp, it may be advisable to disperse the oil above the reef. This may increase oil exposure of the corals but it will prevent oil from becoming incorporated into the mangrove sediments from where it will seep out over the years, forming a chronic pollution source for both the mangrove and coral reef ecosystems. Several elements of the net environmental benefit analysis for a particular area are spill specific and can only be answered at the time of the spill. The Oil Spill Response Team must ensure that the following questions are considered so that they can achieve a successful Oil Spill Emergency Plan:

- How effective is dispersant application likely to be, in view of factors such as the type and amount of oil spilled?
- Will the dispersed oil be diluted to low impact levels and, if so, how long will it take to reach these levels?
- What is the estimated time scale for the recovery of the habitat?

In some cases, the use of more than one response technique will be advisable to reduce risks and any negative consequences (IMO, 1995).

4 PREPLANNING AND METHODOLOGY

After the management of offshore oil platform and in particular the oil spill response team are satisfied that they fully understand the problems associated with oil spills and that they are familiar with all the possible oil spill response options and methods they should now go one step further toward formulating their plan by identifying the following points:

- The Tier response level
  This concept is very important to rationalize the required resources. In general most companies adopt three tiers response system upon which the company will be responsible for combating Tier 1 response level; in case of the OOP field it was decided that Tier 1 is 100 metric ton. Tier 2 will be handled using external resources of neighbouring companies and facilities and Tier 3 can be the subject of the National Oil Spill Contingency plan and even regional plans. The company must seek the approval of government authorities for the proposed size of spill selected as Tier 1.
- The response area
  That is to say the geographical area that the company is responsible to cover and protect, it is obvious that oil slicks can drift to areas far from the source of the spill and it could become Tier 2 or Tier 3 operations. In case of the OOP field an area of 36 kilometres was selected with the OOP platform is located at its centre.
- Sensitivity mapping
  This is a qualitative and semi-quantitative method where all the fauna and flora in the field will be determined. This should include but not limited to mammals (e.g. dolphins), Sea birds, pelagic fish, benthic fish, benthic organisms (e.g. molluscs, crustacean), cephalopods (e.g. octopus and squids) as well as coral colonies. This survey is time consuming and expensive and require the use of divers, marine vessels at various time of the year. The information was later plotted on a field map using AutoCAD software.
- Meteorological and oceanographic data
  These include previous record of weather data for a minimum 10 years period (if not available data from neighbouring facilities can be used). In case of the OOP field 15 years collected data were

tabulated and using simple statistical methods the 3 main predominant wind directions in the field were identified and were used to constitute the wind rose.

- Pre-incident contamination levels
  Where 37 points evenly distributed across the response area were sampled (special procedures for sample collection was used) to determine the existing levels of hydrocarbon contamination across the field. The sample was then analyzed at a professional third party laboratory and the results (baseline pollutant distribution map) were plotted on the field map using AutoCAD software. This information is important for determining the extent of future spills as well as deciding on actions that could be required if the existing levels are high which could be an indication of a previous spill or extensive malpractices in the field.

- Crude oil composition analysis (physical & chemical characteristics)
  The OOP field produces oil from 3 different formations each with its own characteristics (in-field pipelines); however the 3 types combine in the Main Oil Line (MOL). This step is important not only to decide the most effective chemical dispersant but also to understand the behaviour of the cruse in different weather conditions e.g. if the wax content is high. The OOP crude is also known to have a very high concentration of hydrogen sulphide which is very critical as it can affect the safety and health of the oil spill combating team (Hanna, 2003).

- Risk assessment of the OOP operations
  The risk assessment study included the different offshore platform's activities that may result in an oil spill. These activities include operations such as production, drilling, well maintenance and well workover, as well as marine operations such as anchor handling during barge and rig moves in addition to vessel positioning operations.
  The degree of risk for each activity was calculated as follows:
  - At first a list of all oil spill scenarios was made, for which the cause of the oil spill, the amount expected to be spilled before control and the frequency of occurrence(F) of such oil spills were identified;
  - Secondly, The potential oil spill scenarios were incorporated into the Risk Matrix according to their severities (S) and frequencies (F) in order to decide their risk categories (i.e. High Risk, Medium Risk, or Low Risk);
  - The Risk Factor was calculated by applying the following equation:
    Risk (R) = Severity (S) × Frequency (F), according to this equation the highest Risk Factor will have a value of 25.
    Risk Factor values between 1 and 7 represent a Low risk operation or activity. Risk Factor values between 8 and 15 represent medium risk operations that will need to be reduced further if necessary. The Medium Risk activities fall within the ALARP (As Low As reasonably Practicable) cat-egory. High Risk activities will have a risk value between 16 and 25 and these can not be accepted by management or operators and they should be brought down within the ALARP category.
    From the above it was concluded that the highest oil spill risk in the "OOP" platform's operations may be caused either by mechanical damage to the 18″ Main Oil line (MOL), or by a blow out from the drilling operations, it was calculated that both of these two scenarios have a medium degree of risk (within the ALARP category).
  - Measures that must be taken to reduce the risk in these scenarios were defined. The amount of spilled oil expected from these worst scenarios was used later to run a trajectory model using the OILMAP software in order to predict the fate, route and destination of the oil slick. (Hanna, 2003).

- Trajectory modelling
  Using the Oil Map 97 software 3 scenarios were run (using the 3 main wind directions mentioned above). The results of the trajectory modelling are essential for pre planning the oil combating activities and the location of the oil spill combating equipment. The results are also useful in deciding the possibility of affecting other remote sensitive areas (which may result in going to

Tier 2 or 3 levels depending on the size of the spill and the success of the spill control operations) (Hanna, 2003).

5 EMERGENCY RESPONSE PLANNING FRAMEWORK

This framework is developed to ensure that the OOP Oil Spill Emergency Response Plan (OOSERP) will be in compliance with:

- The requirements of the Abu Dhabi Oil Spill Contingency Plan (ADOSCP) developed by Abu Dhabi National Oil Company (ADNOC) in 2002;
- Federal Law No.24, act 199 of the United Arab Emirates (U.A.E.) developed by the Federal Environmental Agency (FEA) as well as;
- The requirements of the International Maritime Organization (IMO) as described in the IMO' s Manual on Oil Pollution, section 2, Contingency Planning.

5.1 *Relationships with other plans*

The level of emergency preparedness activities related to discharges of oil is dependant on several factors such as the sensitivity of the area surrounding the offshore facility, the remoteness of the location, the prevailing weather conditions, the type of oil expected to be spilled and it is also directly related to the expected amount of spilled oil. The ADOSCP recognizes three levels of response planning according to the amount of spilled oil as follows:

- The National Emergency Plan (NEP) for the United Arab Emirates which is still under development by the FEA;
- The Area Contingency Plan (ACP) which is covered by the ADOSCP developed by ADNOC; and
- The Facility Spill Response Plan (FSRP). The facility response plan is covered by the OOSERP which is the subject of this framework.
- The ADOSCP states that "The owner or operator of a facility that has the risk of spilling more than 10,000 Imperial Gallons (IG) of oil is responsible for the facility's oil spill response planning, coordination, and responses (ADOSCP, 2002).
- The plan should cover all activities and operations handled by the operator/owner. This should include:
  - Oil storage tanks greater than 10,000 IG capacity;
  - Pipelines transferring liquid hydrocarbons;
  - Tank vessels.

The OOP is transporting its entire crude oil production through its Main Oil Line MOL to an island where it is being stored and exported. The storage and export operations are being done and managed by another operator under a special service agreement. The service agreement placed the responsibility for managing oil spills from the storage and export operations under the responsibility of the second operator. Therefore, the OOP's Oil Spill Emergency Response Plan (OOSERP) will be only required to handle oil spills resulting from field activities (production, drilling and wellhead maintenance) as well as accidental releases from the MOL or any of its in-field sub sea pipelines. The OOSERP should include the following information as required by the Abu Dhabi Oil Spill Contingency Plan (ADOSCP):

- Initial response actions – responsibilities;
- Reporting of oil spills;
- Safety of personnel;
- Key facility information;
- Incident Command System (ICS) & On-site response organization;
- On-site equipment;
- Facility specific response strategies Tier 1 Response;

- Sensitive resources/facilities in the immediate vicinity and response priorities;
- Response options i.e. containment and recovery , chemical dispersant, shoreline clean-up or monitor and weight;
- Accessing additional resources if needed (Mutual Aid plans – Tier 2 & 3);
- Waste disposal;
- Two scenarios describing the most probable spills (based on actual Risk Assessment)

The OOSERP development must follow a logic sequence where team work plays an important part, the elements of the OOSERP.

5.2 *How to prepare an effective Oil Spill Emergency Response Plan?*

Having discussed almost all the elements and factors necessary for formulating a successful emergency response plan it is important for the field management to have a road map that will ensure that they will achieve the required results, availability of optimal resources with minimum obstacles and delays.

This road map is in the form of management steps requirements that should be integrated with the company HSE Management System (but can be also done independently and later integrated with the HSE Management System if this system is not available at the time the plan is being developed). These steps can be summarized as follows:

1. The company should have an oil spill prevention and mitigation policy that is signed by the most senior manager and is frequently reviewed and updated.
   It must be understood by all concerned parties that the task of emergency response planning requires team work approach as well as total management commitment. It must be understood that this task requires considerable financial resources, time and personnel; therefore the company should be committed to make these resources available. This fact should be clearly stipulated in the policy.
2. The Oil Spill Emergency Response Planning Team should be formed and assigned the task of developing the plan. This should be official and authorized by the most senior manager (GM or above depend on the nature of the organization).
3. All members of the team must be fully aware of the information mentioned under items no. 2 (understanding the problem) and no. 3 (Identifying the available oil spill response options mentioned above. This can be achieved through training and group education.
4. Risk Assessment studies should be made to establish the most credible scenarios.
5. The company should decide on the Tier response levels and the response area and approval of the concerned government authorities.
6. Sensitivity maps should be developed.
7. Weather and oceanographic data must be collected and summarized.
8. Establish the field Baseline Pollutant distribution map (sample, analyze and plot).
9. Trajectory models must be developed.
10. Decide on the best response method for each scenario. Have your plans reviewed by the concerned authorities.
11. Identify the required equipments (i.e. containment, recovery and waste management equipments), personnel and training required for Tier 1, arrange for budget and purchase.
12. Analysis of the different petroleum products handled in the field (e.g. crude oil and Diesel fuel) to select the most effective and environmentally acceptable chemical dispersant. Approval of authorities is a must and clear procedures and policy for using dispersant must developed, documented and published. Decide on the quantities and equipment required for application during Tier 1. Prepare budget and purchase.
13. Decide on the resources required for at least Tier 2 and ensure that agreement in place to cover your requirements (mutual aid or third party agreements).
14. Test & review your plans through drills and exercises these can be Communication drills (no mobilization of equipment), Table Top exercises, Partial Mobilization drills or Major drills

with actual mobilization of resources. Involve other companies and specialists in the evaluation of drills. Exchange information with similar and neighboring companies.

15. Systematically and frequently (according to a fixed schedule) audit your plan, your communication routes and your resources including personnel preparedness and equipment availability.

REFERENCES

ADNOC 2002. Abu Dhabi Oil Spill Contingency Plan. Abu Dhabi: ADNOC.

Hanna, Ahmes L. 2003. Emergency Response Planning for Oil Spills from Offshore Platforms. Alexandria: Arab Academy for Science & Technology and Maritime Transport.

IMO 1988. Manual on oil Pollution Section IV, Combating Oil Spills. London: IMO.

IMO 1995. Guideline on Oil Spill Dispersant Application. London: IMO.

ITOPF 2002. Use of Oil Spill Dispersants, Technical Information, Paper No.4. London: ITOPF.

Reclaiming the Desert: Towards a Sustainable Environment in Arid Lands – Mohamed (ed.)
 ISBN 0 415 41128 9

Effect of salt on the degradation of MSW in the bioreactor landfills

S. Alkaabi & P.J. Van Geel
Department of Environmental Engineering, Caleton University

M.A. Warith, Ph.D.
Department of Environmental Engineering, Reyerson University

ABSTRACT: Bioreactor landfills require sufficient moisture to optimize the biodegradation processes and methane generation. In arid regions this is problematic given the lack of fresh water supplies, and hence the focus of this study was to address the use of saline water for controlling the moisture content in bioreactor landfills. Four bioreactors and Eight biochemical methane potential (BMP) were used to study the impact of salinity on biodegradation of MSW. The salt concentration was 0%, 0.5%, 1%, and 3% (w/v), respectively. In eight BMP flasks, two inoculums, anaerobic digested sludge (ADS) and salt-acclimatized anaerobic digested sludge (AAS), and the same four salt concentrations were assessed. The rate of methane produced in the bioreactors after 214 days was 0.2, 0.165 and 0.095 (L/d/kg dry waste) in R1, R2 and R4 respectively. Similarly, the BMP was 66.1, 34.1, 22.7, and 18.7 ml/d with ADS and 66.2, 35.1, 28.37, and 21 ml/d with AAS as the inoculums.

1 INTRODUCTION

Landfills have been widely used for municipal solid waste (MSW) disposal all over the world. Landfill disposal is the most commonly used waste management method in many countries. The conventional landfill (dry tomb) is designed to minimize problems associated with the production of leachate and gas emissions by controlling the water flow using barrier liners and low permeability covers. The low moisture content in the conventional landfill can result in a slow degradation rate, which extends landfill life and leads to long term monitoring. The cost of monitoring and long term care of the landfill is very high. Leachate that is produced must be collected and stored or treated if it is to be discharged from the site. In addition, the landfill gases must be controlled to reduce the problems associated with gaseous emissions and odor problems (Rinhart 1996).

The process based approach that promotes contaminant biodegradation and that uses the landfill space as a treatment method rather than a storage method is the landfill bioreactor. The bioreactor not only enhances the degradation but also stabilizes the landfill as quickly as possible. Acceleration is accomplished through the addition of water in order to bring the moisture content to an approximate range of 35 to 50 percent (weight basis). The bioreactor landfill thereby avoids the problem of future liner failure and leachate contamination because the waste decomposes within 5 to 10 years, which is considerably less than the typical service life of these systems (Pacey 1989; Warith 2003).

Bioreactor landfills and accelerated waste degradation systems have several advantages over conventional landfills. Increased methane production can result when biological processes in the landfill are accelerated. Methane is a very valuable energy source and bioreactor landfills have the potential to produce high amounts of the gas in very short periods of time. Methane gas could then be collected and could lead to the improvement of the economic viability of landfill gas recovery and utilization. Leachate recirculation not only reduces the cost of treatment, but also lowers the strength of leachate, which lessens its potential for soil and groundwater contamination, a process referred to as "in situ treatment". Finally, bioreactor landfills have great economic benefits due to the reduction of cost associated with avoiding long term monitoring and maintenance and delaying final utilization of the site after closure (Rinehart et al. 2002).

The bioreactor landfill has great advantages in the degradation of MSW, which could be employed in arid and semi-arid regions. However, it requires sufficient moisture to initiate the biodegradation processes, which is problematic in arid regions that face shortages of fresh water supplies. It may therefore be useful to examine the utilization of saline water for controlling the moisture content in the landfill with respect to arid and semi-arid regions. To date, experimental studies on the use of saline water for the biodegradation of MSW in landfills have not been reported.

No studies have been conducted to determine the impact of salinity on the biodegradation of MSW. The objective of this paper is to study the effect of saline water on the biodegradation of municipal solid under different operational conditions.

2 MATERIALS AND METHODS

In order to study the impact of saline water on the degradation of MSW in an anaerobic bioreactor landfill, four laboratory scale bioreactor cells were used. Table.1 shows the experimental variables and waste characteristic. In the first set of experiments, the first reactor (R1) functioned as the control reactor, using water (without salt), whereas the second (R2), third (R3) and fourth (R4) reactors were run at 0.5%, 1.0% and 3.0% (w/v) salt content respectively. Figure.1 shows the configuration of the lab scale bioreactors. Each bioreactor was made from a PVC pipe (diameter 0.3 m and height 1.25 m) with covers equipped with several connector pipes, valves and rubber seals to prevent gas leakage from or into the reactor. At the bottom of the reactor, a 70 to 100 mm thick layer of gravel acted as a drainage layer to collect leachate at the base of the bioreactor cell. A geotextile covered the gravel layer to prevent migration of fines into the drainage area. The next layer consisted of 1.0 m of MSW. Three separate ports on the top of each reactor were used as an inlet for the leachate recycle or for infiltration water, to measure the total biogas produced and to inject air in the aerobic stage.

The reactors were operated under aerobic conditions through air injection for an initial period of sixteen days to enhance the acidogenic stage and decrease the organic content of the MSW. A perforated PVC pipe with a diameter of 20 mm was used as an air distribution system in the aerobic reactors and was connected to the laboratory air compressor. Following the initial aerobic stage (16 days), all reactors were operated under anaerobic conditions for 214 days.

2.1 *Biochemical methane potential*

In the second set of experiments, the biochemical methane potential (BMP) assay was used to determine the methane yield of an organic material during its anaerobic decomposition. BMP tests were carried out to determine the amount of methane that produced from leachate generation at different conditions of salt. The BMP assays were conducted in two sets, the first set consisted of four bottles with 250 ml of leachate from R1 (TVS = 8,000 mg/l) and was run at 0%, 1%, 2% and 3% (w/v) salt concentration with addition of 20% (v/v) anaerobic digested sludge (TVS = 1.43%)

Table 1. Waste characteristics.

Parameters	R1	R2	R3	R4
Density (kg/m^3)	478.57	490	516	516.64
Initial moisture content, % (wt. basis)	38	39	44	36
Mass of waste (kg)	33.5	34.3	36.12	36.165
Salt concentration, % (w/v)	0	0.5	1	3

as an inoculum. Where the second set was run under the same conditions, but with the addition of anaerobic digested sludge acclimatized (TVS = 1.2%) with salt as an inoculum. For the acclimatized inoculum, the anaerobic digested sludge was initially fed with leachate containing a low concentration of salt (e.g 0.2% w/v) and then the salt concentration was increased gradually until it reached (3% w/v).

2.2 *Experimental method and operations*

All reactors were loaded with shredded MSW. The composition of solid waste used in the experiments was 60% (w/w) organic waste, 20% (w/w) paper waste, 15% (w/w) plastics and 5% (w/w) textiles. The waste was placed and compacted manually in 0.1 to 0.15 m lifts. The average bulk density of the solid waste was in the range of 470–520 kg/m^3. This technique was repeated until the thickness of solid waste in each reactor reached a height of 1.0 m. The volume of the solid waste in the anaerobic bioreactor was approximately 0.07 m^3 and the mass was in the range of 33–36.2 kg per reactor.

The leachate generated was collected at the bottom of the reactor and allowed to gravity drain into a storage tank. The leachate was recycled daily in the aerobic stage and three times per week in the anaerobic stage. The volume of leachate recycled was approximately 10% (v/v) of the MSW in the bioreactor (Chugh et al. 1998). The pH was adjusted to neutral through the addition of buffer ($NaHCO_3$) to control the pH of the leachate before it was recycled back to the reactor. The salt concentration in the recycled leachate was maintained at a constant for each operating condition by adding NaCl to the leachate recycle.

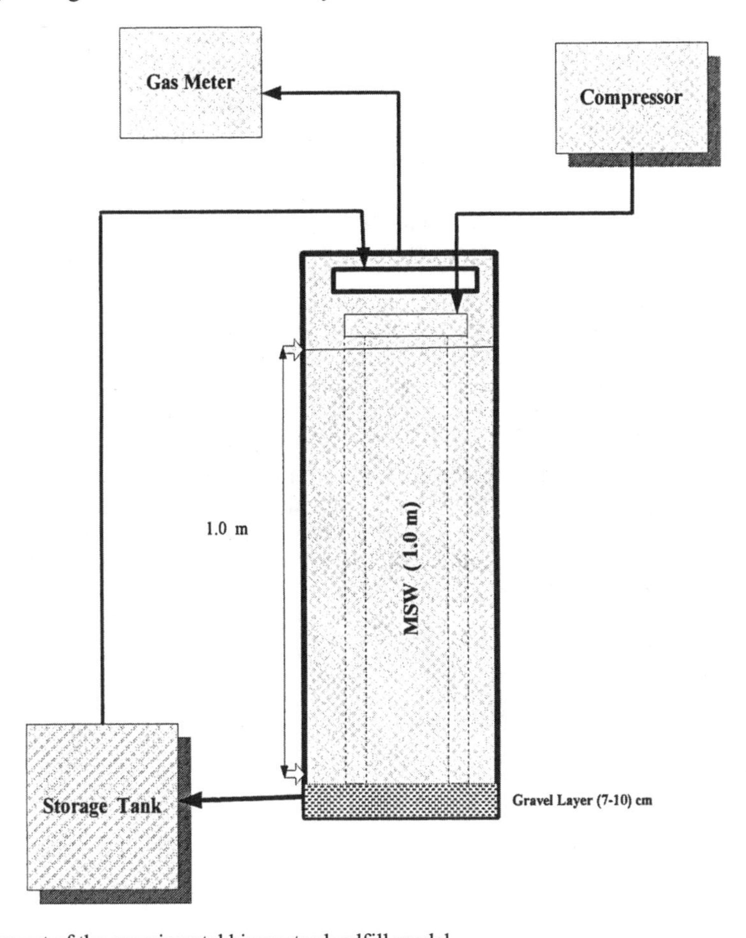

Figure 1. Layout of the experimental bioreactor landfill model.

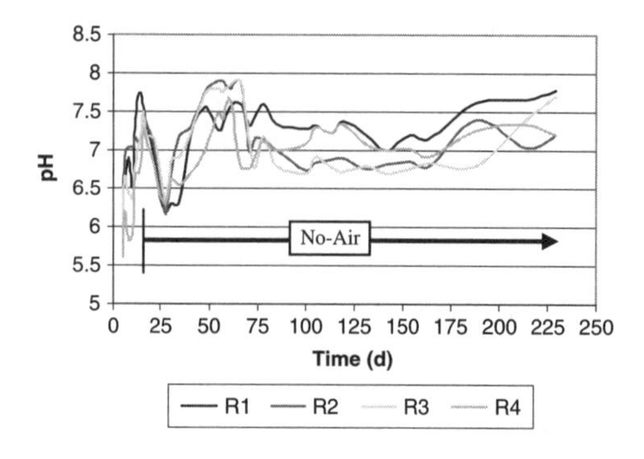

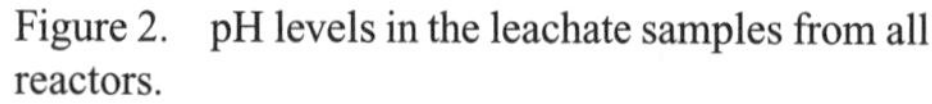

Figure 2. pH levels in the leachate samples from all reactors.

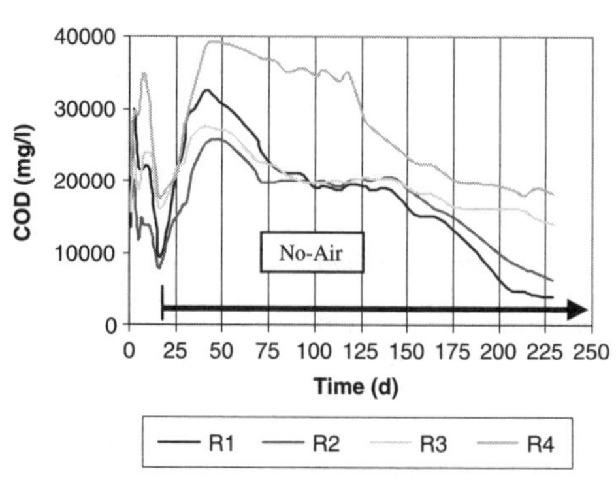

Figure 3. COD concentration profile.

Air was supplied to the four bioreactors to maintain aerobic conditions and to satisfy the oxygen demand for 16 days (Brummeler et al. 1990). Anderson (1990) presented the suitable aeration rate for the municipal solid waste containing a moisture content of 50–70% in the range of 0.6 to 1.8 m^3 air/(d·kg waste). Based on that and on the stoichiometric demand (Haug 1993), 1.3 m^3 air $d^{-1}kg^{-1}$ MSW (dry) was supplied.

2.3 *Analytical analysis*

Leachate samples were analyzed for pH, chemical oxygen demand (COD), volatile fatty acids (VFA), total volatile solids (TVS) and salinity. The pH was measured using a pH meter immediately after sampling to avoid any change due to CO_2 stripping. The COD, TVS and salinity were measured according to standard methods (Arnold E. G. et al. 1992). The VFA concentrations were measured using a simple titration method presented by Anderson et al. (1993). The volume of the biogas produced was measured using a wet tip meter and the biogas composition was analyzed using a gas chromatograph.

3 RESULTS AND DISCUSSION

3.1 *pH of the leachate from reactors*

The initial pH of the leachate samples from R1, R2, R3 and R4 reactors were 6.5, 6.3, 6.1, and 5.6, respectively. Then it increased due to the buffering and decreasing of VFA concentration in the aerobic stage as shown in Figure 2. When the reactors shifted to anaerobic conditions, the pH decreased sharply from 7.4 to 6.3 in all reactors due to the accumulation of VFA. As the methane production increased on day 41, the VFA concentration decreased and the pH increased and stabilized around 7 for all reactors. Agdag et al. (2005) found that the initial pH of leachate from a reactor operated with the addition of buffer was 5.3 and it reached approximately 7 after 20 days. Warith et al. (1998) found that the initial pH of the leachate sample with buffer addition was 6 and it stabilized at approximately 7 after 25 days.

The low value of pH at the beginning of the anaerobic stage is due to the hydrolysis of the solid waste and complex dissolved organic compound into organic acids, which the acidogenic bacteria convert into hydrogen, CO_2 and acetic acid.

3.2 *COD concentration of leachate produced from reactors*

The variation in the COD concentration for all the reactors is shown in Figure 3. Initial COD concentrations of leachate from R1, R2, R3 and R4 were 13600, 15650, 15700 and 20700 mg/l,

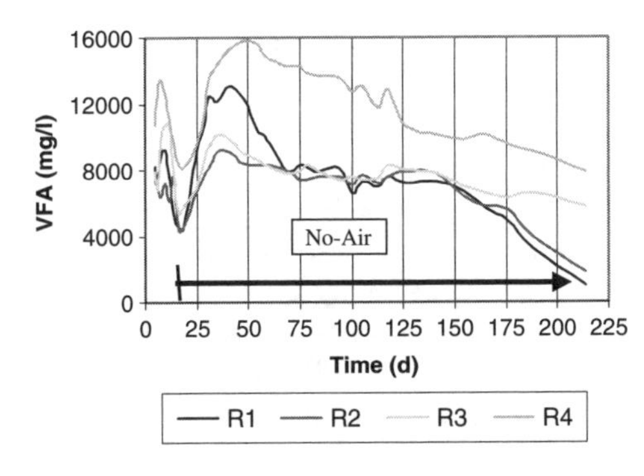

Figure 4. VFA concentration profile.

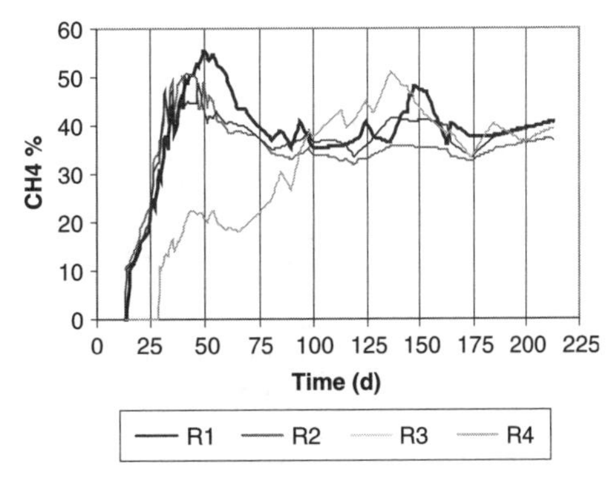

Figure 5. Methane concentration.

respectively. In the aerobic stage, the salinity had less effect on the COD profile of R1, R2 and R3 in comparison to R4. In R4, with high salt concentration, initially the COD stabilized above 30000 mg/l for the first 10 days and then decreased to 17600 mg/l at the end of the aerobic stage.

In the anaerobic stage, there was a rapid increase in the COD concentration for all reactors due to the accumulation of VFAs, as a result of hydrolysis and acidogenic bacteria. The peak value of the COD concentration in R1, R2, R3 and R4 was 32400, 25300, 27600 and 39200 mg/l, respectively on day 40. The increase was followed by a decrease in the COD concentration in reactors R1, R2 and R3 due to an increase in the daily methane production. For R4, the COD concentration stabilized above 35000 mg/l until day 118, and then descended as the daily methane production increased and reached its peak after day 118.

As mentioned, the COD concentration in R1, R2 and R3 decreased until it stabilized around 20000 mg/l from day 75 to 150. During this period, the daily methane production rate also decreased. The COD concentration of R1, R2 and R3 continued to decline until it reached 3900, 6250, 14000 mg/l, respectively at the end of the experiments. For R4, the COD concentration decreased until it reached 20000 mg/l at the end of the experiment.

The COD concentrations in R1, R2 and R3 indicated similar trends over time with limited impact due to the salt concentration. At the end of the experiments, the COD concentration was lowest in the reactor with no salt and greatest in the reactor with the highest salt concentration. During both the aerobic and anaerobic stage, the COD concentration for R4 reached a maximum value and leveled off before a declining. This trend was not present in the other three reactors indicating a lag in the acidogenic and methanogenic breakdown of the COD. For the anaerobic stage, this lag period is clearly illustrated by the delay in the production of methane for R4.

3.3 *VFA concentration of leachate produced from reactors*

The concentration of VFA is shown in Figure 4. The VFA concentration followed a similar trend with time as the COD concentration in all reactors. Initially, it decreased in the aerobic stage and then it increased at the beginning of the anaerobic stage as a result of the end product from hydrolysis and acidogenic bacteria. Then, it fell as the daily methane production increased because the VFA is the substrate for methanogenic bacteria.

3.4 *Methane gas production and concentration*

The daily methane production and concentration (expressed as a percentage of the biogas) of methane in all reactors are shown in Figures 5 and 6. There was a leak in R3, hence, the volume of biogas produced was not measured. The methane production started on day 15 for R1, R2 and R3 while it started on day 30 for R4. The maximum daily production was recorded as 10.6 L for R1

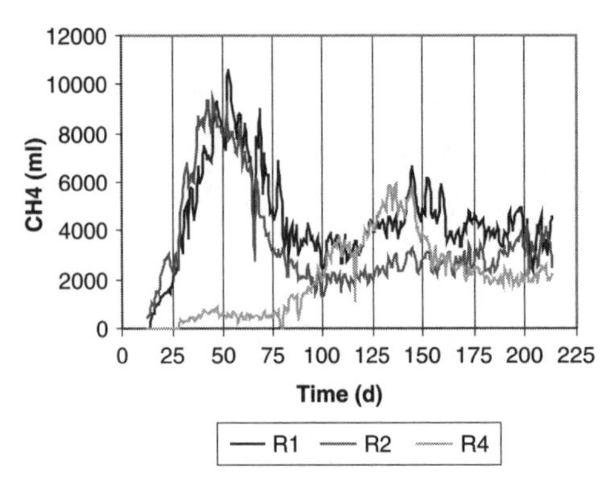

Figure 6a. Methane daily produced.

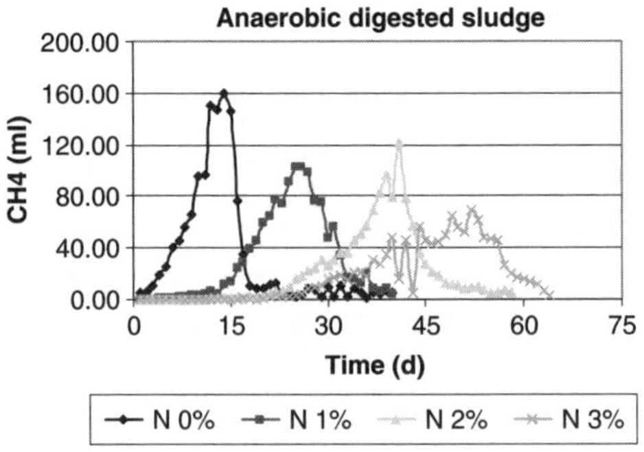

Figure 6b. Methane daily produced.

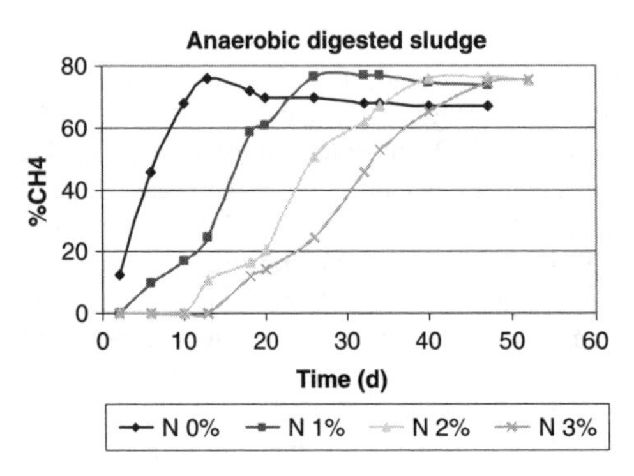

Figure 7. Methane concentration.

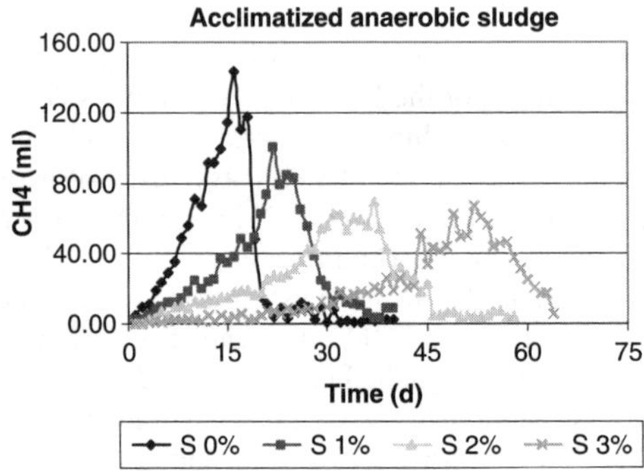

Figure 8. Methane daily produced.

on day 50, 9.3 L for R2 on day 45, and 5.9 L for R4 on day 137. The slight delay for R1 compared to R2 may have been due to higher VFA concentrations which may have slightly inhibited the methanogenic bacteria, where the delay for R4 was due to salt inhibition. The high salt concentration in R4 inhibited the methanogenic bacteria and as a result it took more than 100 days to be acclimatized with the salt. The highest methane concentration was recorded in R1 and it was 53.6%, where the peak of methane concentration in R2 and R3 was 50.2% and 47.4% respectively. In R4, the peak was reached after 118 days and it was 50.2%.

Ağdağ et al. (2005) found the methane concentration in the control (no sludge addition) reached its peak (62%) on day 50. Also Sponza et al. (2005) found the peak of methane concentration was 56% in the shredded reactor on day 50.

The rate of methane produced over the anaerobic stage of the experiment was 0.2, 0.165 and 0.095 (L/d/kg dry waste) in R1, R2 and R4 respectively. The ratio of daily methane produced from R2 and R4 with respect to R1 was 0.783 and 0.445 respectively.

3.5 *BMP assays*

The results of daily methane produced and methane concentration are shown in Figures 6 and 7 for BMP inoculums with anaerobic digested sludge (ADS), and in Figures 8 and 9 for anaerobic acclimatized sludge (AAS). For the BMP assay inoculated with ADS the lag phase of methane production increased as the salt concentration increased. The lag time was 0, 2, 10 and 14 days for 0%, 1%, 2% and 3% (w/v) salt concentration, respectively. On the other hand, the lag time for the

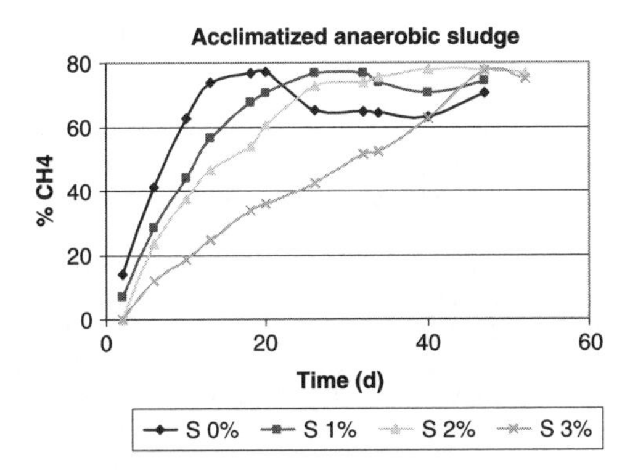

Figure 9. Methane concentration.

BMP assay inoculated with AAS was 0, 0, 2 and 4 days for 0%, 1%, 2% and 3% (w/v) salt concentration, respectively. The average rate of methane production was 66.1, 34.1, 22.7 and 18.7 ml/d for the ADS inoculated BMP assays while it was 66.2, 35, 28.37 and 21 ml/d for the AAS inoculated BMP assays for 0%, 1%, 2% and 3% (w/v) salt concentration, respectively.

The BMP assays indicated similar results as in the bioreactor columns. The increase in salt concentration causes an increase in the lag phase of methane production and a decrease in the daily methane produced.

4 CONCLUSIONS

The results showed that high salt content has a negative impact on the degradation of MSW in anaerobic bioreactors. The result of COD showed that the reactor with salt content higher than 1% (w/v), as in R4, required longer time to reach low COD concentration compared to the others.

The rate of methane produced in the bioreactors after 214 days was 0.2, 0.165 and 0.095 (L/d/ kg dry waste) in R1, R2 and R4 respectively. Similarly, the BMP was 66.1, 34.1, 22.7, and 18.7 ml/d with ADS and 66.2, 35.1, 28.37, and 21 ml/d with AAS as the inoculums.

Similar effects were observed on the rate of methane produced and methane concentration. The increase in salt content caused a reduction of the rate of methane produced as shown in R4 and the BMP results. Also the increase in salt content caused a longer lag phase for the methane produced. The BMP with acclimatized anaerobic sludge decreased the lag phase of methane produced.

From all of that, designers of bioreactor landfills should consider that the degradation of MSW using saline water requires longer time to stabilize. Also, they should consider the effect of saline water on the rate of methane produced if they plan to recover energy from methane.

REFERENCES

Anderson J. G. 1990. Treatment of wastes by composting. In microbiology of landfill sites (E. Senior, ed.). Boca Raton, FL, U.S.A.: CRC Press, Inc., p 59–80.

Anderson G. and Yang G. 1993. Determination of bicarbonate and total volatile acid concentrations in anaerobic digestion using a simple titration, Water Environmental Research Vol. 64 : 53–59.

Arnold E. G., Lenore S. C. and Andrew D. E. 1992. Standard methods for the examination of water and wastewater, 18th edition.

Ağdağ O. N. and Sponza D. T. 2005. Co-digestion of industrial sludge with municipal solid wastes in anaerobic simulated landfilling reactors. Process Biochemistry 40: 1871–1879.

Brummeler E.T. and Koster I.W, 1990. Enhancement of dry anaerobic batch digestion of the organic fraction of MSW by an aerobic pretreatment step. Biological Waste Vol. 31: 199–210.

Chugh S., Chynoweth D. and Clarke W. Pullammanappallil P. and Rudolph V. 1998. Effect of recirculated leachate volume on MSW degradation, Waste Management Research Vol.16: 564–573.
Haug R.T. 1993. The practical handbook of compost Engineering, Lewis Publishers.
Pacey J. 1989. Enhancement of degradation: Laboratory scale experiments. Sanitary Landfilling: Process, Technology, and Environmental Impact, Toronto, Academic Press. pp 103–119.
Reinhart D. 1996. Full scale experiences with leachate recirculating landfills: case studies, Waste Management & Research Vol.14: 347–365.
Reinhart D., McCreanor P. and Townsend T. 2002. The bioreactor landfill: its status and future, Waste Management Research Vol. 20: 172–186.
Sponza D. T. and Agdag O. N. 2005. Effects of shredding of wastes on the treatment of municipal solid wastes (MSWs) in simulated anaerobic recycled reactors, Enzyme and Microbial Technology 36: 25–33.
Warith M. and Sharma R. 1998. Technical review of methods to enhance biological degradation in sanitary landfill, Water Quality. Res. J. Canada, Vol. 33, No. 3: 417–437.
Warith M. A. 2003. Solid Waste Management: New trends in landfill design, Emirates Journal of Engineering Research, Vol.8, No. 1: 61–70.

Reclaiming the Desert: Towards a Sustainable Environment in Arid Lands – Mohamed (ed.)
© 2006 Taylor & Francis Group, London, ISBN 0 415 41128 9

Bioremediation of crude oil contaminated UAE soils: Challenges and advances

F. Benyahia[1], M. Abdulkarim[1], A. Zekri[1], A.M.O. Mohamed[2] & O. Chaalal[1]
[1]Department of Chemical and Petroleum Engineering, [2]Department of Civil and Environmental Engineering, College of Engineering, United Arab Emirates University

ABSTRACT: Biological treatment of crude oil contaminated desert soils was shown to be an effective method when using biopile systems. Because of the poor nutrient and microbial content of the desert soils, bioaugmentation was shown to be essential in the effective treatment of crude oil contaminated desert soils. This investigation has shown the benefits of staged studies involving soil characterization, oil evaporation and bioremediation using engineered biopile systems. Biopile systems were also shown to be superior to landfarming practices that are being phased out in Western countries because of the health hazards associated with them.

1 INTRODUCTION

1.1 *Inland oil spills in the Arabian Gulf area*

The Arabian Gulf region is an arid geographic zone well known for its vast reserves of crude oil and natural gas. The wealth generated by the hydrocarbon sales around the world has given some countries in this region a unique opportunity to develop their economies and transform desert areas into green heavens that became synonymous with prosperity and tourist attractions. Surrounding these green areas are some of the largest crude oil explorations in the world. It is well known that minor oil spills are expected to occur occasionally. However, given the massive scale of exploration and crude oil transportation by pipelines, the probability of a major inland oil spill is not negligible. When such events occur, the prospects of underground aquifer contamination, atmospheric pollution and long term soil contamination are indeed of great concern. The environmental impact of such industrial disasters can potentially, adversely affect nearby green heavens and cause health hazards. The ecological disaster that followed the Iraki invasion of Kuwait and involved massive crude oil spills was described by Al-Daher, et al. 1998. When such large scale crude oil soil contaminations occur, bioremediation was shown to be an effective treatment process (Balta, M.T. et al. 1998). Other natural processes such as phytoremediation for treatment of contaminated soils have been documented (Pletsch, M. 1999). However, their effectiveness in treating large areas of desert soils has yet to be confirmed since it is suspected that plants uptake of soil hydrocarbons will lead to release of these through its leaves into the atmosphere, thus transferring the problem from soil to atmosphere.

Landfarming is a bioremediation technique that has received a great deal of scientific coverage and has been employed over a number of decades by oil companies to treat hydrocarbon contaminated soils (Del Arco, J.P. and Franca, F.P. 2001; 1999). However, a recent study has shown the health hazards associated with landfarming (Hijazi, R.F., et al. 2003) and in fact this practice has been banned in the United States and other Western nations (Ward, O. et al. 2003). A better alternative to landfarming is biopile systems. These systems are essentially "engineering" bioremediation processes that offer more control and produce less atmospheric emissions when properly designed (Kodres, C.A. 1999; Benyahia et al. 2005). Although biopile systems have a higher cost associated with their design and operation compared to landfarming treatment methods, the environmental and health benefits far outweigh such cost difference.

The United Arab Emirates is one of the Arabian Gulf states that have institutionalized environmental protection in its industrial activities and encouraged environmental research aimed at environmental protection and human well-being. The present study is part of a research program supported by the United Arab Emirates University aimed at understanding phenomena occurring after major oil spills on UAE soils, developing effective contaminated soil treatment technologies and offering strategies and guidelines to the relevant industrial sector.

1.2 *Specificities of inland oil spills in desert regions*

Inland crude oil spills have several distinctive characteristics: difficulty to recover spilled hydrocarbons, interaction of spilled oil with soil, partial evaporation of crude oil fractions and effect of temperature and sun radiation fluctuations between day and night. Unlike marine oil spills that have been exposed to significant media attention, inland oil spills receive fewer and lighter media coverage although the environmental impact is no less damaging. In addition, traditional marine oil spill counter-measures such as natural or chemically aided dispersion are not options to consider for inland oil spills.

Because of the complications cited above, a more comprehensive and systematic approach is required to mitigate environmental impacts of major inland oil spills. In desert regions, the soil physical, chemical and biological make up has a profound effect on any anticipated treatment technology. The extreme climatic conditions and day/night temperature variations, coupled with wind speeds tend to complicate the situation. One can clearly see the challenge ahead in the treatment of contaminated desert soils.

1.3 *Important issues in bioremediation of contaminated soils*

When a major crude oil spill occurs inland, the spilled oil will flow onto the surface of the ground and starts to percolate into the soil whilst being slowly absorbed to form a saturated mass of soil. Thereafter, a crude oil pool forms and immediately, partial evaporation begins. The extent of such evaporation depends on the prevailing weather conditions as indicated above.

In order to develop reliable and effective soil bioremediation processes, a systematic experimental procedure has to be adopted and implemented in order top extract useful and dependable design information. The first step in such procedure invariably includes full soil characterization, followed by oil evaporation studies and bioremediation tests exploiting the virtues of hydrocarbon degrading microbial colonies (indigenous or added and in the presence or absence of additives such as macro/micronutrients, surfactants). Vogel T. (1996) discussed comprehensively the issues that concern bioaugmentation, biostimulation and the associated engineering process design matters.

Details of the experimental procedure cited above and employed in this work will be presented in the following sections.

2 EXPERIMENTAL

2.1 *Soil characterization*

The soil selected in this investigation was collected from an area near to an oil field, south of Abu Dhabi (Sahel). Its texture was determined by the standard ASTM classification of clay, silt and sand content. The sieve weight fractions of each grain size were subsequently applied to the soil characterization triangle shown in Figure 1 to determine the soil texture.

The texture of Sahel soil was thus determined as "sandy loam" with an average particle size of 150 microns. The soil bulk density was 1.6 g/ml and the clean soil pH and conductivities were measured as 7.8 and 118 μS respectively. The soil field capacity (water absorption to saturation) and crude oil absorption to saturation were experimentally determined as 23% w/w and 17% w/w

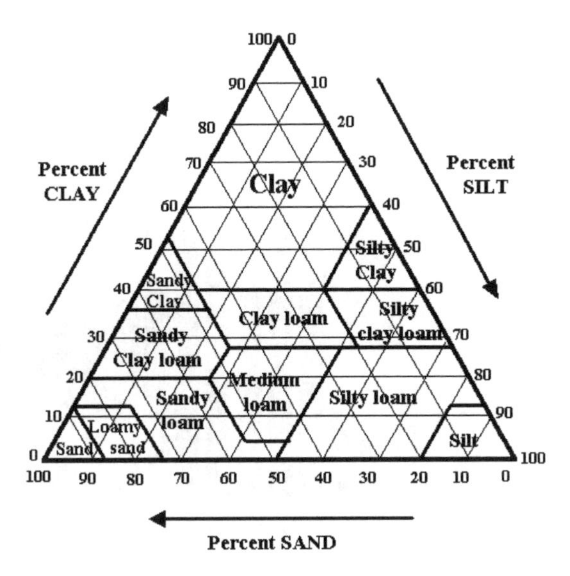

Figure 1. Soil characterization triangle.

respectively. The soil chemical analysis revealed that it was poor in nitrogen, phosphorus and potassium. The soil bacterial count was also found to be low (Image analysis method of soil extracted biomass).

2.2 *Oil evaporation studies*

The oil evaporation studies were conducted on the open roof of the Chemical Engineering department building. They consisted of exposing to the elements accurately weighed amounts of crude oil and synthetic crude oil contaminated soil samples, and weighing these hourly for periods extending to 72 hours. The oil and contaminated soil samples temperatures were also measured hourly. The prevailing weather conditions were recorded (temperature, wind speed, cloudiness and humidity). The crude oil employed in this study was obtained from Bu Hasa oil field, south of Abu Dhabi. Its composition is shown in Figure 2 and suggests that it is a somewhat intermediate crude oil with a significant "heavies" content.

The crude oil evaporation pattern will be presented in the results section.

2.3 *Biopile systems*

The biopile systems employed in this investigation were similar to those described by (Benyahia et al. 2005). They consisted of an air flow metering system, CO_2 traps before and after the biopile cell, an air washing/humidifying bottle and VOC adsorption column. The CO_2 traps consisted of tall 1 L graduated bottles fitted with a sampling port. The air flowrate was set to 400 cc/min for all systems (flow controller supplied by Roxpur Instruments, UK).

Backflow liquid trap bottles were added between caustic traps and after the biopile cell to prevent accidental contamination. Details are shown in Figure 3.

In this investigation, 4 biopile trains were employed. The content of the biopile cells in these trains is shown in Table 1. The amounts of soil and initial crude oil were 1850 g and 277.5 g respectively in all 4 cases. A biological product called Amnite P300 (supplied by Cleveland Biotech Ltd, UK) has been considered for application in some runs aimed at investigating bioaugmentation. It consisted of a mix of *Pseudomonas* Putida (Gram negative) and *Bacilli* Subtilis (Gram positive) bacteria acclimatized on hydrocarbon carbon sources, and immobilized on bran cereal to facilitate extended storage and convenience of use. The amount of Amnite P300 was 55.5 g in each of biopile B and D. The concentration of biomass in Amnite P300 was around 9×10^7 CFU/g of bran

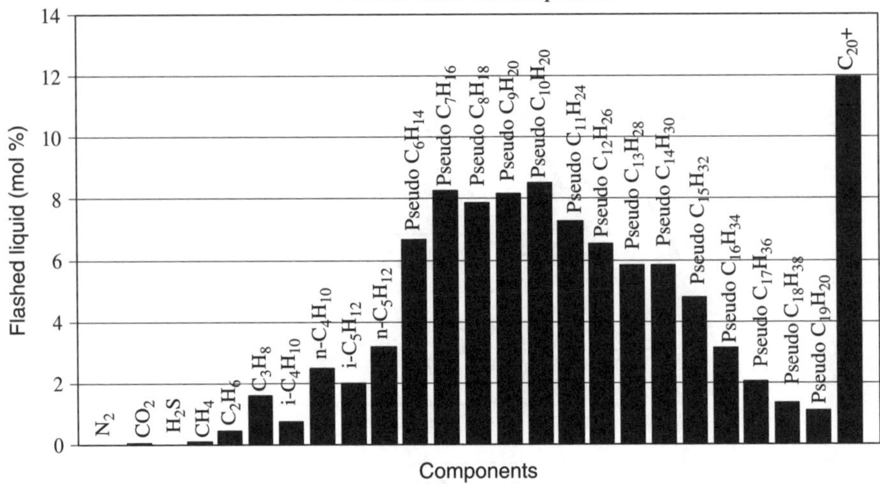

Figure 2. Crude oil composition.

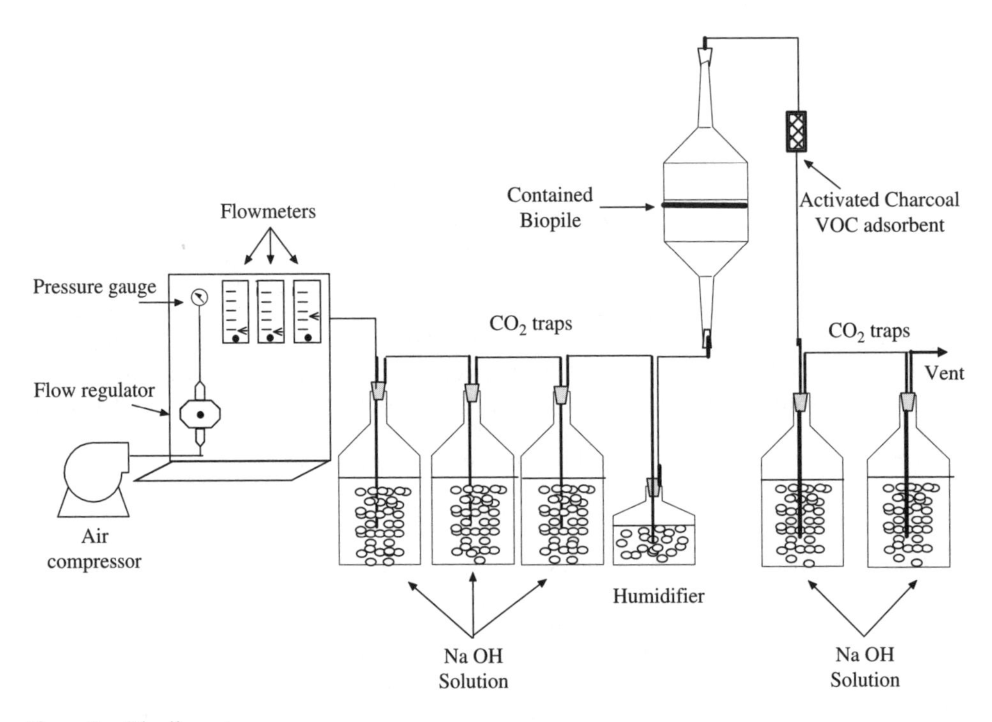

Figure 3. Biopile system.

(source: Cleveland Biotech Ltd, UK). All biopile systems had Nitrogen, Phosphorus and Potassium added in the form of urea (N) and K_2HPO_4 (P and K). Amounts of 9.74 g of K_2HPO_4 and 23.8 g of urea were added in all biopile systems.

The soil biopiles were thoroughly blended after adding the relevant ingredients prior to being loaded into the respective biopile cell. The biopile cells were ceramic inverted cones fitted with perforated bases and lined with geotextile cloth to prevent sand from escaping through the perforations, but allowing free flow of gases and vapor.

Table 1. Biopile systems formulations.

Biopile system label →	A	B	C	D
Biopile cell content ↓				
Clean soil	✓	✓	✓	✓
Crude oil	✓	✓	✓	✓
Amnite P300		✓		✓
N-P-K	✓	✓	✓	✓
Tween 80				✓
$HgCl_2$	✓			

Biopile system A served as a control run for biological activity and therefore had $HgCl_2$ biocide added to kill indigenous bacteria present in the soil. Biopile system C served to evaluate indigenous bacteria biological activity whilst biopile systems B and D served to evaluate bioaugmented biological activity without and with surfactant (Tween 80) respectively. Tween 80 has been added as a solution with 10 times the critical micelle concentration (0.27 mMol). Therefore biopile D served as an evaluation of surfactant aided bioavailability of nutrients.

2.4 *Bioremediation monitoring system*

The biological activities in biopile systems A to D were monitored by means of evolved CO_2 measurements at the biopile cell effluent. In order to accomplish this task, CO_2 present in the feed air supply was absorbed in sodium hydroxide traps preceding the biopile cell and CO_2 produced by bacterial activity was absorbed in sodium hydroxide traps placed after the VOC's capturing column, as shown in Figure 3. Daily samples of the NaOH solutions containing carbonates representing CO_2 were collected and titrated with hydrochloric acid solutions of 1 and 0.01 M strengths. The titration was conducted automatically using a pH meter endowed DOSIMAT titrator (Metrohm, Switzerland) and a titroprocessor to pinpoint end points. To enhance the measurements, two color indicators were also employed (phenolphthalein and methyl orange). This technique was extremely reliable and produced very accurate and repeatable results. Such CO_2 generation data served to log cumulative CO_2 production from which daily CO_2 generation rates could be extracted.

3 RESULTS AND DISCUSSION

3.1 *Soil texture and physico-chemical characteristics*

The sandy loam texture of the soil employed indicated that it is permeable and tends to dry quickly. Its field capacity also indicated the level of moisture not to be exceeded to avoid water saturation and therefore block gas transfer. In our work, there were two sources of moisture: from the saturated air feed to the biopiles and from the biological activity of the biopile system where it is sufficient. Obviously the former source was the main one and could reach 30% of the soil field capacity under the prevailing air flow conditions. Due to the poor nutrient content of the soil, additional Nitrogen, Phosphorus and Potassium were added to stimulate bacterial growth.

3.2 *Oil evaporation trends and modeling approach*

Although the bioremediation tests were conducted in the laboratory at constant temperature, it was deemed instructive and for the purpose of aiding large scale biopile design, to investigate oil evaporation under conditions that are close to a real life situation. The runs were conducted during the summer when temperatures were relatively high during the day and still above laboratory conditions during the evening. The nocturnal and diurnal temperature variations over a period of 55 hours is shown in Figure 4. The typical sinusoidal pattern with a large amplitude has a very significant bearing on the oil evaporation pattern. The hourly weights of oil samples and contaminated soil exposed to the elements provided data that enabled the calculation of the amounts of crude oil evaporated every hour. The pattern of crude oil evaporation and the sample temperatures can be seen in Figure 5. A similar pattern was observed for oil evaporation from contaminated soil (not shown in this article).

The oil loss trend followed a damped oscillatory pattern that followed the temperature variations. The damping can be explained in terms of losses of lighter fractions present initially in the crude oil. This interesting oil evaporation pattern is easily amenable to mathematical analysis and simple empirical models were fitted to experimental data (Abdulakrim, M. et al. 2005).

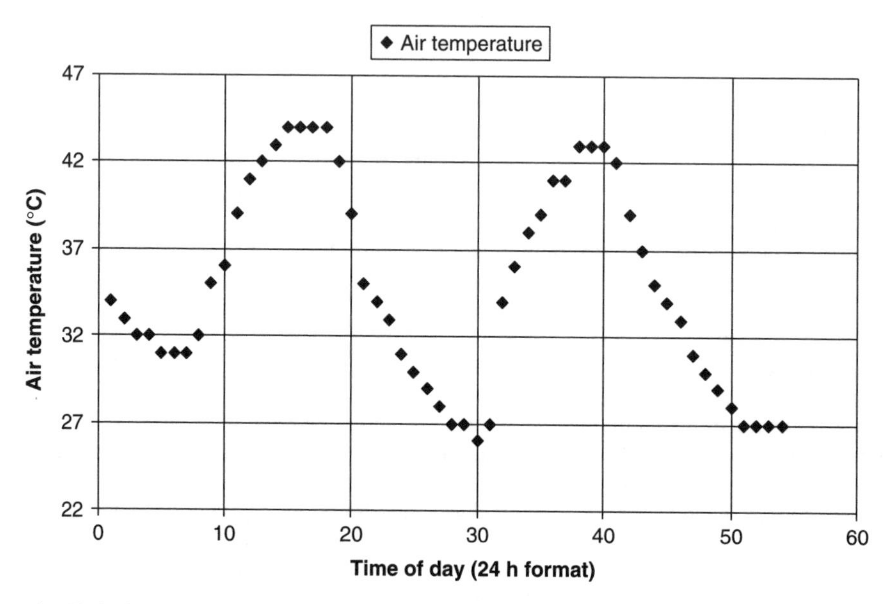

Figure 4. Typical summer air temperature variation in the UAE.

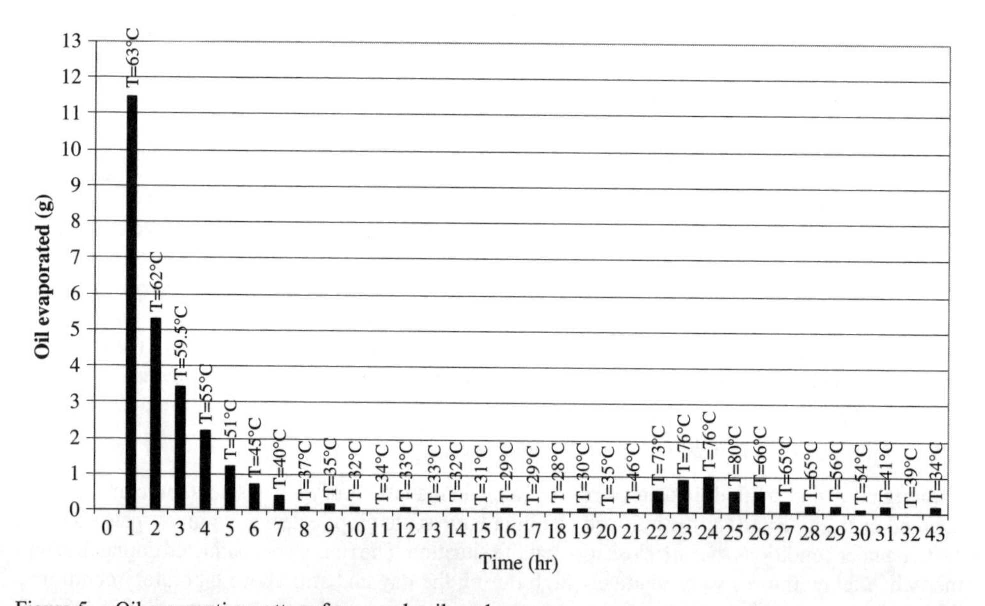

Figure 5. Oil evaporation pattern from crude oil pool.

The high oil sample temperatures recorded were due to the solar radiation during the day and black body nature of crude oil that tends to absorb solar thermal energy. The oil evaporation studies showed that up to 25% of the original amount of crude oil could evaporate within 24 hours (essentially light fractions).

3.3 *Trends in carbon dioxide evolution in various biopile systems*

One of the prime objectives of this investigation is to evaluate biological activities of biopile systems that have been biostimulated, bioaugmented or had a surfactant added to enhance bioavailability of nutrient. Details of such systems have been presented in section 2.3.

The carbon dioxide that evolved from biopile systems A, B, C and D is shown in Figure 6. It is represented by the cumulative CO_2 generation curves. The trends depicted in systems B and D are typical sigmoids and have previously been reported (Benyahia F., et al. 2005). On the other hand, the control system A in which $HgCl_2$ has been added shows a negligible amount of CO_2 that probably indicates that the biocide action of $HgCl_2$ has not been 100% effective. System C which represents the biological activity of indigenous bacteria in the presence of N-P-K nutrients through the curve just above that of system A, suggests that either the initial concentration of biomass was insufficient or a strong form of inhibition by crude oil components applied.

Indeed, the amount of CO_2 in system C marginally exceeded that of system A. This result is a confirmation of the poor nature of the UAE soils in terms of hydrocarbon degrading bacteria and essential nutrients.

On the other hand, system B which was bioaugmented by means of Amnite P300, displayed a healthy respiration rate that exceeded that of system D which had an anionic surfactant added (Tween 80). The effect of the surfactant can clearly be seen in Figure 6: production of CO_2 started earlier that any other system and the cumulative amount of CO_2 also exceeded all systems up to 2400 hours (100 days), after which biological activity subsided.

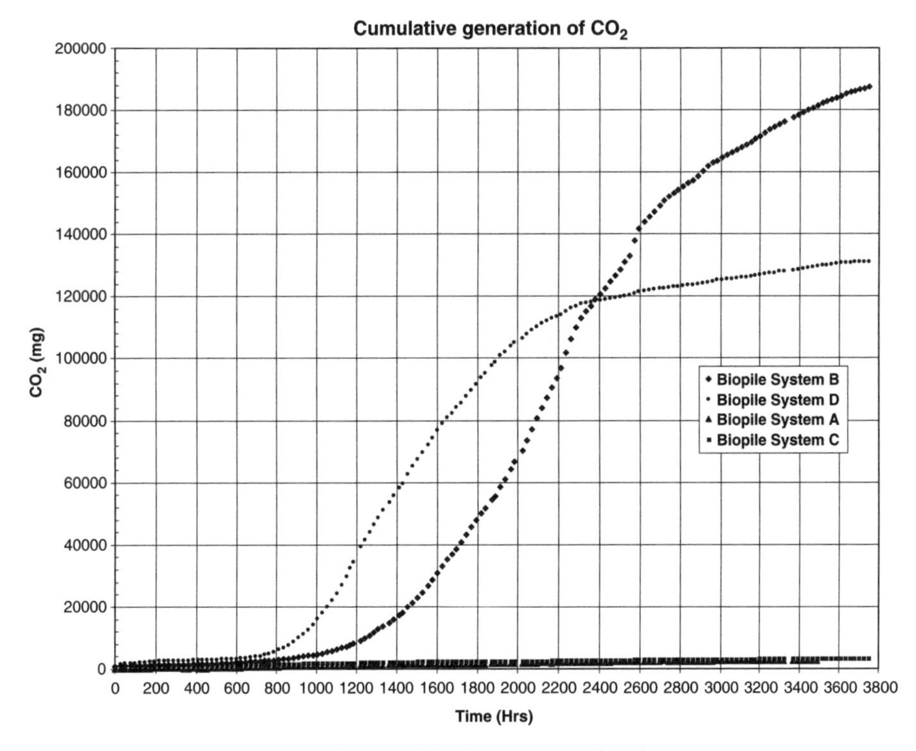

Figure 6. Cumulative CO_2 generation from the biopile systems employed.

The CO_2 produced in systems B and D also included a small amount produced by the hydrolysis of urea that was added as a sources of Nitrogen. This was confirmed by the detection of small amounts of ammonia in the biopiles exhaust and the high pH values of the soil after treatment. The urea hydrolysis is thought to have been catalyzed by the enzyme urease produced by the bacteria in the soils.

4 CONCLUSIONS

The success towards design of bioremediation systems to treat crude oil contaminated soils crucially depends on the outcome of a series of staged experiments involving soil characterization, oil evaporation and effects of biological and chemical additives to the contaminated soils.

The texture, chemical analysis and microbiological character of soils will dictate the level of humidification, biostimulation and bioaugmentation required.

Biopile systems have been proved to be effective, controllable and environmentally friendly techniques to treat contaminated soils. Extensive laboratory tests have shown that such systems can be optimized for treatment efficiency and design cost. Ultimately, it is expected that the current landfarming practices would be phased out because of their poor treatment efficiency and displacement of part of the problem from soil to atmosphere.

This investigation has clearly shown the benefits of bioaugmentation in the bioremediation of crude oil contaminated desert soils. The addition of an anionic surfactant enhanced the bioavailability of nutrients and resulted in a faster response to the biological treatment of crude oil contaminated desert soil.

ACKNOWLEDGEMENTS

The authors would like to acknowledge the technical assistance of Ahmed Embaby and Bassam Hindawi in this work. The financial support of the UAE University research sector to this work under grant 03-7-12/02 is also gratefully acknowledged.

REFERENCES

Abdulkarim, M., Benyahia, F., Zekri, A., Chaalal, O. and Hasanain, H. 2005. Insight into crude oil evaporation modeling and environmental impact assessment following inland oil spills in the Arabian Gulf region. *7th World Congress of Chemical Engineering,* Glasgow (UK). ISBN 0 85295 494 9.

Al-Daher, R., Al-Awadi, N. and Al-Nawawy. A. 1998. Bioremediation of damaged desert environment using the windrow soil pile system in Kuwait. *Environment International* vol 24 (1/2) pp175–180.

Balta, M.T., Al-Daher, R., Al-Awadhi, N., Chino, H. and Tsuji, H. 1998. Bioremediation of oil contaminated desert soil: the Kuwaiti experience. *Environment International* vol 24 (1/2) pp163–173.

Benyahia, F., Abdulkarim, M., Zekri, A., Chaalal, O. and Hasanain, H. 2005. Bioremediation of crude oil contaminated soils: a black art or an engineering challenge?. *Process Safety and Environmental Protection* vol 83 (B4) pp364–370.

Del Arco, J.P. and de Franca F.P. 1999. Biodegradation of crude oil in sandy sediment. International *Biodeterioration and Biodegradation* vol 44 pp87–92.

Del Arco, J.P. and de Franca F.P. 2001. Influence of oil contamination levels on hydrocarbon biodegradation in sandy sediment. *Environmental Pollution* vol 110 pp515–519.

Hejazi, R.F., Husain. T. and Khan, F. 2003. Landfarming operation of oily sludge in arid region: human health risk assessment. *Journal of hazardous materials* vol B99 pp287–302.

Kodres, C.A. 1999. Coupled water and air flows through a bioremediation soil pile. *Environmental Modelling and Software* vol 14 pp37–47.

Pletsch, M., Santos de Araujo, B. and Charlwood B.V. Novel biotechnological approaches in environmental remediation research. 1999. *Biotechnology Advances* vol 17 pp679–687.

Vogel, T.M. 1996. Bioaugmentation as a soil bioremediation approach. *Current opinion in biotechnology* vol 7, pp311–316.

Ward, O., Singh, A. and van Hamme, J. 2003. Accelerated biodegradation of petroleum hydrocarbon waste. *J. Ind. Microbiol. Biotechnol.* vol 30 pp260–270.

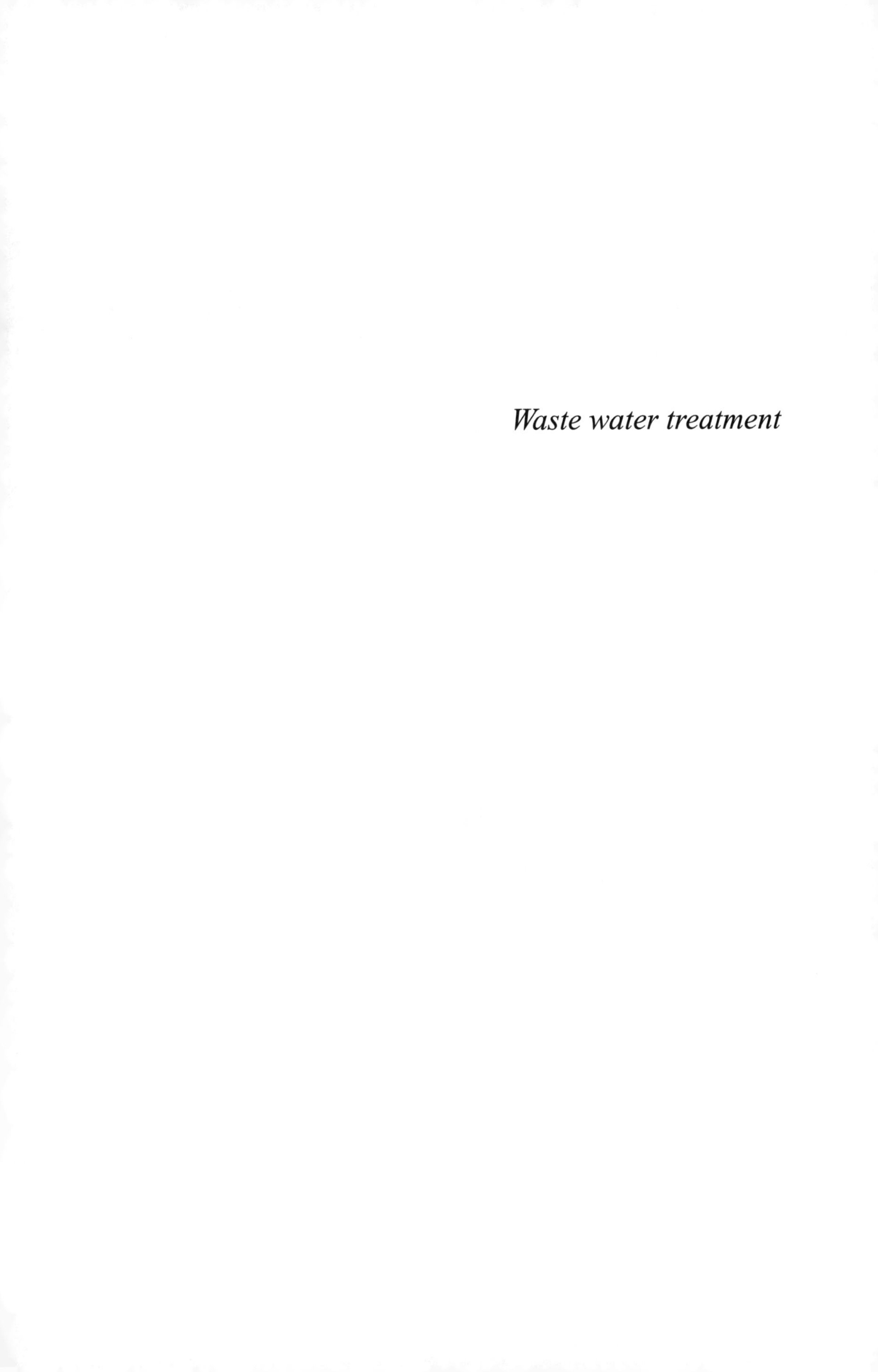

Waste water treatment

Reclaiming the Desert: Towards a Sustainable Environment in Arid Lands – Mohamed (ed.)
© 2006 Taylor & Francis Group, London, ISBN 0 415 41128 9

Studies leading to a pilot process for the removal of dibenzothiophene from solutions

K.Y. Mataqi
Kuwait Institute for Scientific Research, Al-Shuwaikh, State of Kuwait

ABSTRACT: The separation of dibenzothiophene (DBT) from hydroxybiphenyl (HBP) mixtures was attained by the adsorption of HBP with alumina and the elution of DBT with toluene. Alumina (5 g) can adsorb HBP up to 40 mM. The HBP was removed, in a controlled way, by elution with dichloromethane. DBT was extracted from DBT-toluene mixture with deionised water. A pilot system was designed after combining the above steps and characterized by: (1) no significant increase in the conversion of DBT to HBP when the air distribution (10%) was throughout the chemostat rather than as one central supply. (2) as the number of immobilised cells added to the system increases the conversion of DBT to HBP increases. To use all the DBT added to the system the DBT has been recycled. But when the recycled DBT compared with the freshly DBT (6.0 mM) added to the system in the conversion to HBP the fresh DBT showed a constant conversion of HBP, whereas, the conversion to HBP was reduced when the recycled DBT was added.

1 INTRODUCTION

Early in the development of the petroleum industry neither the petroleum technologists nor microbiologists were aware of a requirement for knowledge of microbial physiology and biochemistry. Microbial applications were introduced as the industry developed and became not only more complex but more scientifically sophisticated (Davis, 1967).

Sulphur dioxide released into the atmosphere from the combustion of both crude oil and coal remains one of the most prominent and intractable issues of environmental concern, as it is a major contributor to the air pollution. It is harmful to humans, as it is a respiratory irritant, and in the presence of water generates acid rain, which attacks buildings, causes pipe corrosion, soil pollution, kills trees, and pollutes lakes (Chang, et al., 2000).

The compound dibenzothiophene (DBT) has become an accepted model representative of the recalcitrant thiophenic structures found in coal and crude oils (40% and 60% of the total organic sulphur respectively). Furthermore, DBT seems to be the most difficult of all organosulphur compounds to remove by hydrodesulphurization and it is the major polluting component of coal and crude oils. So DBT has received most attention in recent biodesulphurization studies (Oldfield, et al., 1997; Kilbane and Bielaga, 1988).

HBP is the final desulphurization product when DBT is the sulphur source for growth and in many reports end products can inhibit the utilization of the initial substrate (Omri, et al., 1992;Chang, et al., 2000). So any process developed for the removal of DBT from solution requires the removal of HBP in order to optimize the desulphurisation. Therefore, some requirements must be satisfied before developing such a system. Firstly, a method for rapid and accurate quantification of DBT and HBP will be developed using gas liquid chromatograph (GC) with a specific internal standard (benzylphenylsulphide).

Microorganisms used in any industrial process need to be separated from the product, either by filtration or centrifugation. Thus the removal of the organisms can be the first step of the downstream processes required to purify the product. A good solution is to immobilize the cell/enzyme to enable easy recovery of the cell and reuse of immobilized cell/enzyme.

Immobilization is defined as the ability to physically confine or localize in a defined region or space cells or enzyme molecules whilst retaining the catalytic activities, for repeated and continuous use. There are three major techniques used for cell immobilization: adsorption, entrapment, and coupling. The mechanisms for each technique are quite different (Corcoran, 1983).

2 AIM

To utilize immobilized cells to develop a prototype process to remove sulphur from DBT.

3 MATERIALS AND METHODS

3.1 *Bacterial strains*

Bacterial strains used were from the biotechnology department in Kuwait Institute for Scientific Research.

3.2 *Minimal medium*

Content	gl^{-1}
KH_2PO_4	0.5
K_2HPO_4	4.0
NH_4Cl	1.0
$MgCl_2 \cdot 6H_2O$	0.1
Trace element solution	10 ml
Vitamin solution	1.0 ml

The pH was adjusted to 7.5 with KOH (2 M) and then sterilized by autoclaving at 121°C for 15 min. Glucose (22 mM) was sterilized and added aseptically to the cooled minimal medium to give a final concentration of 20 mM. A 0.5% w/v solution of the sulphur source (DBT) was prepared in ethanol and filter-sterilized using a 0.22 μm Millipore filter. The sterilized DBT (33 mM) were transferred aseptically into the medium to give a final concentration of 31 mM. Agar plates were prepared as above by adding agar (1.5% w/v) to the above minimal medium prior to sterilization. After cooling to 45°C in a water bath, the plates were poured and dried overnight at 37°C.

3.3 *Trace element solution*

The trace element solution contained the following:

Content	gl^{-1}	Content	gl^{-1}
NaCl	1.00	$CuCl_2$	0.05
$CaCl_2$	2.00	$NaMO_4 \cdot 2H_2O$	0.10
$FeCl_2 \cdot 6H_2O$	0.50	$Na_2WO_4 \cdot 2H_2O$	0.05
$ZnCl_2$	0.50	Concentrated HCl	10 ml
$MnCl_2 \cdot 4H_2O$	0.50		

The salts were dissolved in the order shown above. After the addition of the first component the pH was adjusted to 6.0 with KOH (2 M). This was repeated after each compound was added. The solution was autoclaved at 121°C for 20–30 min. When the sterilized trace element solution had cooled 10 ml were transferred aseptically into the medium.

3.4 *Vitamin solution*

The vitamin solution contained the following:

Content	$mg\,l^{-1}$
Ca-pantothenate	400
Inositol	200
Niacin	400
P-aminobenzoate	200
Vitamin B12	0.5

The pH was adjusted to 7.5 with KOH (2 M) and autoclaved at 110°C for 20–30 min. When the sterilized vitamin solution had cooled 1.0 ml was transferred aseptically into the medium.

3.5 *The modified minimal medium*

This medium is the same as the one described above but the final concentration in the medium for glucose was lowered to 10 mM and the DBT lowered to 6.0 mM.

3.6 *Use of Gas Chromatograph-Mass Spectroscope (GC-MS)*

Samples (3 ml) of culture filtrates were extracted three times with methylene chloride (0.1% and 0.3 ml). The mixture was allowed to stand and the supernatants containing DBT were collected carefully and pooled. Gas Chromatograph-Mass Spectroscope (GCMS) analysis was carried out in a Shimadzu model 6000 with a HP5972 series mass selective detector with a 5890 series II Gas chromatograph. The scanning mass range used was 10.00 to 700 amv.

The column (3.6 m by 0.3 mm diameter) was packed with silicone (0.1 μm thickness of the film). The silicone column was washed with methylene chloride (0.1%). The carrier gas (nitrogen) was set at a flow rate of 1 ml min^{-1} at 250°C. DBT extracted from the samples was injected (0.25 ml) in triplicate onto the column with a temperature programme at 260°C (injector and detector) throughout the experiment. To measure the DBT in the samples, known amounts of DBT (0–10 mM) were injected onto the column and the signals recorded and used to produce a standard curve. The peak areas are proportional to concentration of DBT added (see Appendix A). The concentration of the DBT in the steady-state samples was determined using the standard curve.

3.7 *Use of gas chromatograph with an internal standard*

DBT concentration was measured using a gas chromatograph equipped with a flame ionization detector (Shimadzu GC-14A), coupled to a computing integrator (Shimadzu C-R5A). The glass column (1 × 2.0 m × 5 mm o.d. × 3 mm i.d.) was packed with 5% Carbowax 20 M on Chromosorb W-AW 60–80 mesh to fit GC-14A and held at 150°C. Helium (2.0 kg cm^{-Z}) was used as the carrier gas. Benzylphenylsulphide (20 mM) dissolved in ethanol was the internal standard.

A ratio (Q value) was calculated where Q was defined as the peak area for DBT divided by the peak area for internal standard of constant value benzylphenylsulphide (20 mM). A standard curve was produced by preparing known amounts of DBT and plotting the Q values for each value against the concentration of DBT in the solution (see Appendix A).

Unknown samples (1 μl) were added to an equal volume of benzylphenylsulphide (20 mM) and subjected to the same procedure. The concentration of DBT was determined from the Q value standard curve.

3.8 *Estimation of hydroxybiphenyl (HBP) chemical assay*

Two methods were used (i) microplate titre and (ii) standard assay. In both procedures Gibbs reagent was added to a solution expected to contain hydroxybiphenyl at pH 8.0. After mixing the amount of blue-complex formed was measured at $O.D._{610}$.

$$\text{HBP} + \text{Gibbs reagent} \xrightarrow[\text{pH8.0}]{} \text{blue colour}$$

3.9 *Reagents*

- $NaHCO_3$ (38.5 mM) was dissolved in 100 ml of distilled water.
- 2,6 dichloroquinone-4-chloramide (Gibbs reagent) (1 mM) was dissolved in 100 ml of ethanol.

3.10 *Use of microtitre plates*

Flasks (250 ml) containing minimal medium (50 ml) with glucose as carbon source (20 mM), DBT as sulphur source (31 mM) at pH 7.5 were inoculated with one loopful of a liquid culture of the strains to be examined and incubated at 30°C in a shaking water bath for 2 days. A sample (1 ml) of the culture was transferred into an Eppendorf tube and centrifuged at 1000 g for 5 min. The supernatant (150 μl) was transferred to the microtitre plates. Sodium bicarbonate (30 μl) was added and mixed for 3 min to adjust the pH to 8.0. Gibbs reagent (20~t1) was added and after 3 min. mixing, the solution was observed for a blue-coloration. The blue-complex is an indication that HBP was present.

3.11 *Standard assay*

Flasks (250 ml) containing minimal medium (50 ml) with glucose as carbon source (20 mM) and sodium sulphate as sulphur source (0.2 mM) were inoculated with one loopful of a liquid culture of the strain to be examined and incubated at pH 7.5 and at 30°C with shaking for 2 days. The $O.D._{610}$ of the overnight culture was measured and the volume required to inoculate a similar culture to give initial $O.D._{610}$ of 1.0 was calculated.

Flasks (250 ml) containing the same medium (50 ml) but with DBT as sulphur source (31 mM) were inoculated with the volume of inoculum calculated, flasks were incubated at pH 7.5 and at 30°C in shaking water bath for 1 hour. A sample (2 ml) of the culture was transferred into an Eppendorf tube and centrifuged at 1500 rpm for 5 min. The supernatant (1 ml) was transferred to a test tube, $NaHCO_3$ (200 μl) was added and mixed for 3 min. to adjust to pH 8.0. Gibbs reagent (133 μl) was added and mixed for 3 min.

The $O.D._{610}$ was measured using a spectrophotometer at 610 nm with distilled water as the blank. The O.D. measured was compared with a standard curve prepared with known amount of HBP over the range of 1 to 10 μmoles.

3.12 *Use of gas chromatograph with an internal standard*

The procedure was the same as for quantification of DBT.A standard curve was produced by preparing known amounts of HBP and plotting the Q values for each value against the concentration of HBP in the solution (see Appendix A).

Unknown samples (1 μl) were added to an equal volume of benzylphenylsulphide (20 mM) and subjected to the same procedure. The concentration of HBP was determined from the Q value standard curve.

4 IMMOBILIZATION OF MICROBIAL CELLS

4.1 *Entrapment of cells in beads*

4.1.1 *By barium alginate*

Cells (150 ml) cultured in modified minimal medium were concentrated by centrifugation at 10,000 g for 10 minutes. The precipitate was washed with phosphate free modified minimal medium and recentrifuged, resuspended in 5 ml of this modified medium and mixed with 2.0% (w/v) sodium alginate in 1:3 (v/v) cell:alginate ratio. The beads were formed by carefully dropping the cells-alginate mixture from a syringe (5 mm diameter) close to surface of a $BaCl_2$ (0.1 M) solution in an evaporating dish. The beads were left in $BaCl_2$ (0.1 M) solution for 1–2 hours to harden before they were washed twice with saline (0.9%) to remove the excess Ba^{+2} ions. The beads were kept in a flask (250 ml) with modified minimal medium (50 ml) overnight at 30°C before use.

5 RESULTS

5.1 *Development of a method to quantify dibenzothiophene and hydroxybiphenyl using gas chromatograph with an internal standard*

To quantify the concentrations of DBT and HBP an internal standard must be selected. Therefore, a range of possible standards (benzylphenylsulphide, propanol, hexane, pentane and cyclohexane) of equal concentrations (20 mM) were dissolved in ethanol (5 ml) before being injected (1.0 μl) onto the GC column using the described standard conditions. The retention times for those compounds together with the different compounds which required quantification (seen in bold Table 1) were recorded. The internal standard chosen because of it's retention time being significantly different from that of ethanol, toluene, DBT or HBP was benzylphenylsulphide. To quantify DBT and HBP in unknown solutions, samples with known amounts of DBT or HBP together with a concentration of the internal standard (20 mM) were injected (1.0 μl) onto the GC column using the described standard conditions. The areas of the peaks were measured for DBT and HBP and divided by the areas of the internal standard peaks. This gives Q values for the peaks. A standard curve of Q values versus actual amounts of DBT or HBP in the samples was drawn. This standard curve can be used to quantify any solution containing DBT and/or HBP as long as a constant amount of internal standard is added (see Appendix A) and the conditions maintained.

Table 1. The retention time for different compounds with gas chromatograph.

Compound	Retention time (minutes)
Benzylphenylsulphide	1.4
Toluene*	4.6
Propanol	4.5
Hexane	12.4
Pentane	12.3
Cyclohexane	12.6
Dibenzothiophene*	12.5
Hydroxybiphenyl*	4.7
Ethanol*	0.4

The gas chromatograph was equipped with a flame ionization detector (Shimadzu GC-14A), coupled to a computing integrator (Shimadzu C-R5A). The glass column (1 × 2.0 m × 5 mm o.d. × 3 mm i.d.) was packed with 5% Carbowax 20 M on Chromosorb W-AW 60–80 mesh to fit GC-14A and held at 150°C. Helium (2.0 kg cm^{-2}) was used as the carrier gas at a flow rate of 1 ml min^{-1} at 250°C.
* Compounds requiring quantification.

5.2 *The assessment of the capacity of alumina to bind hydroxybiphenyl*

To determine whether alumina can adsorb HBP and then assess the capacity of alumina for this ability, flasks (150 ml) with a known amount of alumina (5 g) and toluene (10 ml) were shaken (50 rpm) at room temperature to make a slurry. Different concentrations of HBP, in duplicate, (1.00, 5.00, 10.0, 15.0, 20.0, 40.0 or 50.0 mM) were dissolved in ethanol (10 ml) before being added individually to the flasks. The flasks were shaken gently (50 rpm) at room temperature for 10 minutes to allow HBP to be adsorbed onto the alumina. The contents of the flasks were allowed to settle for 1 hour. Samples (lml) of the supernatant were collected and tested for the presence of HBP using the Gibbs reaction. When no blue colouration was developed it was assumed that all HBP added had been adsorbed. Table 2 shows that 5 grams of alumina adsorbed up to 40 mM HBP. When 50 mM HBP was added to the alumina when the supernatant was mixed with the Gibbs reagents a blue colouration developed indicating HBP was present in the supernatant.

To confirm this, a column (80 cm × 3.0 cm diameter) with a tap was packed with a slurry of alumina (5 g) in toluene (10 ml). Different concentrations of HBP (20, 40, 60 or 80 mM) dissolved in ethanol (5 ml) were individually added and allowed to drain into the column. Before the top of the column became dry and cracks and/or bubbles occurred toluene was added to elute the column. Samples (1 ml) were collected at known time intervals from the column and tested for the presence of HBP. No HBP was detected when the concentration of HBP added was between 20 and 40 mM, but HBP was detected when the concentration was between 60 and 80 mM. This confirmed that 5 grams of alumina can adsorb 40 mM of HBP.

5.3 *The use of dichloromethane to remove hydroxybiphenyl from an alumina column*

To use an alumina column continuously to separate the HBP produced from DBT, a method is required for the controlled removal of HBP from the column. A column was packed carefully with alumina (5 g) in deionised water (10 ml). Different concentrations of HBP, in duplicate (1.0, 3.0, 6.0, 10, 20, 30 or 40 mM) dissolved in ethanol (5 ml) were added individually onto the top of the column and allowed to drain into the column. Dichloromethane (10 ml) was added onto the column to elute the HBP from the alumina. Samples (0.2 ml) were taken at known time intervals to measure the concentration of HBP using the GC. The results are shown in Table 3.

The results showed that dichloromethane elutes HBP with 100% recovery over the range of concentrations tested.

5.4 *The removal of toluene from dibenzothiophene with deionised water*

To remove toluene from DBT/toluene mixture, deionised water (20 ml) in a separating funnel (150 m1) was utilized using liquid/liquid extraction. Different concentrations of DBT, in duplicate

Table 2. The capacity of alumina to bind hydroxybiphenyl.

Hydroxybiphenyl	Gibbs Assay
1.0	−ve
5.0	−ve
10	−ve
15	−ve
20	−ve
40	−ve
50	+ve

Alumina (5 g) in toluene (10 ml) was added to flasks (150 m1) to make slurry with gentle shaking (50 rpm) at room temperature.
Different concentrations of HBP, in duplicate, (1.0, 5.0, 10, 15, 20, 40 or 50 mM) were dissolved in ethanol (10 ml) before being added individually to the flasks.
−ve = no colour (i.e. no HBP), +ve = blue colour (i.e. HBP present).

(1.0, 3.0, 6.0, 10 or 20 mM) in toluene (5.0 mM): (10 ml) were added individually into the deionised water with gentle shaking (50 rpm). The mixture (DBT/toluene mixture and the deionised water) was left for 1 hour to settle at room temperature. Two layers were formed and samples (0.2 ml) from each layer were taken and analyzed for DBT and toluene using the GC. The results are shown in Table 4. The results showed that two different layers formed in the separating funnel. The top layer was toluene (5.0 mM) (10 ml) and the bottom layer was the solution of deionised water (20 ml) with the different concentrations of DBT used.

5.5 *The time required to wash out dibenzothiophene from the bioreactor*

To measure the time required to wash out a known concentration of DBT added to the bioreactor, Ba^{+2}-alginate beads (190 beads) were prepared with a mixture containing cells (2.1×10^{11}cells ml^{-1}). The beads were incubated as a batch culture in a bioreactor (working volume 800 ml) containing the standard modified minimal medium at optimum physiological conditions; pH 7.5, 30°C, and oxygen tension 10% air saturation. The beads were incubated for 10 hours before the pump associated with a reservoir of the same medium (with no added DBT) was switched on at the chosen dilution rate ($0.13\,h^{-1}$). The culture was allowed to grow as a continuous culture for a further 8 hours. Samples (1 ml) were taken at known time intervals from the bioreactor culture filtrate to measure the concentration of DBT using the GC. The results are shown in Table 5.

5.6 *The effect of recycling dibenzothiophene back into the system*

To study the effect of recycling the DBT, after separation from toluene, back into the system, Ba^{+2}-alginate beads (308 beads) were prepared with a mixture containing a known cell count

Table 3. The removal of different concentrations hydroxybiphenyl by dichloromethane from the column.

HBP*	HBP**
1.0	0.99
3.0	2.98
6.0	5.97
10.0	10.0
20.0	19.99
30.0	29.97
40.0	40.0

Alumina (5 g) in toluene (5 mM) (10 ml) was added to a column.
Different concentrations of HBP, in duplicate, (1.0, 3.0, 6.0, 10.0, 20.0, 30.0 or 40 mM) were dissolved in ethanol (5 ml) before added to the column.
* HBP = hydroxybiphenyl measured using GC before the addition to the column.
** HBP = hydroxybiphenyl measured using GC after eluted from the column.

Table 4. The removal of toluene from different concentrations of dibenzothiophene by deionised water.

	Toluene (mM)					DBT (mM)				
Top layer	5.0	5.0	5.0	5.0	5.0	–	–	–	–	–
Bottom layer	–	–	–	–	–	1.0	3.0	6.0	10	20

A mixture of toluene (5.0 mM) (10 ml) with different concentrations of DBT in duplicate (1.0, 3.0, 6.0, 10 or 20 mM) and deionised water (20 ml) were added into a separating funnel and allowed to separate into two distinct layers.
The DBT and toluene measured using the GC.

Table 5. The time required to wash out dibenzothiophene from bioreactor.

Time (hour)	DBT (mM)
0.00	6.00
*10.0	3.20
12.0	1.60
14.0	0.65
16.0	0.00
18.0	0.00

DBT = Dibenzothiophene.
Beads containing immobilized cells (2.1×10^{11} cells ml^{-1}) were incubated in bioreactor culture containing standard modified minimal medium minus DBT pH 7.5, at 30°C and oxygen tension 10% air saturation.
DBT was measured using the GC.
*The pump switched on at $D = 0.13\,h^{-1}$ with the same medium but with no DBT added.
The results showed that it requires 16 hours to wash out a known concentration of DBT (6 mM) from a bioreactor with a working volume of 800″ and a cell content of 2.1×10^{11}cells ml^{-1}.

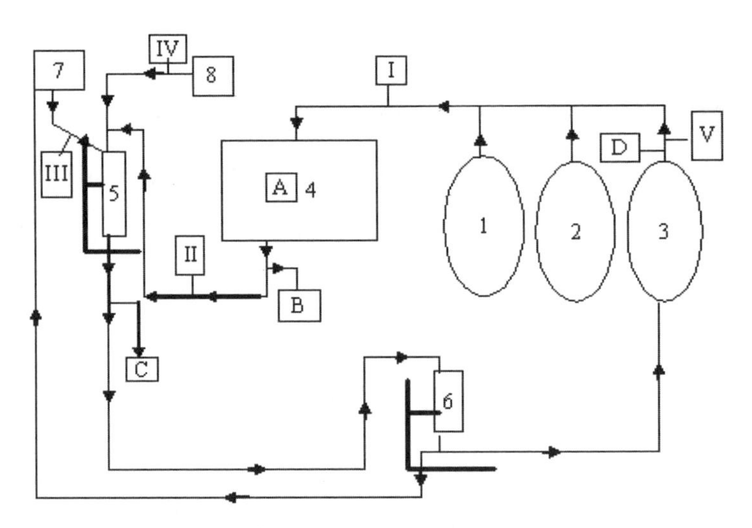

Figure 1. Schematic diagram of a bioreactor system.

(4.4×10^{12} cells ml^{-1}). The beads were incubated as a batch and then as a continuous culture in a bioreactor at the optimum physiological conditions; pH 7.5, 30°C and with oxygen tension 10% air saturation. The system was built up as described above and is shown diagrammatically in Fig. 1. To keep the concentration of DBT (6 mM) constant in the bioreactor, DBT (fresh or recycled) was added to the bioreactor during the incubation period (9 days). To maintain the concentration of DBT in the bioreactor at 6 mM the flow rates of the fresh and the recycled DBT were different (0.13 and $0.27\,h^{-1}$ respectively). This is because the concentration of the recycled DBT is less than 6 mM. The pumps associated with the reservoirs of toluene and dichloromethane were switched on and off for eluting DBT and HBP respectively at known. Samples (1 ml) were taken at known time intervals from the culture filtrate of the bioreactor, from the eluant of the column, and from the separating funnel and analyzed for DBT and HBP using the GC. The results are shown in Table 6.

Table 6. The effect of using the recycled dibenzothiophene on the system.

Time (hour)	DBT^1 (mM)	DBT^2 (mM)	DBT^3 (mM)	HBP^1 (mM)	HBP^2 (mM)	HBP^3 (mM)	HBP^4 (mM)
0.00	6.00	0.00	0.00	0.00	0.00	0.00	0.00
*9.00	3.25	0.00	0.00	2.66	2.70	0.00	0.00
♣♦12.0	2.30	2.00	1.65	3.60	3.65	0.00	3.55
*24.0	2.40	2.10	1.60	3.55	3.60	0.00	3.60
*36.0	2.30	2.15	1.55	3.65	3.65	0.00	3.55
*48.0	2.35	2.10	1.60	3.60	3.65	0.00	3.60
72.0	**2.80**	**2.70**	**2.40**	**1.85**	**1.80**	**0.00**	**1.70**
*96.0	2.35	2.05	1.70	3.55	3.60	0.00	3.50
120	**3.30**	**3.25**	**3.10**	**1.45**	**1.50**	**0.00**	**1.30**
* 144	2.70	2.45	2.20	3.25	3.15	0.00	3.15
168	**5.20**	**5.05**	**4.85**	**1.35**	**1.40**	**0.00**	**1.30**
♥*192	2.70	2.55	2.35	3.20	3.25	0.00	3.10

DBT^1 = Dibenzothiophene measured from the culture filtrate of the bioreactor, DBT^2 = Dibenzothiophene measured from the alumina column, DBT^3 = Dibenzothiophene measured after the removal of toluene, HBP^1 = Hydroxybiphenyl measured from the culture filtrate of the bioreactor using Gibbs assay, HBP^2 = Hydroxybiphenyl measured from the culture filtrate of the chemostat using the GC, HBP^3 = Hydroxybiphenyl measured from the alumina column using the GC, HBP^4 = Hydroxybiphenyl measured from the alumina column after switching the dichloromethane pump on.

The culture (*Rhodococcus erythropolis*) was grown in the standard modified minimal medium pH 7.5 containing glucose (10 mM) as a carbon source and DBT (6 mM) as sole source of sulphur. The culture was maintained at 30°C and 10% air saturation.

*The pump turned on at $D = 0.13\,h^{-1}$ containing the fresh medium plus fresh DBT (6 mM).

The numbers in bold are the results of the fresh medium plus the recycled DBT/deionised water mixture while the fresh DBT was switched off.

♣ The pump of toluene switched on for 10 minutes.

♦ The pump of the dichloromethane switched on for 10 minutes.

The cell count = 4.4×10^{12}cells ml^{-1} immobilised as Ba^{+2} - alginate in bead form (308 beads).

♥ The experiment stopped at 216 hours because the beads collapsed.

No HBP was produced from DBT during the first 6 hours of incubation. At 9 hours of incubation the DBT^1 concentration was 3.25 mM and HBP^1 production 2.66 mM was equal to the DBT removed from the bioreactor. During the period from 24 to 48 hours of incubation and with the addition of the same medium and the fresh DBT (6 mM) the conversion rate of DBT to HBP was constant (3.6 ± 0.05 mM) (see Table 6). Whereas when the same medium and the recycled DBT/deionised water mixture was added into the chemostat vessel the conversion rate was reduced to 1.4 mM after 168 hours of incubation (see in bold Table 6). This reduction may be due to the fact that the recycled DBT is added with deionised water and this reduces the concentration of nutrient in the bioreactor. After 120 hours of incubation the conversion rate of DBT to HBP was reduced with the addition of either the fresh DBT or the recycled DBT. The experiment was stopped at 216 hours because the beads had collapsed.

5.7 *The effect of different dilution (growth) rates on the conversion of dibenzothiophene to hydroxybiphenyl*

To investigate the limitations of the pilot process, Ba^{+2}-alginate beads (210 beads) were prepared with a mixture containing a cell count of 2.0×10^{10} cells ml^{-1}. The beads were incubated as a continuous culture in a bioreactor (working volume 800 ml) containing the standard modified minimal medium at the optimum physiological conditions; pH 7.5, 30°C, and oxygen tension 10% air saturation. The system was built up as described above. The dilution rate (D) was progressively

Table 7. The effect of different dilution (growth) rate on the conversion of dibenzothiophene to hydroxybiphenyl.

D (h^{-1})	Time (hour)	DBT[1]	DBT[2]	DBT[3]	HBP[1]	HBP[2]	Conversion of DBT to HBP* (%)
0.13	12	2.65	2.60	2.60	3.30	3.25	43.74
	24	2.70	2.70	2.65	3.25	3.25	
	36	2.75	2.65	2.60	3.25	3.20	
	48	2.70	2.70	2.60	3.30	3.20	
	12	2.40	2.40	2.40	3.55	3.50	
0.15	24	2.50	2.45	2.40	3.50	3.50	40.70
	36	2.40	2.35	2.35	3.50	3.45	
	12	2.20	2.10	2.10	3.75	3.75	
0.17	24	2.20	2.15	2.10	3.80	3.80	38.68
	36	2.15	2.10	2.10	3.80	3.70	
	48	2.15	2.10	2.10	3.80	3.70	
	12	4.00	4.00	3.95	1.90	1.90	
0.20	24	4.00	4.00	4.00	1.95	1.90	16.64
	36	4.00	4.00	3.95	1.90	1.90	

DBT[1] = Dibenzothiophene measured from the culture filtrate of the bioreactor, DBT[2] = Dibenzothiophene measured from the alumina column, DBT[3] = Dibenzothiophene measured after the removal of toluene, HBP[1] = Hydroxybiphenyl measured from the culture filtrate of the bioreactor, HBP[2] = Hydroxybiphenyl measured from the alumina column. The culture (*Rhodococcus erythropolis*) was grown in the standard modified minimal medium pH 7.5 containing glucose (10 mM) as a carbon source and DBT (6 mM) as sole source of sulphur. The culture was maintained at 30°C and 10% air saturation, and at different dilution (growth) rate (0.13, 0.15, 0.17, and 0.20 h^{-1}).
The pump of the feed line switched on after 5 working volumes.
The pump of toluene switched on for 10 minutes.
The pump of the dichloromethane switched on for 10 minutes.
The cell count = 2.0×10^{10} cells ml^{-1} immobilised as Ba^{+2} - alginate in bead form (210 beads).
*The conversion percentage of DBT to HBP = The total HBP produced during the period (mM) as a percentage of DBT (mM) added to the culture vessel during the same period.

increased from 0.10 to 0.20 h^{-1}. Samples (lml) were taken at known time intervals from the culture filtrate of the bioreactor, from the eluant of the column, and from the separating funnel and analyzed for DBT and HBP using the GC. The results are shown in Table 7.

As reported in the previous experiment the optimum dilution (growth) rate was 0.13 h^{-1} when the HBP production was the highest (3.25 ± 0.05 mM) and the conversion percentage of DBT to HBP of 43.74%.

6 CONCLUSION

The separation of DBT from HBP by the use of alumina column initially eluted with toluene was important for the development of a pilot plant process. The removal of the HBP bound to the alumina column was controlled by the rate of elution with dichloromethane. Finally, the DBT was shown to be removed from the DBT-toluene mixture by extraction with deionised water. With no pilot system reported in the literature for the removal of DBT from liquid media these steps (the growth in bioreactor, the separation of DBT from HBP using alumina column and the extraction of DBT from toluene with deionised water in a separating funnel) (see Figure 1) were combined to develop a complete treatment for the removal of a known concentration of fresh DBT with immobilized cells in carbon-limited minimal medium. DBT was recycled back into the system in a controlled way with high recovery. The limitations of this system were reported in this study showing that the percentage conversion of DBT to HBP (43.74%) was achieved with the optimum dilution rate ($0.13\,h^{-1}$) (see Table 7). Furthermore the optimum physiological conditions reported could account for the removal of 21 mM DBT from the medium. The use of a larger alumina column and further modification of pump rates may well increase the amount of DBT that can be removed from the system. This approach using immobilized wild type organisms has resulted in a higher conversion rate of DBT to HBP than that obtained by the use of genetically engineered organisms.

7 FUTURE WORK

There are a number of ways to continue the work started in this project e.g.:

1. Scale up the system for field work.
2. Other organic sulphur sources could be examined such as: light gas oil, diesel, etc.
3. A wider range of microbes to be examined using the different organic sulphur sources.

APPENDIX A

Q value Against Concentration of Hydroxybiphenyl

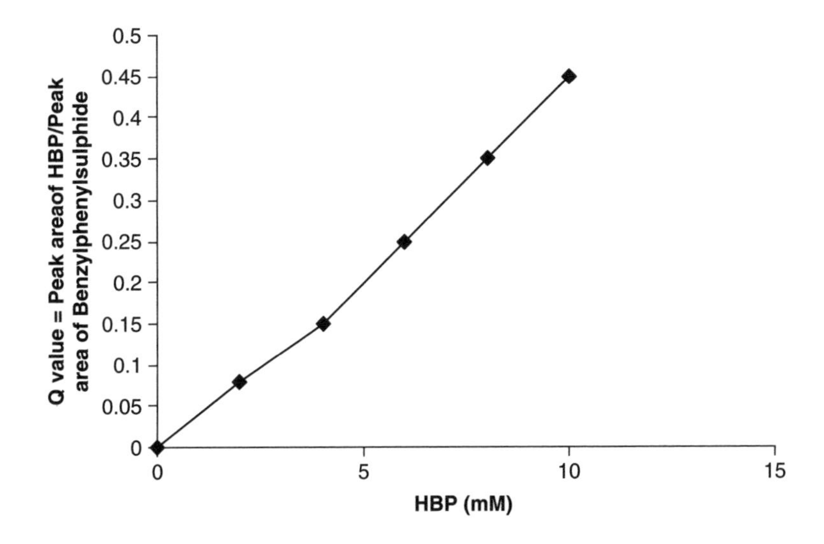

REFERENCES

Chang, J. H., Chang, Y.K., Ryu, H. W. and Chang, H. N. 2000. Desulphurisation of Model and Diesel Oils by Resting Cells of *Gordona sp. FEMS Microbiology Letters.* 182: 309–312.

Corcoran, E. 1983. *The Production and use of Immobilized Living Microbial Cells.* In: Topics in Enzyme and Fermentation Biotechnology (Wiseman, A., ed), Vol. 10: pp. 12*50,* Wiley, New York.

Guiseley, K. B. 1989. Chemical and Physical Properties of Algal Polysaccharides used for Cell Immobilisation. *Enzyme and Microbial Technology.* ll: 706–716.

Kilbane, J. J. and Bielaga, B. A. 1988. Toward Sulphur-Free Fuels. *American Chemical Society.*69: 317–330.

Oldfield, C., Wood, N. T., Gilbert, S. C., Murray, F. D. and Faure, F. R. 1997. Desulphurisation of Benzothiophene and Dibenzothiophene by Actinomycete Organisms belonging to the genus *Rhodococcus* and Related Taxa. *Microbiology.* 74: 119–132.

Omori, T., Monna, L., Saiki, Y., and Kodama, T. 1992. Desulphurisation of Dibenzothiophene by *Corynebacterium* sp. Strain SY1. *Applied and Environmental Microbiology.* 58: 911–915.

Reclaiming the Desert: Towards a Sustainable Environment in Arid Lands – Mohamed (ed.)
 ISBN 0 415 41128 9

A new thermodynamic model for mixed solvent electrolyte solutions

N. Boukhalfa
Department of Industrial Chemistry, University of Biskra, Algeria

A.H. Meniai
Department of Industrial Chemistry, University of Constantine, Algeria

E.A. Macedo
Laboratory of Separation and Reaction Engineering, University of Porto, Portugal

ABSTRACT: The approach tested in this work consists in adapting the Pitzer model, initially designed for aqueous solutions of electrolytes, to the case of solutions with a mixed solvent. This modified model was applied successfully to correlate the activity coefficients of the electrolytes: CsCl, KCl, LiCl, RbCl, NaCl, and NaBr in methanol-water mixed solvent solutions at 25°C. The experimental data for these systems are from literature. The model parameters are optimized by using the Marquardt-Levenberg minimization method. We have elaborated a computer code in FORTRAN language for this purpose. The results are very encouraging; the new model can correlate the experimental data for all systems studied with good precision.

1 INTRODUCTION

Activity coefficients are one of the most important thermodynamic properties used to analyze problems associated with phase equilibrium of systems in diverse areas such as environmental chemistry, industrial chemistry, biochemistry and many others.

Many petroleum problems would be more easily solved if a reliable and accurate model representative of phase equilibrium of systems including, along with hydrocarbon elements, salts and water and even further additives such as alcohols, was available. Methanol is widely used in petroleum industry to facilitate transport in certain circumstances.

In the present work, we introduce a new approach based on a modification of the Pitzer model (Pitzer 1973, Pitzer & Mayorga 1973, 1974), in order to adapt it to solutions with mixed solvent (formed of water+methanol). The central concept of the modification concerns the definition of a solvent equivalent to the superposition of water and methanol in order to meet the requirements of the Pitzer Model witch is designed for water as single solvent.

We have applied the new approach to correlate the activity coefficients of the electrolytes: CsCl, KCl, LiCl, RbCl, NaCl, and NaBr in methanol-water mixed solvent solutions at 25°C. The experimental data of CsCl, KCl, LiCl and RbCl are those of Koh et al. (1985). For NaCl, we have used the recent data reported by Basili et al. (1996) and for NaBr we have used the experimental data of Han and Pan (1993).

Note that for the experimental data reported by Koh et al. (1985), it is the ratio of activity coefficients of electrolytes in mixed solvents system with respect to aqueous system which is measured.

2 THERMODYNAMIC MODEL

2.1 *Pitzer model*

The Pitzer's equations for the activity coefficient of the electrolyte MX in a solution of molality m and ionic strength I is given by:

$$\ln \gamma_{MX} = |z_M z_X| f^{\gamma} + m\left(\frac{2 \nu_M \nu_X}{\nu}\right) B_{MX}^{\gamma} + m^2 \left[\frac{2(\nu_M \nu_X)^{3/2}}{\nu}\right] C_{MX}^{\gamma} \tag{1}$$

where:

$$f^{\gamma} = -A_{\varphi}\left[\frac{I^{1/2}}{1+b\, I^{1/2}} + \frac{2}{b}\ln\left(1+b\, I^{1/2}\right)\right] \tag{2}$$

$$B_{MX}^{\gamma} = 2\,\beta_{MX}^{(0)} + \frac{2\,\beta_{MX}^{(1)}}{\alpha^2 I}\left[1 - e^{-\alpha I^{1/2}}\left(1 + \alpha I^{1/2} - (1/2)\alpha^2 I\right)\right] \tag{3}$$

$$I = \frac{1}{2}\sum_i m_i\, Z_i \tag{4}$$

I = ionic strength; m_i = ionic molality; Z_i = ionic charge number; γ_{MX} = electrolyte activity coefficient; μ = stoichiometric number; and A_{φ} = Debye-Huckel coefficient.
The optimal values for α and b are 2.0 and 1.2 respectively (Pitzer 1973).

For 1-1 electrolyte, equation (1) can be written as:

$$\ln \gamma_{MX} = f^{\gamma} + m\, B_{MX}^{\gamma} + m^2\; C_{MX}^{\gamma} \tag{5}$$

For electrolytes whose concentrations do not exceed 2 m, the parameter C_{MX}^{γ} may be neglected (Pitzer 1973).

At 25°C, the Debye-Huckel coefficient is written as:

$$A_{\varphi} = 272.058\, d^{1/2}\, D^{-3/2} \tag{6}$$

where d = solvent density; D = solvent dielectric constant. The values used in this work are those of Koh et al. (1985).

From the above, the Pitzer parameters are: $\beta_{MX}^{(0)}$, $\beta_{MX}^{(1)}$ and C_{MX}.

2.2 *Modification of the Pitzer model*

The central concept of the modification of the Pitzer model concerns the substitution of a single equivalent solvent in place of the mixed solvent. This is because it is useful to define a solvent equivalent to the superposition of water and methanol in order to meet the requirements of the Pitzer Model which is designed for water as single solvent. As a consequence, Pitzer's equation may be applied to mixed solvent systems by using the appropriate values for the density and the dielectic constant in equation (6).

3 RESULTS AND DISCUSSION

3.1 *Minimization*

The Modified Pitzer Model was applied to correlate the mean ionic activity coefficients of CsCl, KCl, LiCl, RbCl, NaCl, and NaBr in water-methanol mixed solvent solutions, using the experimental data available in literature. The model parameters are optimized by minimizing the following objective function:

$$OBJ = \sum_{i}^{NP} \left(\ln \gamma_{MX}^{cal} - \ln \gamma_{MX}^{\exp} \right)^2 \tag{7}$$

where NP = number of data points; cal = calculated value; exp = experimental value.

The standard deviation σ is given by:

$$\sigma = \left[\frac{\sum_{i}^{NP} \left(\ln \gamma_{MX}^{cal} - \ln \gamma_{MX}^{\exp} \right)^2}{NP} \right]^{1/2} \tag{8}$$

We have elaborated a computer code in FORTRAN language based on the Marquardt-Levenberg minimization method (Marquardt 1963) in order to optimize the parameters of the new model.

3.2 *Results*

The Figures 1–6 illustrate the results. It can be seen that the approach proposed in this work shows good fit with the experimental data for all systems studied.

The Marquardt-Levenberg minimization method used to optimize the model parameters is very accurate and converges after a few iterations.

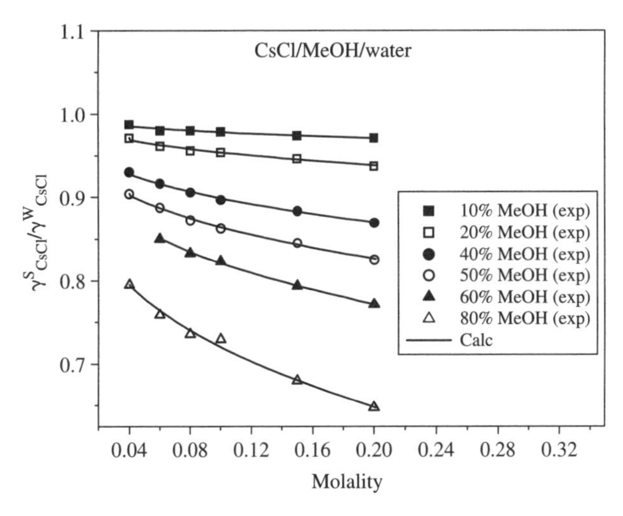

Figure 1. Ratio of activity coefficient of CsCl in water-methanol mixed solvent solution. Experimental data are those reported by Koh et al. (1985).

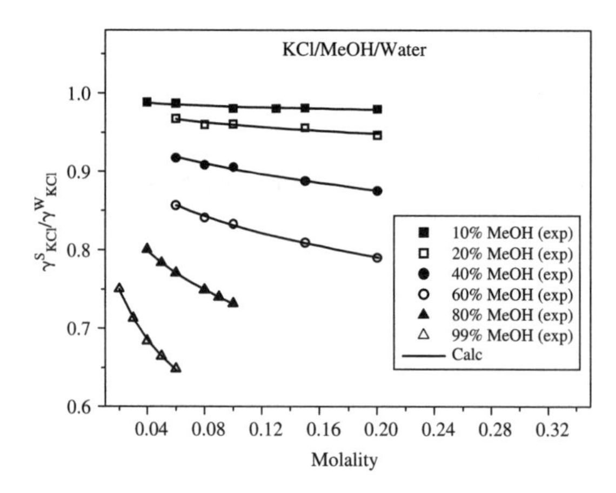

Figure 2. Ratio of activity coefficient of KCl in water-methanol mixed solvent solution. Experimental data are those reported by Koh et al. (1985).

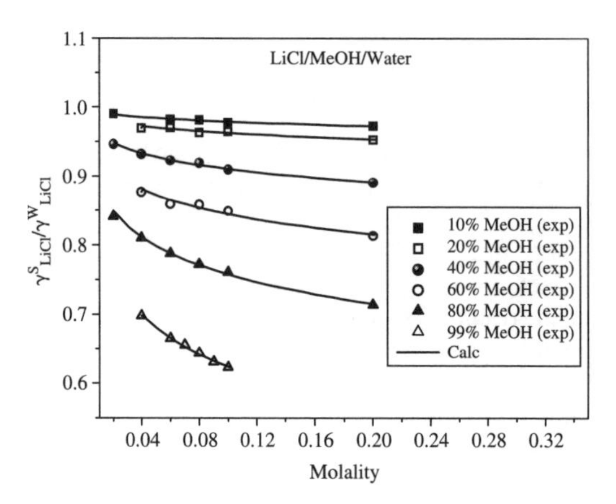

Figure 3. Ratio of activity coefficient of LiCl in water-methanol mixed solvent solution. Experimental data are those reported by Koh et al. (1985).

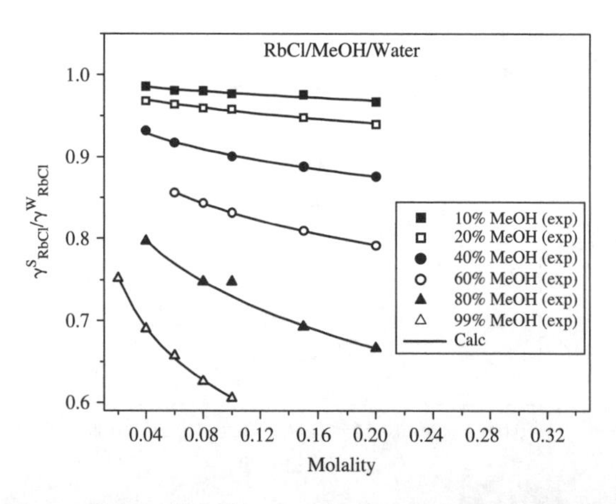

Figure 4. Ratio of activity coefficient of RbCl in water-methanol mixed solvent solution. Experimental data are those reported by Koh et al. (1985).

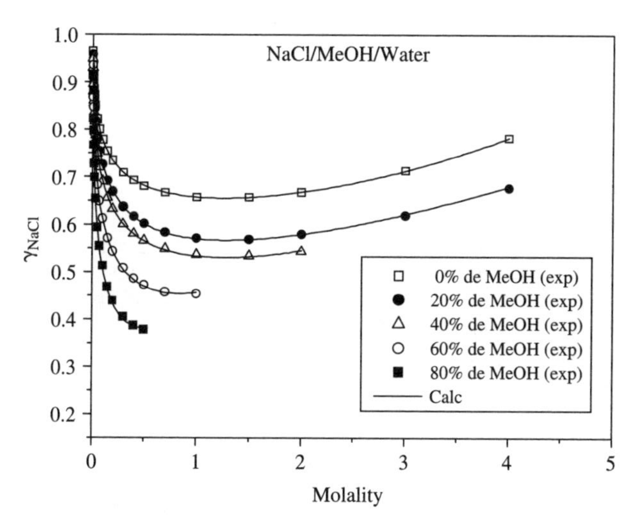

Figure 5. Activity coefficient of NaCl in water-methanol mixed solvent solution. Experimental data are those reported by Basili et al. (1996).

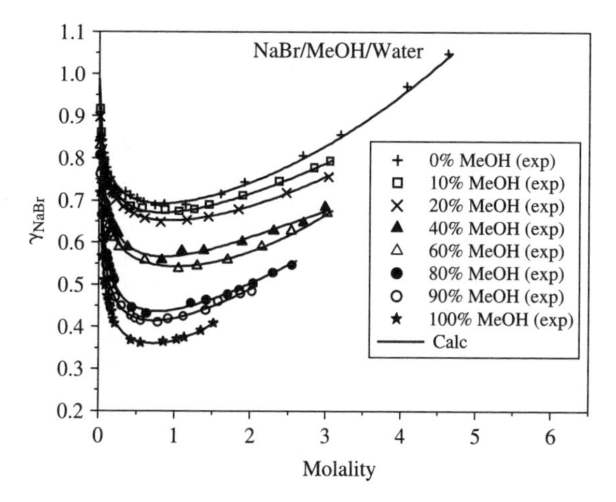

Figure 6. Activity coefficient of NaBr in water-methanol mixed solvent solution. Experimental data are those reported by Han & Pan (1993).

The model parameters optimized are presented in Table 1. The standard deviations obtained with the model proposed in this work are also included in Table 1.

As it can be seen, the standard deviations indicate a good precision of the new model.

4 CONCLUSION

In this work, we have applied a new approach to correlate the activity coefficients of the electrolytes: CsCl, KCl, LiCl, RbCl, NaCl, and NaBr in methanol-water mixed solvent solutions at 25°C. The Marquardt-Levenberg minimization method is used to optimize the parameters of the new model. The results obtained by the new approach in the case of the systems studied seem very promising.

Table 1. Values of $\beta^{(0)}_{MX}$, $\beta^{(1)}_{MX}$ and C_{MX} and the standard deviation σ for CsCl, KCl, LiCl, RbCl, NaCl, and NaBr in methanol-water mixtures.

Electrolyte	%MeOH	m. max	$\beta^{(0)}_{MX}$	$\beta^{(1)}_{MX}$	C_{MX}	$10^3\ \sigma$
CsCl	10	0.2	0.02199	0.046	–	1.64
	20	0.2	0.001	0.09399	–	1.66
	40	0.2	−0.027	0.25799	–	1.49
	50	0.2	−0.063	0.395	–	1.89
	60	0.2	−0.1	0.51599	–	1.59
	80	0.2	−0.2347	1.0036	–	7.33
KCl	10	0.2	0.05099	0.25	–	1.41
	20	0.2	0.027	0.31499	–	2.06
	40	0.2	−0.01899	0.497	–	1.98
	60	0.2	−0.026	0.711	–	1.52
	80	0.1	−0.15399	1.177	–	0.81
	99	0.06	−0.10499	1.689	–	1.95
LiCl	10	0.2	0.15299	0.29899	–	0.97
	20	0.2	0.135	0.41299	–	2.19
	40	0.2	0.1459	0.5419	–	1.58
	60	0.2	0.1369	0.824	–	5.31
	80	0.2	0.09099	1.314	–	4.4
	99	0.1	0.071	2.012	–	2.37
RbCl	10	0.2	0.014	0.209	–	1.8
	20	0.2	0	0.268	–	1.56
	40	0.2	0.01099	0.377	–	1.6
	60	0.2	−0.028	0.66	–	1.17
	80	0.2	−0.139	1.072	–	2.1
	99	0.1	−0.19799	1.909	–	2.34
NaCl	0	4	0.07336	0.2877	0.00293	0.86
	20	4	0.10674	0.08209	−0.0053	1.97
	40	2	0.1136	0.2785	–	3.81
	60	1	0.2236	−0.02018	–	5.02
	80	0.5	0.5593	−0.7906	–	9.8
NaBr	0	4.62	0.10268	0.28505	0.00046	7.84
	10	3.05	0.1199	0.2747	−0.0083	7.2
	20	3.03	0.1032	0.3933	0.00118	3.7
	40	2.99	0.1843	0.1544	−0.0259	8.7
	60	3.02	0.1342	0.7726	0.0085	5.6
	80	2.54	0.2878	0.5371	−0.0384	9.4
	90	2.02	0.3149	0.948	−0.0407	7.5
	100	1.51	0.31183	1.55086	–	9.55

As a consequence, it seems possible and beneficial to extend this approach to the study of other thermodynamic properties, such as the effects of salts on phase equilibria. We propose also to integrate this new approach into thermodynamic models representing multicomponent systems.

REFERENCES

Basili, A. Mussini, P.R. Mussini, T. & Rondinini, S. 1996. Thermodynamics of the cell: [Na_xHg_{1-x}|NaCl(m)| AgCl|Ag] in (methanol+water) solvent mixtures. Journal of Chemical Thermodynamics. 28: 923–933.

Han, S. & Pan, H. 1993. Thermodynamics of the Sodium Bromide Methanol Water and Sodium Bromide Ethanol Water two ternary systems by the measurements of electromotive force at 298.15°K. Fluid Phase Equilibria. 83: 261–270.

Koh, D.S. Khoo, K.H. & Chee-Yan Chan. 1985. The Application of the Pitzer Equations to 1-1 Electrolytes in Mixed Solvents. Journal of Solution Chemistry. 14(9): 635–651.
Marquardt, D.W. 1963. An Algorithm for least squares estimation of nonlinear parameters. J. Soc. Ind. Appl. Math. 11: 431.
Pitzer, K.S. 1973. Thermodynamics of electrolytes. I. Theoretical basis and general equations. The Journal of Physical Chemistry. 77(2): 268–277.
Pitzer, K.S. & Mayorga, G. 1973. Thermodynamics of electrolytes. II. Activity and osmotic coefficients for strong electrolytes with one or both ions univalent. The Journal of Physical Chemistry. 77(19): 2300–2308.
Pitzer, K.S. & Mayorga, G. 1974. Thermodynamics of electrolytes. III. Activity and osmotic coefficients for 2-2 electrolytes. The Journal of Solution Chemistry. 3(7): 539–546.

Reclaiming the Desert: Towards a Sustainable Environment in Arid Lands – Mohamed (ed.)
© 2006 Taylor & Francis Group, London, ISBN 0 415 41128 9

Characterization and removal of phenol from wastewater generated at a condensate crude oil refinery

W. Al Hashemi
Emirates National Oil Company, Dubai, UAE

M.A. Maraqa
Department of Civil and Environmental Engineering, UAE University, Al Ain, UAE

M.V. Rao
Department of Chemistry, UAE University, Al Ain, UAE

ABSTRACT: Phenols are present in discharge effluents of many heavy industries such as refineries. Emirates National Oil Company (ENOC) processes condensate oil and produces wastewater that is treated at the Refinery Wastewater Treatment Plant (ENOC-RWTP). Characterization of phenol level at ENOC-RWTP and assessment of the effectiveness of employed pollution control technologies in reducing phenol level at the treatment plant have been conducted in this study. It was found that the main sources of phenol in the received waste streams at ENOC-RWTP are the tank water drain (average 11.8 mg/l), the desalter effluent (average 1.4 mg/l) and the neutralized spent caustic (average 234 mg/l). However, there are large fluctuations from the average phenol level within each waste stream. Also, the level of phenol and its derivatives (substituted phenols) in these streams vary significantly with phenol, m-p-cresols, o-cresol, tri and tetra-chlorophenols and to a lesser extent 4-chloro-3-cresol are common among these streams. The study further showed that the most effective process employed in the reduction of phenols within the plant is the sequencing batch reactors. The study further showed that the relationships between phenol level in the discharged treated effluent and levels of COD, BOD_5 and sulfides are weak but statistically significant.

1 INTRODUCTION

The development of petroleum and petrochemical industry in the Arabian Gulf region has been growing rapidly. In the last 50 years, exploration, drilling, extraction, refining and chemical engineering activities of oil and gas industry have all become an essential component in the economy of many of the Arabian Gulf countries. This speedy development has resulted in many changes such as landscape, economy, human development and interactions with other regions of the world as well as having impacts upon the environment and society. In fact "one of the most serious challenges facing the modern Middle East is the protection of its environment and the need to balance sustainable development with environmental security" (Morris, 2001).

Water pollution is a serious concern in the UAE. This critical problem is made even more serious with the fact that water is very scarce in the region. Like most industries, oil refineries generate enormous quantities of wastewater. Such wastewater may contain several pollutants including phenols. Because of its toxicity (Shumway, 1966; Mitrovic et al., 1968; European Inland Fisheries Advisory Commission, 1973; Haley et al., 1987) phenol became a wastewater quality parameter that the regulators closely look at in the effluents of heavy industry such as refineries. For example, Jebel Ali Free Zone Authority in Dubai stipulates a maximum phenol level of 0.1 ppm in industrial effluents discharged to marine environment (PCFC, 2005).

In oil refineries the typical ranges of phenol concentration is 50 to 185 ppm for catalytic cracking wastewater to sour water streams, respectively (Abuzaid, 1995). Similar levels of phenols in waste streams of a full fledged oil refinery have also been reported by Patterson (1998). The author reported that the level of phenols typically falls in the range of 80–185 ppm for raw outlets sour water, 40–80 ppm for general waste stream, about 80 ppm for post-stripping, 40–50 ppm for catalytic cracker, and 10–100 ppm for general wastewater. It should be mentioned that industrial wastewaters originating from oil production processes may contain phenols in high concentrations that exceed sometimes 10 g/l (González et al., 2001).

Several studies have been reported in the literature on the use of biodegradation for removal of phenols from wastewater (Sokót, 1998; Shin et al., 1999; Onysko et al., 2002; González et al., 2001; Rao and Viraraghavan, 2002; Alemzadeh et al., 2002; Klimenko et al., 2002; Goudar et al., 2000). In many of these studies, Pseudomonas putida (Pp) strain-type bacteria have been found to be effective for the degradation of phenols (Sokót, 1998; Onysko et al., 2002; González et al., 2001). The efficiency of biological methods employed for the treatment of phenol-contaminated wastewater varied with operating process variables such as pH, temperature, dissolved oxygen, and nutrients. It was further found that anaerobic degradation is typically more effective in removing phenol than degradation under aerobic conditions (Shin et al., 1999; Alemzadeh et al., 2002).

Phenol levels in most of the previously mentioned studies ranged between 100–1000 mg/l. At high phenols levels, however, the inhibitory nature of phenol has been observed. Goudar et al. (2000), for example, used Pseudomonas putida and Trichosporon cutaneum strains for phenol degradation in batch experiments with an initial phenol concentration above 100 mg/l. The authors found no phenol degradation at concentrations $>$1300 mg/l. Also, the lag phase increased with the increase in initial phenol levels for concentrations $<$1300 mg/l.

Most of the studies conducted on the removal of phenol from wastewater were limited to laboratory conditions. Transfer of laboratory results to filed conditions may not be straightforward, especially when processed wastewater is generated at condensate crude oil refineries. These refineries commonly receive crude oil from various sources, leading to fluctuations in waste characteristics. To our knowledge, no investigation has been made to characterize and assess the effectiveness of the commonly employed treatment processes for the removal of phenols from condensate refinery wastewater in the UAE. So, the main objective of this study was to investigate the efficiency of treatment processes employed at ENOC-RWTP in reducing phenol concentration. Another objective was to identify and quantify individual phenols in waste steams that contain these chemicals.

2 SAMPLE COLLECTION AND ANALYSIS

2.1 *Site description*

ENOC is the first oil refinery in the Emirate of Dubai, UAE. The company is a condensate splitter refinery with various plants including Merox sulfur removal and sulfur recovery units. Due to the nature of condensate, the plant has little environmental emissions when compared to crude oil refineries which have other units such as hydro-crackers, vacuum distillation, and coking units. The crude received by ENOC refinery are at times varying in basic quality due to oil nature, and use of different injected anti-corrosive chemicals and drilling mud. Wastewater generated from oil processing at ENOC is treated at the refinery wastewater treatment plant before discharge into the harbor of Jebel Ali Free Zone.

The wastewater treatment plant receives incoming wastewater from 18 streams. These streams include: wet slop, sour-water-stripper bottoms (SWS bottoms), desalter effluent, boiler blow-down, cooling water blow-down, utility water caustic soda pumps flushing, LPG vaporizer polluted condensate, fuel oil heater polluted condensate, Merox plants caustic soda, caustic heater condensate, pretreated neutralized spent caustic wastewater, oily water from Merox plants, pad drainage and storm water run-off, naphthenic neutralized spent caustic wastewaters, tank drain water, storage area stormwater, and sanitary wastes. It is important to note that the refinery processes are extremely complex and that the flowing streams come with different constituents and different flow rates.

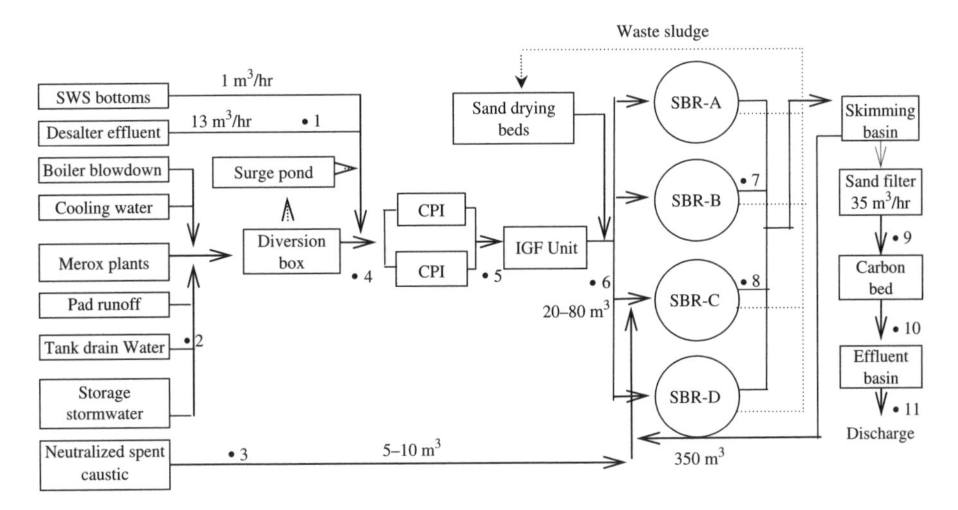

Figure 1. Waste streams and process flow diagram of ENOC-RWTP.

As shown in Fig. 1, many of the streams are first directed into a diversion box which feeds into corrugated plate induced separators (CPI) for oil removal. The CPI enhances coagulation of emulsified oil particles and floatation of these particles. After having floatable oil removed by the CPI, the flow goes into induced gas floatation (IGF) units, where air and a polyelectrolyte/strong cationic additive polymer are injected to further reduce the oil content. The outlet from the IGF is directed to a biological sequencing batch reactor (SBR) for further removal of dissolved organic matter.

The SBR is a fill-and-draw type system involving a single complete-mix reactor in which all the sets of the activated sludge process occur. Mixed liquor remains in the reactor during all cycles, thereby eliminating the need for separate secondary sedimentation tanks. It should be noticed that the pH is controlled in the SBRs to 7–9 to achieve proper operation of the biological system downstream. Decanted water from the SBRs is directed to a skimming basin that acts as a settling tank of escaped active sludge and is also used to remove some of the floatable solids. The effluent is then pumped through a sand filter which helps in reducing escaped suspended solids. Effluent from the filter goes into an activated carbon bed which acts as a polishing step for the removal of remaining organic substances. Effluent of the carbon bed is then sent to a final storage pond before it is discharged into the harbor.

2.2 *Sample collection and analysis*

Eleven locations within the treatment plant were sampled as shown in Fig. 1. Three of these locations (1–3) are of the waste streams including desalter water, drain water, and neutralized spent caustic. Location 4 represents an average characteristic of all waste streams excluding the neutralized spent caustic (which is sent to a separate storage tank and then added directly under controlled conditions to some SBRs). This location is referred to as raw mixed inlet. The other 7 sampling locations (4–11) are located at the outlets of unit processes within the treatment plant.

There are 4 SBR units (SBR-A through SBR-D Fig. 1) at ENOC-RWTP. SBR-A and B receive treated waste streams leaving the IGF unit, while SBR-C and D receive a mixture of treated wastewater from the IGF unit, neutralized spent caustic, and treated wastewater from the skimming pond. Two locations at the outlet of the SBRs were thus chosen (7 and 8). The other sampling locations (9–11) include the outlet of the sand filter, the outlet of the activated carbon reactor, and the effluent pond.

Phenol in the waste streams is determined as both total phenol (using the spectrophotometric and the HACH methods) and as individual phenols. Individual phenols in the waste streams are characterized by both HPLC with Fluorescence and Photodiode array detectors (US EPA 555) and GC with ECD (as pentafluorobenzyl derivatives, US EPA 604) methods. Several other water parameters have

Table 1. Sampling dates and locations.

Location	30/12/03	6–7/1/04	14/1/05	8/2/04	14/2/04	3/3/05	4/3/05	5/3/05
1	X	–	X	–	X	X[a]	X[a]	X
2	X	–	X	–	X	X[a]	–	–
3	X	–	X	–	X	X[a]	–	–
4	X	X	X	X	X	X[b]	X[b]	X[b]
5	X	X	X	X	X	X[b]	X[b]	X[b]
6	X	X	X	X	X	X[b]	X[b]	X[b]
7	X	X	X	X	X	X[b]	X[b]	X[b]
8	X	X	X	X	X	X[b]	X[b]	X[b]
9	X	X	X	X	X	X[b]	X[b]	X[b]
10	X	X	X	X	X	–	–	–
11	X	X	X	X	X	–	–	–

[a] Analysis included quantification of individual phenols.
[b] Samples were taken every 6 hours from 3–5/3/05.

been tested including: temperature, pH, total dissolved solids (TDS), dissolved oxygen (DO), oil and grease (O&G), chemical oxygen demand (COD), 5-day biochemical oxygen demand (BOD_5), total organic carbon (TOC), and sulfides. Testing procedure was based on the well-established methods by the American Public Health Association (APHA) and the US EPA. Sample analysis for all parameters, except individual phenols, was conducted at the ENOC-RWTP site. Individual phenols were analyzed, however, at the Central Laboratories Unit at the UAE University.

Samples were collected from the sampling locations on different days as shown in Table 1. Samples of the continuous-flow units (CPI, IGF, sand filter, and carbon bed) are drawn at the same time without consideration of hydraulic retention time. However, testing was further conducted over a 48-hr period using 6-hr samples in order to determine fluctuation in inlet characteristics of these units.

3 RESULTS AND DISCUSSION

3.1 *Troublesome waste streams*

Based on the levels of phenols used in the design of ENOC-RWTP, the streams that are considered most troublesome in terms of phenol loading are the desalter effluent, the neutralized spent caustic, and the tank water drains. Samples from these waste streams were collected for the analysis of phenol and phenol individuals as well as other water characteristics. The results are summarized in Table 2 and are further discussed below.

The desalter effluent water is characterized by heavy brownish light particles in suspension with a relatively pungent mixture of petroleum odor. The desalter effluent water shows a neutral pH of low TDS (Table 2). Meanwhile, the organic matter content (COD, TOC and O&G) is relatively low compared to the other two streams. Also, the phenol level obtained during the analysis (average $\approx$ 1.4 mg/l) is lower than the level found for the other two streams.

Samples from the neutralized spent caustic have dark greenish-yellow color with black floating and settleable particulate matter. This waste stream, which is a mixture of many different compounds is generated in the Merox processes, has a blend of very strong, pungent petroleum and sulfide-type odor. As shown in Table 2, neutralized spent caustic stream is characterized by high TDS and high organic matter content (COD, TOC and O&G). The phenol level in this stream (average 233.5 mg/l) is the highest among the other streams and fluctuates in the value as depicted by the large standard deviation (Table 2).

Samples from the drain water of the condensate tanks have dark yellow to brownish color with black floating and settleable sediments. As is the case with the neutralized spent caustic, this stream has a blend of very strong, pungent petroleum and sulfide-type odor. The tank drain water is generally

Table 2. Average characteristics of the three troublesome waste streams.[a]

Sample	pH	Temp. (°C)	TDS (mg/l)	COD (mg/l)	Phenol (mg/l)	DO (mg/l)	O&G (mg/l)	TOC (mg/l)
Desalter effluent (Location 1)	7.0 (0.6)	39.1 (17.7)	129 (138)	754 (152)	1.4 (1.1)	3.7 (1.5)	<5 (NA)	118 (5.7)
Neutralized spent caustic (Location 3)	7.0 (0.2)	23.9 (0.4)	35954 (28763)	11093 (5941)	233.5 (127.2)	2.1 (1.1)	87.3 (28.0)	5212 (NA)
Tank drain (Location 2)	5.8 (0.2)	23.5 (0.4)	28860 (38565)	34608 (49238)	11.8 (10.2)	1.3 (0.9)	125.8 (96.2)	22215 (23992)

[a] Values in parenthesis represent the standard deviation.

discolored due to presence of high iron. This stream also has higher organic matter than that associated with the neutralized spent caustic. However, the phenol level (average 11.8 mg/l) is almost 20 times lower than what has been found in the neutralized spent caustic stream but again fluctuates widely.

3.2 *Phenols characterization in waste streams*

Finger prints of phenols in the main troublesome streams are shown in Fig. 2. Analysis of individual phenols in the desalter effluent indicates that straight phenols and cresols are the highest in the stream at nearly equal levels (Fig. 2). Also, 2,4,6-trichlorophenol, 2,4-dichlorophenol, 2,4,5-trichlorophenol, o-cresol + 2-cyclohexyl-4,6-dinitrophenol, 2,3,4-tertrachlorophenol, and 2,4-dimethylphenol all exist in nearly equal concentrations. Furthermore, 4-chloro-3-cresol and 2,6-dichlorophenol could be present in this waste stream as well. It is evident, however, that the levels of phenol derivatives are less than the levels in the drain water and the neutralized caustic streams.

The level of individual phenols in the tank drain varies significantly with certain types of phenols such as cresols, 2,4-dimethylphenol and 2,6-dichlorophenols. Furthermore, m,p-cresol seems to be the highest in concentration recorded after straight phenol and shows similar peak concentrations as o-cresol + 2-cyclohexyl-4,6-dinitrophenol. These compounds are probably found due to added chemicals during exploration and pre-treatments which are undertaken on the crude condensates prior to arrival at the refinery. Other nitro and chloro-phenol are evident in this stream but at trace levels.

Most of the phenol compounds present in the neutralized spent caustic waste stream are similar to those found in the tank draw water. These include 2,4,6-trichlorophenol, m,p-cresol, 2,3,4-tetra-chlorophenol, o-cresol + 2-cyclohexyl-4,6-dinitrophenol and straight phenol. However, in this waste stream, o-cresol + 2-cyclohexyl-4,6-dinitrophenol and m,p-cresol were detected in concentrations nearly equal to that of straight phenol. The detected phenol compounds in this waste stream are probably product of the Merox and subsequent neutralization reactions that take place in the plant operations. These are found due to possibly the violent catalysis oxidation reactions and subsequent acid-base reactions that take place which are undertaken on the final raw products prior to final product storage and export from the refinery. Again, straight phenol and m-cresol seem to be the highest in concentration among other phenols.

3.3 *Changes in phenol level in time and throughout processes*

Fluctuations in phenol levels and other wastewater parameters were investigated by taking samples from the inlet of unit processes every 6 hrs for 48 hrs. Figure 3 shows variations of phenols in the incoming and treated effluent of each unit process. Although results of the HACH method were presented in Fig. 3, those by the spectrophotometric method (not shown here) were identical. As the figure shows, the level of phenol in the incoming flow to the unit processes varies over time from about 0.2 to about 1.3 mg/l, except for the wastewater entering the sand filter and the carbon bed where the level retains a value of about 0.1 mg/l.

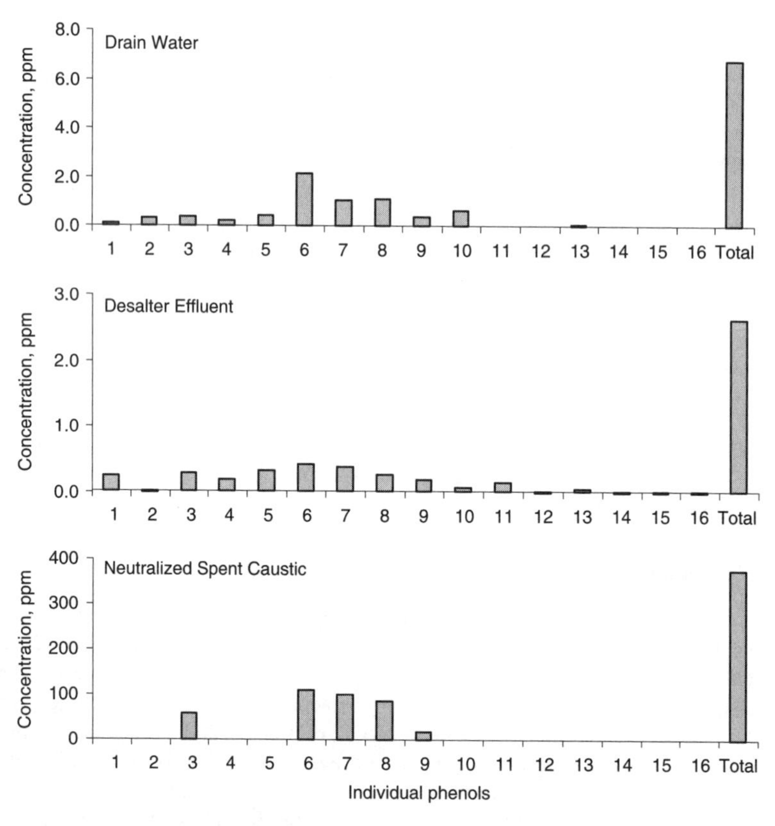

Figure. 2. Characterization of phenols in the troublesome waste streams. 1 = 2,4-dinitrophenol, 2 = 2-methyl-4,6-dinitrophenol, 3 = 2,4,6-trichlorophenol, 4 = 2,4-dichlorophenol, 5 = 2,4,5-trichlorophenol, 6 = phenol, 7 = m,p-cresol, 8 = o-cresol + 2-cyclohexyl-4,6-dinitrophenol, 9 = 2,3,4-tetrachlorophenol, 10 = 4-chloro-3-cresol, 11 = 2,4-dimethylphenol, 12 = o-chlorophenol, 13 = 2,6-dichlorophenol, 14 = o-nitrophenol, 15 = p-nitrophenol, and 16 = pentachlorophenol.

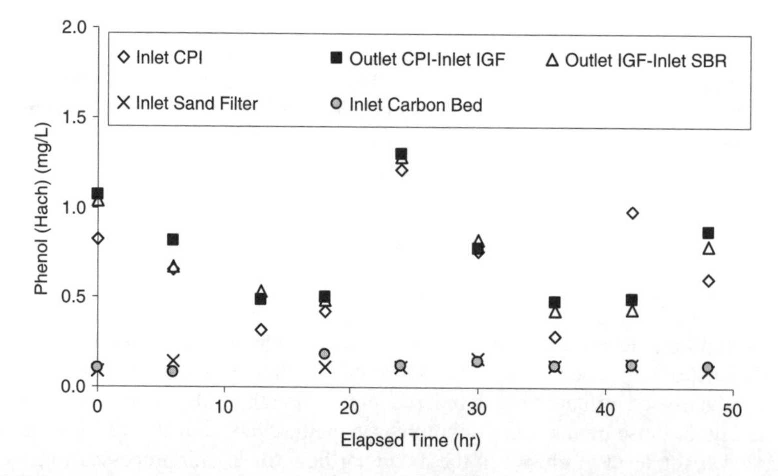

Figure 3. Changes in phenol level at the inlet of unit processes at ENOC-RWTP. Zero time corresponds to 00 hr on 3/3/2005.

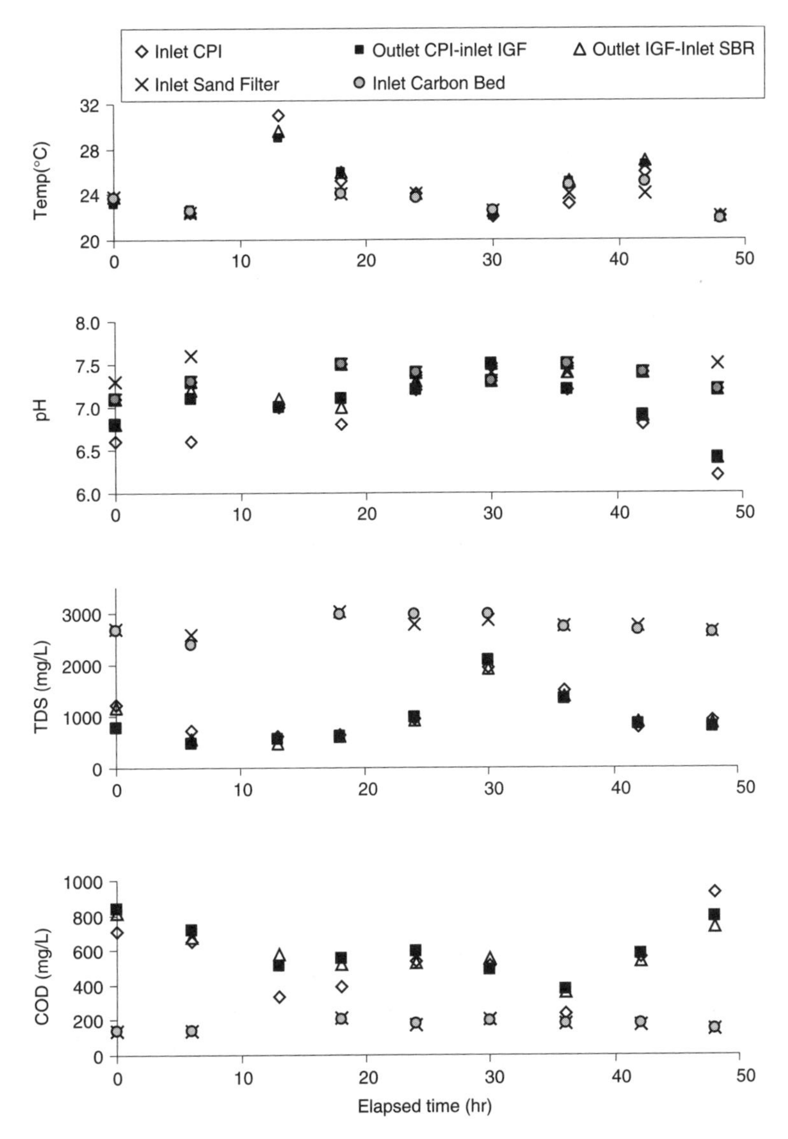

Figure 4. Changes in temperature, pH, TDS and COD at the inlet of unit processes at ENOC-RWTP. Zero time corresponds to 00 hr on 3/3/2005.

Figure 4 shows changes in other wastewater parameters (temperature, pH, TDS, and COD) in the inlet of unit processes at ENOC-RWTP. Changes in temperature (22–28°C) reflect the effect of ambient conditions; with lower temperatures during night and higher ones during daytime. Temperature, however, does not differ much among different processes at a certain time during the day. The pH, on the other hand, falls in the range of 6.5 to 7.5 with no particular changes with time or process. The pH at the inlet of the sand filter and carbon bed remains almost constant (pH 7.2) due to homogenization of the incoming streams in the SBR.

TDS variations at the outlet of the CPI, IGF and SBR seem to be related to variation in the TDS of the inlet to the treatment plant especially those of the tank water drain. Moreover, TDS at the outlet of the SBR, sand filter and carbon bed does not fluctuate much but is at a higher level compared to

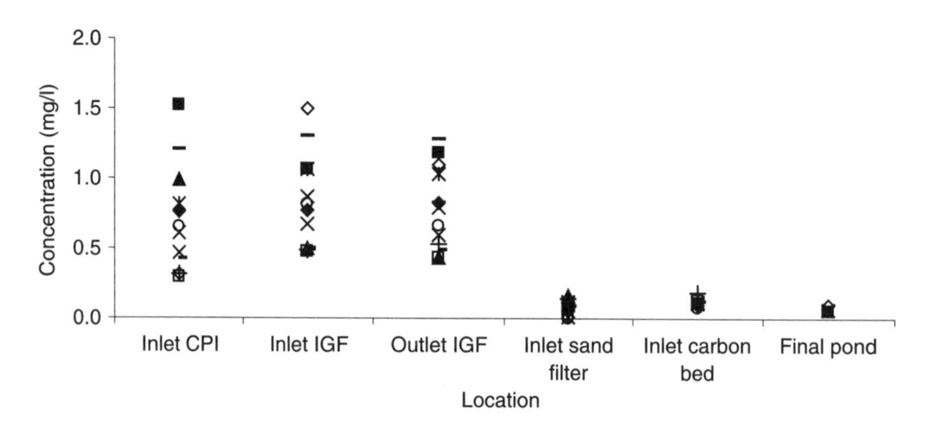

Figure 5. Phenol changes across unit processes at ENOC-RWTP. Different symbols represent different data sets taken on different times.

that of the CPI and IGF. This is due to homogenization of waste streams in the SBR process. However, higher TDS at the outlet of the SBR is probably due to dosing of nitrogen and phosphate compounds through addition of urea and phosphoric acid in addition to processing neutralized spent caustic waste that is characterized by a high TDS.

Figure 4 shows significant variations in the COD of the influent of the CPI, IGF and SBR, with no particular trend of increase or decrease in COD within the monitoring duration. Inlet IGF has generally higher COD than that of inlet CPI due possibly to dissolution (emulsification) of organic matter in the CPI that was originally not dissolved when the stream entered the CPI. The figure also reveals that the SBR process is the most effective in removing COD as demonstrated by the significant drop in COD at the inlet of the sand filter as compared to the levels at the outlet of the IGF. COD also changes temporally at the inlet of a certain process as well as among different processes. Such behavior is similar to that observed for phenol (Fig. 3).

Figure 5 shows phenol removal across the unit processes at ENOC-RWTP. The figure depicts the effect of the SBR on phenol reduction. Moreover, the CPI and IGF processes do not have any significant removal effect as similar phenol levels, compared to the entering waste stream, are observed in the outlet of these processes. The observed slight increase in phenol in the inlet of the carbon bed, compared to the level entering the sand filter, could be due either to experimental errors or time lag effect in collecting the samples. The carbon bed has a positive effect on removal of phenol as depicted by the decrease in phenol level in the effluent pond in comparison to that of the inlet of the carbon bed.

3.4 *Relationship between phenol and other water parameters*

Previous records available at ENOC-RWTP for the effluent (treated wastewater) allow comparison between parameter values. It should be noted that previous data represent monthly averages from daily final effluent pond readings. Figure 6 shows effluent phenol levels plotted versus BOD, COD, and sulfide. The figure also shows the best fit line between phenol and the other water parameters. The best linear equation along with the coefficient of determination (r^2) and confidence interval (CI) of the slope are also presented in Table 3. The CI for the slope is determined according to the method described by Anderson (1987).

As shown in Table 3, the established relationships are statistically significant although their associated r^2 values are relatively low due to the scatter of the data as depicted in Fig. 6. The relationships presented in Table 3 could be used to estimate effluent BOD, COD and sulfide based on an already determined effluent phenol level. Meanwhile, by meeting a phenol value of 0.1 mg/l or less in the final effluent prior to discharge, one can estimate based on these relationships whether the effluent limit of BOD_5, COD and sulfide will be met or not.

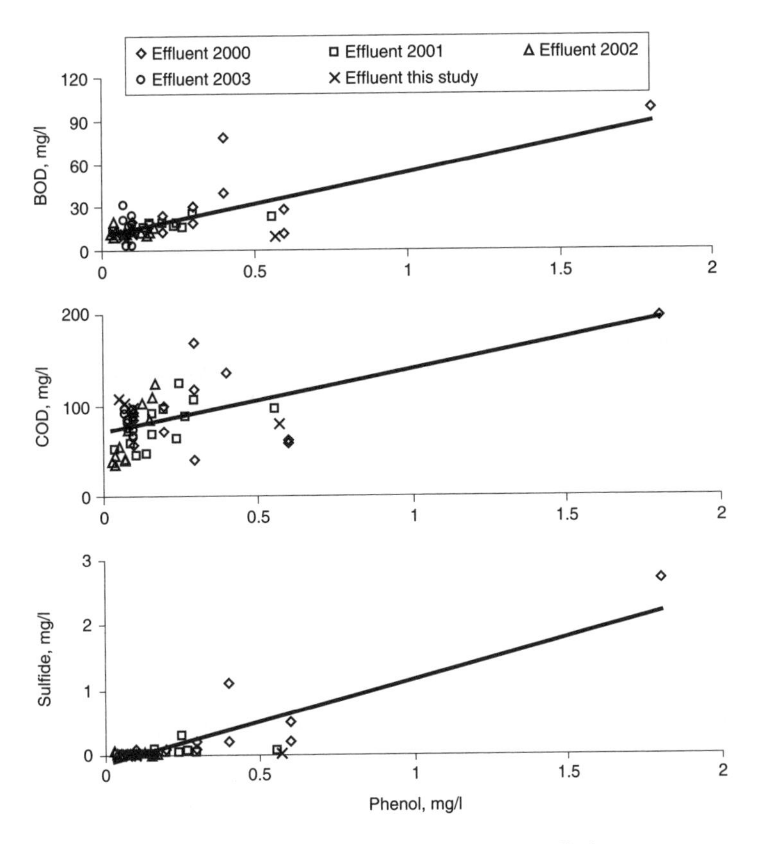

Figure 6. Correlation between phenol and other water parameters in the discharge water.

Table 3. Relationships between phenol and other parameters of treated wastewater.

Relationship	Equation	r^2	Confidence interval (CI)
Phenol- BOD_5	$BOD_5 = 43.7 \times \text{phenol} + 9.7$	0.56	43.7 ± 15.3
Phenol-COD	$COD = 68.6 \times \text{phenol} + 71.3$	0.25	68.6 ± 37.2
Phenol-Sulfide	$\text{Sulfide} = 1.29 \times \text{phenol} - 0.13$	0.76	1.29 ± 0.38

4 CONCLUSION

The following conclusions are drawn based on the data collected in this study and those collected from the refinery testing records:

1. Phenol in refinery wastewaters at ENOC-RWTP comes from the tank water drain, the desalter effluent, and the neutralized spent caustic waste streams. The latter has the highest phenol levels of 234 mg/l but large deviations from the average phenol level exist within each stream.
2. The concentration of phenols in the troublesome waste streams varies between certain types of phenols with straight phenol and m-p-cresol are the most dominant in the waste streams. Other phenols that are common among the waste streams include o-cresol, tri and tetra-chlorophenols and to a lesser extent 4-chloro-3-cresol.
3. The existence of 7 out of 16 phenol compounds in the different waste streams indicates that the source of phenol is probably the processed condensate crude oil and is not a byproduct of oil processing at the refinery.

4. The waste stream at the inlet of unit processes at ENOC-RWTP varies in its phenol and other water characteristics over time but variations become less pronounced after the SBR.
5. The SBR is the most effective process within the treatment plant for the reduction of phenol, organic loading, and sulfides.
6. A significant relationship exists between phenol concentration in the discharged effluent and other water parameters such as COD, BOD_5 and sulfides.

ACKNOWLEDGMENT

The authors are grateful to the Emirates National Oil Company (ENOC) for facilitating the work presented in this study. Thanks are also to the Central Laboratories Unit at the United Arab Emirates University for conducting the analysis of the individual phenols.

REFERENCES

Abuzaid N.S., 1998. Phenolic wastewater; characteristics, environmental impact and treatment, Research Institute, King Fahd University of Petroleum and Minerals, Dhahran, Saudi Arabia.

Alemzadeh I., Vossoughi F. and Houshmandi M., 2002. Phenol biodegradation by rotating biological contactor, Biochemical Engineering Journal, 11: 19–23.

Anderson R.L., 1987. Practical statistics for analytical chemists, van Nostrand Reinold, New York.

European Inland Fisheries Advisory Commission, 1973. Water quality criteria for European freshwater fish. Report on mono-hydric phenols and inland fisheries. Water Res. 7: 929.

González G., Herrera G., García M.T. and Pena M., 2001. Biodegradation of phenolic industrial wastewater in a fluidized bed bioreactor with immobilized cells of Pseudomonas putida, Bioresource Technology, 80: 137–142.

Goudar C.T., Ganji S.H., Pujar B.G. and Strevett K.A., 2000. Substrate inhibition kinetics of phenol biodegradation, Water Environmental Research, 72(1): 50–54.

Haley T.J. and Berndt W.O, 1987. The Handbook of Toxicology, Hemisphere Publishing Corporation (HPC), New York.

Klimenko N., Winther-Nielsen M., Smolin S., Nevynna L. and Sydorenko J., 2002. Role of the physico-chemical factors in the purification process of water from surface-active matter by biosorption, Water Research, 36: 5132–5140.

Mitrovic V.A., 1968. Some pathologic effects of sub-acute and acute poisoning of rainbow trout by phenol in hard water, Water Res., 2: 249.

Moris M.E., 2001. Water scarcity and security concerns in the Middle East, Emirates Institute Center for Strategic Studies and Research.

Patterson J., 1980. Wastewater Treatment Technology, Ann Arbor Science Publishers.

PCFC, 2005. Environmental Control Rules and Requirements, Fourth Edition, Group Environment, Health & Safety Department, Ports, Customs & Free Zone Corporation, Dubai, UAE.

Onysko K.A., Robinson C.W. and Budman H.M., 2002. Improved modeling of the unsteady-state behavior of an immobilized-cell, fluidized-bed bioreactor for phenol biodegradation, The Canadian Journal of Chemical Engineering, 80: 239–252.

Rao J.R. and Viraraghavan T., 2002. Biosorption of phenol from an aqueous solution by Asperigllus niger biomass, Biosource Technology, 85: 165–171.

Shin H., Yoo K. and Park, J.K., 1999. Removal of polychlorinated phenols in sequential anaerobic-aerobic biofilm reactor packed with tire chips, Water Environment Research, 71(3): 363–367.

Shumway D.I., 1966. Effects of effluents on flavor of salmon fish. Agricultural Experiment Station, Oregon State University, Corvallis.

Sokót W., 1998. Experimental verification of the models of a continuous stirred-tank bioreactor degrading phenol, Biochemical Engineering Journal, 1: 137–141.

US EPA 555, Determination of chlorinated acids in water by high performance liquid chromatography, Method 555.

USEPA 604, Methods for organic chemical analysis Municipal and Industrial wastewater, Method 604 – Phenols.

Reclaiming the Desert: Towards a Sustainable Environment in Arid Lands – Mohamed (ed.)
© 2006 Taylor & Francis Group, London, ISBN 0 415 41128 9

Occurrence and significance of bacteriophages in biological wastewater treatment processes

M.A. Khan
Department of Biotechnology, Manipal Academy of Higher Education (MAHE) Dubai Campus, Knowledge Village, Dubai, UAE

H. Satoh & T. Mino
Institute of Environmental Studies, Graduate School of Frontier Sciences, The University of Tokyo, Hongo, Bunkyo-ku, Tokyo, Japan

ABSTRACT: In this study, the occurrence of bacteriophages in wastewater was demonstrated by isolating host bacteria and their co-occurring bacteriophages from activated sludge treatment process. Bacterial isolates were obtained from an activated sludge process treating urban sewage, and bacteriophages were obtained by plaque assay using the bacterial isolates obtained in this study as the host. The host range test was conducted with a combination of bacteriophage isolates and bacterial isolates. In another experiment the mutagen mitomycin C was used to induce temperate bacteriophages from bacteria isolated from activated sludge. The results of this study demonstrated that 90% of the bacterial isolates from the activated sludge released phage-like particles, suggesting that temperate bacteriophages are very common in activated sludge. In conclusion, the results of this study suggested that lytic as well as temperate bacteriophages are active members of the activated sludge process and could significantly affect the activities of the bacterial populations therein.

1 INTRODUCTION

Bacteriophages are the important member of various natural microbial ecosystems like marine (Fuhrman, J. A. 1999), lake water (Ogunseitan et al., 1990) and soil (Marsh et al., 1994), where quite a high number of viruses have been detected during several researches since last decade. It is believed that the presence of viruses could be a major factor for the bacterial population dynamics of such ecosystems due to various ecological roles played by bacteriophages such as bacterial mortality by lysis (Weinbauer and Suttle, 1996), material flow and horizontal gene transfer (Fuhrman, J. A. 1999). Biological wastewater treatment processes such as the activated sludge process mainly depend on bacterial populations and their activities involved in the processes. However very little attention has been paid to predict role of bacteriophages in the biological wastewater treatment processes, where density of the bacteria is much higher than the natural ecosystems. Past few studies suggested the influence of bacteriophage presence on the treatment efficiency of the activated sludge process (Ogata et al., 1980 and Hantula et al., 1991). However, the exact role played by bacteriophages in the activated sludge processes or their effect on the microbial community is not known at the moment.

Several studies in the past (Ewert and Paynter 1980, Ogata et al., 1980, Hantula et al., 1991) have been conducted on bacteriophages that infect bacteria in activated sludge. These studies reported the isolation of bacteriophages and the increase in the number of phage-like particles in the course of wastewater treatment by activated sludge based on electron microscopic counting. These observations strongly indicate bacteriophages are multiplying themselves in activated sludge, most probably lysing their host bacteria (which are removing pollutants in wastewater). If bacteriophages are an important microbial member in activated sludge, they may probably influence bacterial population structure in activated sludge, and thus affecting the performance of wastewater treatment.

There are two major pathways of bacteriophage replication that can occur when a bacteriophage infects a suitable host. One is the lytic cycle, in which the phage genome replicates immediately after infection and the newly formed viral progeny are released as the host cell bursts. The other is the lysogenic cycle, in which the DNA-containing phage maintains a stable symbiosis with its bacterial host and is the alternative to lytic replication. Lysogeny can be achieved by integration of the viral genome into the host chromosome and transmission of the integrated DNA (termed a prophage) to progeny cells during division. The lysogenic state may continue for many generations until the process of induction, whereby a chemical or physical factor results in activation of the prophage genes and initiation of the lytic cycle. The prophage induction can be caused by physical factors like temperature, pressure, exposure to UV radiation or by chemical mutagenic agents such as mitomycin C, polynuclear aromatic hydrocarbons and polychlorinated biphenyls (PCB's) (Jiang et al., 1998d; Paul et al., 1999). Mitomycin C treatment is most commonly used for the prophage induction (Jiang et al., 1998c). The mutagen mitomycin C causes prophage induction by stimulation of excision-repair mechanisms of the bacteria, which result in excision of the prophage genome from the host chromosome causing initiation of viral replication and host lysis (Paul et al., 1999).

The occurrence of temperate bacteriophages have been suggested in various microbial environment (Marsh et al., 1994, Ogunseitan et al., 1990, Montserrat et al., 1998), however, in activated sludge little importance has been paid to study the temperate bacteriophages. To determine the role of bacteriophages, it is essential to study the occurrence of lytic and temperate bacteriophages in the activated sludge environment.

In this study, the occurrence of bacteriophages was demonstrated by isolating host bacteria and their co-occurring bacteriophages from the pilot-scale anaerobic-aerobic (A/O) sequencing-batch reactor (SBR) treating municipal wastewater. The isolated bacteriophages were characterized by studying their lytic activities on many selected cultivable bacterial hosts. Furthermore to determine whether the bacteria isolated from activated sludge were in lysogenic association with the temperate bacteriophages 1) Various concentrations of mitomycin C was applied to bacteria isolates to induce lysogen to turn lytic, then 2) Phage-like particles released by the bacteria were detected by observation under fluorescent microscope.

2 MATERIALS AND METHODS

2.1 *Activated sludge*

A bench scale sequential batch anaerobic-aerobic activated sludge process treating municipal wastewater was used as the source of host bacteria and bacteriophage. The influent is characterized as follow; COD typically 250 mg/L plus 40 mg/L supplement of acetate, TP typically 3 mgP/L, TN typically 30 mgN/L. The effluent is as follows; COD 20 mg/L, TP less than 0.5 mgP/L, TN 6 mgN/L. The sample for the isolation of bacteria and bacteriophages were taken on the same day. The isolation of bacteria was done immediately. The sample was stored at 4°C for about 2–3 weeks until the isolation of bacteriophages.

2.2 *Isolation of bacteria from activated sludge*

For the isolation of the bacteria, activated sludge samples were taken at the end of either aerobic or anaerobic phases. Ten ml of the sample was diluted with 60 ml of 0.9% NaCl, 1 ml of 10.4% $Na_5P_3O_{10}$ and 0.1 ml of Tween-80 (2% v/v) solution (Merzouki et al., 1999), then sonicated. Dilution series were set up from 10^{-1} to 10^{-8}. One milliliter of each dilution series was spread over solid agar plates of mLB medium containing sodium acetate 5 g/L, tryptone 10 g/L, yeast extract 5 g/L, NaCl 10 g/L, $MgSO_4.7H_2O$ 0.2 g/L, $MnSO_4.7H_2O$ 0.05 g/L, and agar 15 g/L. After incubation at

20°C for 2–5 days, colonies of various colors and morphological characteristics were observed. Pure cultures of host isolates were obtained by picking individual colonies and repeatedly streaking at least three times on their respective fresh agar media.

2.3 *Characterization of bacterial isolates*

The isolates were identified by Biolog identification system (Biolog Inc, Hayward, USA). Gram staining was performed for all the isolated pure cultures to know their morphological characteristics (Jenkins et al., 1993). Nile blue A staining was done to examine polyhydroxyalkanoates (PHA) accumulation of the isolates (Kitamura et al., 1994).

2.4 *Screening test of bacteriophages*

In order to see, if the bacterial isolates have their co-occurring bacteriophages, screening test of bacteriophages was done. To the activated sludge sample, equivolume of 10% beef extract solution was added, and stirred overnight at 20°C. The sample was centrifuged, and filtered by 0.25 μm membrane filters. For some cases, enrichment of bacteriophage with host bacteria was tried. After the elution of bacteriophages with beef extract solution, the elute was filtered by 0.25 μm membrane filter. The plaque formation test was done with the combination of the filtrate and each of the bacterial isolates to observe if plaques are formed or not.

2.5 *Isolation of bacteriophages*

Bacteriophages were isolated by picking them from a single plaque obtained in the screening test above mentioned. To ensure the purity of the phage, isolated plaques were picked and plated with their respective host isolates at least twice. The nomenclature of bacteriophages derives from the nomenclature of their hosts from which the bacteriophage was isolated. For example, a bacteriophage originally isolated and plaque purified on host P26 was designated ϕP26.

2.6 *Plaque formation test*

Each of sample containing bacteriophages was filtered through 0.25 μm membrane filter, 0.1 ml of which was directly mixed with 1 ml of exponentially growing host bacterial culture and incubated for about 30 min at room temperature before mixing with 3 ml of soft agar and poured over the preset bottom agar layer. The plates were incubated upside down at 20°C for about 3–5 days to observe any kind of plaques on the lawn of the host bacterium tested.

2.7 *Host range determination of isolated bacteriophages*

Host range of the isolated bacteriophages was studied by plaque formation test with the isolated bacteriophage to be tested on the lawn of different bacterial isolates.

2.8 *Prophage induction of activated sludge bacteria by mitomycin C*

To each bacteria at their logarithmic growth phase were added mitomycin C (Sigma Chemical Co, St. Louis, MO) at the final concentration of 0.01, 0.05, 0.1 mg/ml respectively. The cultures were incubated for 10–24 hrs with shaking at 30°C. A control for each bacterial culture was incubated under the same conditions except the addition of mitomycin C. Each sample was incubated in dark to prevent photo activation of the host cell DNA repair system. The change in bacterial growth was monitored by measuring cell absorbance at 600 nm every 2 hours for the first 12 hours. Then cultures were incubated overnight for 12–24 hours. All the induced cultures including their controls were centrifuged at 3,000 rpm for 5 min at room temperature. The supernatant was collected for detection of phage-like particles either using plaque formation tests or by direct staining with DAPI under the epifluorescent microscope.

2.9 *Observation of phage-like particles in the mitomycin C exposed bacterial cultures*

To observe the phage-like particles under the fluorescent microscope by direct DAPI staining (Hennes et al., 1995; Hara et al., 1991; Proctor and Fuhrman 1992), about 0.5 ml of the induced culture supernatant was mixed with the 50 mg/ml (final concentration) of freshly prepared DAPI solution. The mixture was incubated for five minutes in the dark at the room temperature (~30°C). After the incubation, DAPI stained phage-like particles were fixed on the glass slides by drying in the air in dark for a few minutes. To the fixed samples was added a drop of 20% glycerol which act as an antifade mounting solution. The slides were covered with a coverslip and viewed at a magnification of ×1,000 with an Olympus BX-51 epifluorescence microscope equipped with a violet filter set for DAPI labeled phage particles with the blue UV excitation. The images of the samples investigated under the fluorescent microscope were taken by a charge coupled device camera (CCD camera) equipped with the fluorescent microscope and saved on the hard disk of a personal computer connected to the CCD camera.

3 RESULTS AND DISCUSSION

3.1 *Isolation of bacteria from activated sludge*

Fifteen bacteria were isolated from the activated sludge. All the isolates formed colonies within 2–3 days of incubation at 20°C. Gram staining of most of the isolates have shown that seven were Gram negative rods, four were Gram positive cocci, and one was Gram positive oval shape bacterium, while two isolates did not show clear gram staining reactions. For one isolate (P-32), Gram staining was not performed. The identification of bacteria by the Biolog system was successful only for five host isolates. The identified isolates belong to the γ-subclass of *Proteobacteria*, and the high G + C group and low G + C group of Gram positive bacteria (Table 1).

3.2 *Plaque formation test*

Of the 15 bacterial isolates described above, 9 supported plaque formation, as shown in Table 2. The remaining six isolates did not support any plaques. The number of plaques on all the infected nine host isolates increased after further incubation at 20°C. The phage titer in unit volume of activated sludge mixed liquor was in the range of 10^3 to 10^4 PFU ml^{-1} for six bacteriophages, as shown in Table 2.

In the present study, nine out of fifteen (60%) isolated bacteria were shown to have their co-occurring bacteriophages. According to Hantula et al. (1991), 15% of the bacterial strains isolated from activated sludge samples were sensitive to their co-occurring phages, which is percentage-wise significantly lower than the result obtained in this study. One of the causes of the difference could be attributed to the difference in method to extract bacteriophages from activated sludge sample. Hantula et al. (1991) tested filtrate of sonicated activated sludge to screen out bacteriophages. On the other hand, elution of bacteriophage with beef extract solution was included in the protocol of the present study. Another possible cause is the difference in the culture media of the bacterial isolates. Hantula et al. (1991) used media containing tryptone, glucose, and yeast extract. In the preliminary work for the present study, we also employed culture medium with glucose, tryptone, yeast extract and so on, but plaque formation was rarely observed.

3.3 *Host range of isolated bacteriophages*

For each bacterial host, one plaque formed in the bacteriophage screening test was picked, and bacteriophage was purified by plaque formation method on the lawn of the original host bacterium.

Table 1. Host bacteria isolated from the bench-scale A/O SBR treating urban sewage.

	Identification by Biolog System			Characteristics			
I.D. No.	Identified species	Similarity (%)	Comments	Gram* stain	Growth rate	Shape	PHA accumulation
P-26	*Escherichia coli*	71.6	γ-subclass of *Proteobacteria*	−ve	fast	small or long rod	–
P-27	Unidentified	<50	–	variable	fast	fibre like	–
P-28	*Cornybacterium liquifaciens* (*Brevibacterium*-like)	53.3	High G + C gp.	+ve	fast	coccus (bi/tri/multi)	–
P-29	Unidentified	<50	–	+ve	fast	coccus	–
P-30	*Photobacterium logei*	64.4	γ-subclass of *Proteobacteria*	−ve	fast	rod reddish colony	–
P-31	Unidentified	<50	–	+ve	fast	coccus (bi/tri/multi)	–
P-32	Unidentified	<50	–	unknown	slow	Not known	–
P-33	*Sphingobacterium mizutaii*	75.2	Low G + C gp.	−ve	fast	rod	–
P-34	Unidentified	<50	–	variable	slow	rod	–
P-35	Unidentified	<50	–	−ve	fast	rod	–
P-36	Unidentified	<50	–	+ve	fast	coccus	–
P-37	*Sphingobacterium sp.*	58.3	Low G + C gp.	−ve	fast	rod	–
P-38	Unidentified	<50	–	+ve	fast	big oval shape	–
P-39	Unidentified	<50	–	−ve	fast	rod	–
P-40	Unidentified	<50	–	−ve	slow	rod	–

* −ve – Gram-negative; +ve – Gram-positive; Variable – not sure; unknown – Gram staining not performed.

Table 2. Bacteriophages isolated from bench-scale anaerobic-aerobic (A/O) SBR.

		Bacteriophage		
Isolating bacterial host	Plaque formation time (days)	I.D.	Plaque size and morphology	Phage titer ml^{-1} (Approx.)[a] PFU ml^{-1}
P26	1	ϕP26	2–3 mm, clear	5.2×10^3
P27	1	ϕP27	1 mm, clear	ND*
P30	1	ϕP30	2–3 mm, clear	3.5×10^4
P35	1	ϕP35	2–3 mm, clear	2.5×10^4
P31	1–2	ϕP31	1–2 mm, clear	ND*
P36	1–2	ϕP36	<1 mm, clear	1.5×10^3
P37	1	ϕP37	1–2 mm, clear	1.5×10^3
P38	1–2	ϕP38	1–2 mm, clear	1.7×10^3
P40	1	ϕP40	1 mm, clear	ND*

[a] Phage titer is sum of the visible approximate number of plaques formed by the phage at least two times plating with its isolating bacterial host; ND*: not determined.

The host ranges of the eight bacteriophages (ϕP26, ϕP27, ϕP30, ϕP35, ϕP36, ϕP37, ϕP38 and ϕP40) isolated were examined on 9 host isolates obtained in this study. The result is summarized in Table 3. Three bacteriophages failed to infect their original host, although they successfully formed plaques in the isolation procedure. Bearing in mind the fact that ϕP36 and ϕP40 formed plaques

Table 3. Host ranges of bacteriophages isolated from bench-scale anaerobic-aerobic (A/O) SBR.

	Host Bacteria								
Bacteriophages	P26, G (−)* rod	P28, G (+)* coccus	P29, G (+) coccus	P30, G (−) rod	P35, G (−) rod	P36, G (+) coccus	P37, G (−) rod	P38, G (+) oval	P40, G (−) rod
φP26	−	−	ND	−	+	−	−	−	−
φP27	−	+	+	−	−	−	+	−	+
φP30	−	−	ND	+	+	−	−	−	−
φP35	−	+	ND	−	+	−	−	−	−
φP36	−	+	−	−	ND	−	−	−	−
φP37	−	+	ND	−	−	−	+	−	−
φP38	−	+	+	−	−	ND	+	+	+
φP40	−	−	−	−	ND	−	−	−	−

+; Plaque formed, −; plaque not formed, ND; not determined, *G (−) – Gram-negative; G (+) – Gram positive.

on their original host in the isolation procedure, all the bacteriophages tested for their host range were shown to have a broad host range. Furthermore, bacteriophages φP27, φP35, φP37 and φP38 formed plaques both on Gram-negative and Gram-positive bacterial hosts.

The host range study of isolated bacteriophages showed that all of bacteriophages tested infected more than one bacterial host. This indicates that bacteriophages with a broad host range are common in the activated sludge system. Hantula et al., 1991 also reported broad-host-range phages from activated sludge, but in their report, 10% of bacteriophages they obtained had a broad host range, which is ten times lower than that obtained in the present study. It is rather difficult to find specific explanation for the difference between their result and ours.

The second feature observed from the host range study was that three phages (φP26, φP36 and φP40) failed to form plaques on their original hosts for which they were first isolated. Although plaque formation was not observed, this does not necessarily mean that infection didn't occur. It is possible that the bacteriophage infected the host bacteria and resulted in lysogenic relationship. But regardless whether bacteriophage and the host started lysogenic interactions or not, the observed phenomena certainly indicate that infectivity of the bacteriophages changed or the resistance of host bacteria to the bacteriophages varied during the course of the study. The chemostat studies by Lenski et al. (1985) and Fuhrman (1999) have proved that resistance of host bacteria against bacteriophage evolves within a short time as a response to the presence of phage by spontaneous single mutations in host bacteria. Also, Lenski et al. (1985) demonstrated that bacteriophages can also modify themselves by mutation and therefore able to infect other host bacteria. The observed phenomena in this study could be explained based on the observations reported in these findings, but since the frequency of the phenomena observed was high, there might be some other reasons than simple mutation in the underlying mechanisms.

The experimental results described above strongly indicate that bacteriophages are an active member in activated sludge microbial ecosystem, and have very close relationship with host bacteria. We obtained bacteriophages and their host bacteria from activated sludge processes treating synthetic wastewater (Khan et al., 2002). In the study too, bacteriophages with very wide host range were found, and some of the bacteriophages were found to lose their plaque formation capability on their original host. It is easily expected that bacteriophages with broad host range have higher chances to survive in activated sludge. Microbial population structure in activated sludge is very complex, and so bacteriophages with narrower host range will have less chance for survival then those with broader host range. In addition, it was observed that some of the bacteriophages easily lost plaque formation capability on their original host. This observation strongly indicates the existence of mechanism that bacteriophage do not completely eliminate host bacteria but are ready to coexist with the host bacteria.

3.4 *Prophage induction of bacteria isolated from activated sludge by mitomycin C*

In total, 14 bacterial isolates obtained from the bench-scale A/O SBR treating real wastewater was tested for the presence of inducible prophage by adding different concentration of mitomycin C. The results of prophage-induction experiments have shown that all fourteen bacterial isolates growth was affected by applying different mitomycin C concentrations as compared to their controls (without mitomycin C).

The release of phage-like particles from bacterial isolates was confirmed by DAPI staining, which was applied directly to the mitomycin C induced and control culture supernatants. The mitomycin-C induction experiments with fourteen isolates have shown two types of observations. Firstly, among a total of 14 isolates tested, 13 produced phage-like particles after the induction, suggesting that over 92% of the bacterial isolates from activated sludge environments were probably lysogens. The lysogens are believed to be common in the environment where density of host bacteria are generally low and bacterial cells exist under the nutrient limited conditions, as observed in the marine environment (Jiang et al., 1998c and Weinbauer et al., 1996). Several studies (Jiang et al., 1994a, 1996b, 1998c, Paul et al., 1999, Weinbauer et al., 1996) on lysogeny in the marine environment indicated that 47% of marine bacteria are lysogenic. However, in the present study, frequency of the lysogens among activated sludge bacteria was found to be quite high as compared to that found in the marine environment. One of the possible reasons is that in activated sludge quite a high number of bacteria, which compete for the available nutrients, prefer to have lysogenic relationship with their interacting phages rather than being lysed by them. Lysogenic relationship of bacteria with phage gives the competitive advantages over their non-lysogenic counterparts (Edlin et al., 1975) by conferring host bacteria advantages such as immunity from infection by similar or related phages and the acquisition of new functions coded by the virus genome (Fuhrman, J. A. 1999). Another reasons could have been that in the present study phage-like particles were observed through DAPI-staining and their visualization under epifluorescence microscopy (Hennes et al., 1995, Hara et al., 1991 and Procter and Fuhrman 1992), however in most of studies on lysogeny the mode of observation of phage-like particles were electron microscopic methods (Jiang et al., 1998d, Chiura, H. X. 1997a, Montserrat et al., 1998), which probably underestimates the lysogens in their marine environment.

Another feature observed during the mitomycin C induction test was that in most of the cases (11 out of 14), induction was also observed in the control cultures (without mitomycin C). However, numbers of phage-like particles observed were quite less in the control as compared to the induced samples. One of the possible reasons for spontaneous induction during growth of bacterial isolates in their control culture could be the change in temperature of the incubation because experiments were not carried out under the strict temperature control. One report demonstrated that even slight temperature increase from 27°C to 30°C resulted in prophage induction (Paul et al., 1999). Our result could be explained by this report as almost all induction experiments were not performed under strict temperature control and most of the time experiments were preformed at the room temperature of ~30°C. However, a slight variation in the temperature could be highly possible during long incubation time of 12–24 hours. It has been suggested that variety of treatments like, most of the DNA-damaging agents, radiation, heat, and pressure, as well as other phages and plasmids (Cochran et al., 1998) could also cause prophage induction in bacterial cultures. Spontaneous inductions have been reported in several bacteria isolated from marine water (Chiura et al., 1999b) and from wine during malolactic fermentation process (Montserrat et al., 1998). This study clearly indicated that bacteria could spontaneously produce phage-like particles during their growth and is a normal natural phenomenon.

In activated sludge the phenomenon of spontaneous induction of phage-like particles could be significant if the important bacterial group contains prophage and under some physical or chemical stimuli they can cause lysis of bacterial cell from within. This could results in the deterioration of the process and gives other bacteria a chance to dominate and so efficiency of the process could be severely affected. One of significance of the released phage-like particles is that they could cause generalized gene transfer within various bacteria (Chiura et al., 1997a).

Although in the present study gene transfer via phage-like particles was not demonstrated but this phenomenon is highly possible in the complex activated sludge environment where quite a high number of bacteria along with their co-occurring lytic and temperate bacteriophages are reproducing.

4 CONCLUSION

In this study, bacteria and their co-occurring bacteriophages were isolated from activated sludge and the host ranges of the bacteriophages were examined on the bacterial isolates. Nine out of fifteen bacterial isolates were found to have their related lytic bacteriophages. All of eight bacteriophages tested for host range had broad host range. To determine the presence of temperate bacteriophage in activated sludge, the mutagen mitomycin C was used to induce temperate bacteriophage from bacteria isolated from activated sludge. The results of this study demonstrated that 13 out of 14 bacteria isolated from the activated sludge released bacteriophage-like particles in their growth medium. The overall experimental results of this study indicated that lytic as well as temperate bacteriophages are an active part of the activated sludge microbial ecosystem, having very close ecological relationship with their host bacteria. It is highly possible that bacteriophage can significantly affect the activities of the bacterial populations in activated sludge system, and thus affects biological wastewater treatment efficiency.

REFERENCES

Cochran, P. K., and J. H. Paul. 1998. Seasonal abundance of lysogenic bacteria in a subtropical estuary. *Appl. Environ. Microbiol.* 64:2308–2313.

Chiura, H. X. 1997a. Generalized gene transfer by virus-like particles from marine bacteria. *Aquat. Microb. Ecol.* 13:75–83.

Chiura, H. X., B. Velimirov, and K. Kogure. 1999b. Virus-like particles in microbial population control and horizontal gene transfer in the aquatic environment. In C. R. Bell, M. Brylinsky, and P. Johnson-Green (eds), Proceedings of the Eighth International Symposium on Microbial Ecology, in press. Atlantic Canada Society for Microbial Ecology, Halifax, Canada.

Ewert, D. L., and M. J. B. Paynter. 1980. Enumeration of bacteriophages and host bacteria in sewage and the activated-sludge treatment process. *Applied & Environmental Microbiology*, 39(3), 576–583.

Edlin, G., L. Lin, and R. Kudrna. 1975. Lambda lysogens of Escherichia coli reproduce more rapidly than non-lysogens. *Nature* (London) 255:735–737.

Fuhrman, J. A. 1999. Review article: Marine viruses and their biogeochemical and ecological effects. *Nature*, 399, 541–548.

Hantula, J., A. Kurki, P. Vuoriranta, and D. H. Bamford. 1991. Ecology of bacteriophages infecting activated sludge bacteria. *Applied & Environmental Microbiology*, 57(8), 2147–2151.

Hara, S., K. Terauchi, and I. Korike. 1991. Abundance of viruses in marine waters: assessment by epifluorescence and transmission electron microscopy. *Appl. Environ. Microbiol.* 57:2731–2734.

Hennes, K. P., and C. A. Suttle. 1995. Direct counts of viruses in natural waters and laboratory cultures by epifluorescence microscopy. *Limnol.Oceanogr.* 40:1050–1055.

Jenkins, D., M. G. Richard, and G. T. Daigger. 1993. Manual on the causes and control of activated sludge bulking and foaming, 2nd ed. Lewis publishers, Inc. Chelsea, Michigan. pp 19–82.

Jiang, S. C., and J. H. Paul. 1994a. Seasonal and diel abundance of viruses and occurrence of lysogeny/bacteriocinogeny in the marine environment. *Mar. Ecol. Prog. Ser.* 104:163–172.

Jiang, S. C., and J. H. Paul. 1996b. Occurrence of lysogenic bacteria in marine microbial communities as determined by prophage induction. *Mar. Ecol. Prog. Ser.* 142:27–38.

Jiang, S. C., C. A. Kellogg, and J. H. Paul. 1998c. Characterization of marine temperate phage-host systems isolated from Mamala Bay, Hawaii. *Appl. Environ. Microbiol.* 64:535–542.

Jiang, S. C., and J. H. Paul. 1998d. Significance of Lysogeny in the Marine Environment: Studies with Isolates and a Model of Lysogenic Phage Production, *Microbial Ecology*, 35:235–243.

Kitamura, S., and Y. Doi. 1994. Staining method of poly (3-hydroxyalkanoic acids) producing bacteria by Nile blue. *Biotechnology Techniques*, 8(5), 345–350.

Khan, M. A., H. Satoh, T. Mino, H. Katayama, F. Kurisu, and T. Matsuo. 2002. "Bacteriophage-host interaction in the enhanced biological phosphate removing activated sludge system". *Water Science and Technology*. 46, No 1–2, pp 39–43.

Lenski, R. E., and B. R. Levin. 1985. Constraints of the coevolution of bacteria and virulent phage: a model, some experiments, and predictions for natural communities. *The American Naturalist*, 125(4), 585–602.

Merzouki, M., J. P. Delgenes, N. Bernet, R. Moletta, and M. Benlemlih. 1999. Polyphosphate-Accumulating and Denitrifying Bacteria Isolated from Anaerobic-Anoxic and Anaerobic-Aerobic Sequencing Batch Reactors, *Current Microbiology*, 38, 9–17.

Marsh, P., and E. M. H. Wellington. 1994. Phage-host interactions in soil. *FEMS Microbiology Ecology*, 15, 99–108.

Montserrat, P. I., B. Albert, and L. F. Aline. 1998. Lysogeny of *Oenococcus oeni* (syn. *Leuconostoc oenos*) and study of their induced bacteriophages. *Current Microbiology*, 36, pp 365–369.

Ogunseitan, O. A., G. S. Sayler, and R. V. Miller. 1990. Dynamic interactions of *Pseudomonas aeruginosa* and bacteriophages in lake water. *Microbial ecology*, 19, 171–185.

Ogata, S., H. Miyamoto, and S. Hayashida. 1980. "An investigation of the influence of bacteriophages on the bacterial flora and purification powers of activated sludge." *Journal of General and Applied Microbiology* 26:97–108.

Paul, J. H., P. K. Cochran, and S. C. Jiang. 1999. Lysogeny and transduction in the marine environment. Microbial Biosystems: New Frontiers: Proceedings of the 8th International Symposium on Microbial Ecology. Bell CR, Brylinsky M, Johnson-Green P (eds). Atlantic Canada Society for Microbial Ecology, Halifax, Canada.

Proctor, L. M., and J. A. Fuhrman. 1992. Mortality of marine bacteria in response to enrichments of the virus size fraction from seawater. *Mar. Ecol. Prog. Ser.* 87:283–293.

Weinbauer, M. G., and C. A. Suttle. 1996. Potential significance of lysogeny to bacteriophage production and bacterial mortality in coastal waters of the Gulf of Mexico. *Appl. Env. Microbiol.*, 62, 4374–4380.

Reclaiming the Desert: Towards a Sustainable Environment in Arid Lands – Mohamed (ed.)
© 2006 Taylor & Francis Group, London, ISBN 0 415 41128 9

Cell immobilization: Applications and benefits in the treatment of ammonia in wastewater

F. Benyahia & A. Embaby
Department of Chemical and Petroleum Engineering,
College of Engineering, United Arab Emirates University

ABSTRACT: Polyvinyl Alcohol (PVA) has been shown to be an excellent immobilizing material for nitrifying biomass. The "freeze-thaw" cross-linking method was found to be superior to any other method reported in the literature and imparted PVA with excellent mechanical properties. The load and pH shock tolerance tests conducted with immobilized cells in bubble columns have shown that the nitrifying biomass recovered its activity in about 10 hours compared to days or weeks with free cells in conventional activated sludge systems.

1 INTRODUCTION

1.1 *The Nitrogen cycle and its impact on wastewater systems*

Nitrogen is an essential nutrient for biological growth, accounting for about 13% of the mass of cell proteins. Its availability to specific micro-organisms depends upon the chemical form it exists in, within the environment. Nitrogen forms vary from gaseous (N_2) to ionic in solutions (nitrate), with several other intermediate forms such as ammonium, nitrite etc... The forms that are ionic have various oxidation states and transformations from one oxidation state to another are brought about by living organisms. Figure 1 depicts the complex Nitrogen cycle with relevant paths from the various forms of Nitrogen, and in particular, the form most likely to be encountered in wastewater streams. Thus ammonium ion NH_4^+ in equilibrium with free ammonia NH_3 is the species concerned in this investigation because of its toxicity, foul smell and other problems such as dissolved oxygen depletion.

Although there are several methods available for the removal of ammonia (or ammonium ion) from wastewater streams, biological oxidation (referred to as nitrification) followed by biological denitrification (leading to nitrogen gas) are the preferred techniques for both technical and economic considerations.

1.2 *Biological nitrification*

Ammonia is oxidized to nitrate in the environment and in biological wastewater treatment by two groups of chemoautotrophic bacteria that operate in sequence. The first group of bacteria in this process of nitrification, represented principally by members of the genus *Nitrosomonas* oxidize ammonia to nitrite which is then further oxidized to nitrate by the second group, usually represented by members of the genus *Nitrobacter*. These oxidation reactions can be represented as (Painter, H.A., 1970):

$$NH_4^+ + 1.5O_2 \xrightarrow{\text{Nitrosomonas}} NO_2^- + 2H^+ + H_2O + \text{energy} \qquad (240 - 350\,\text{kJ}) \qquad (1)$$

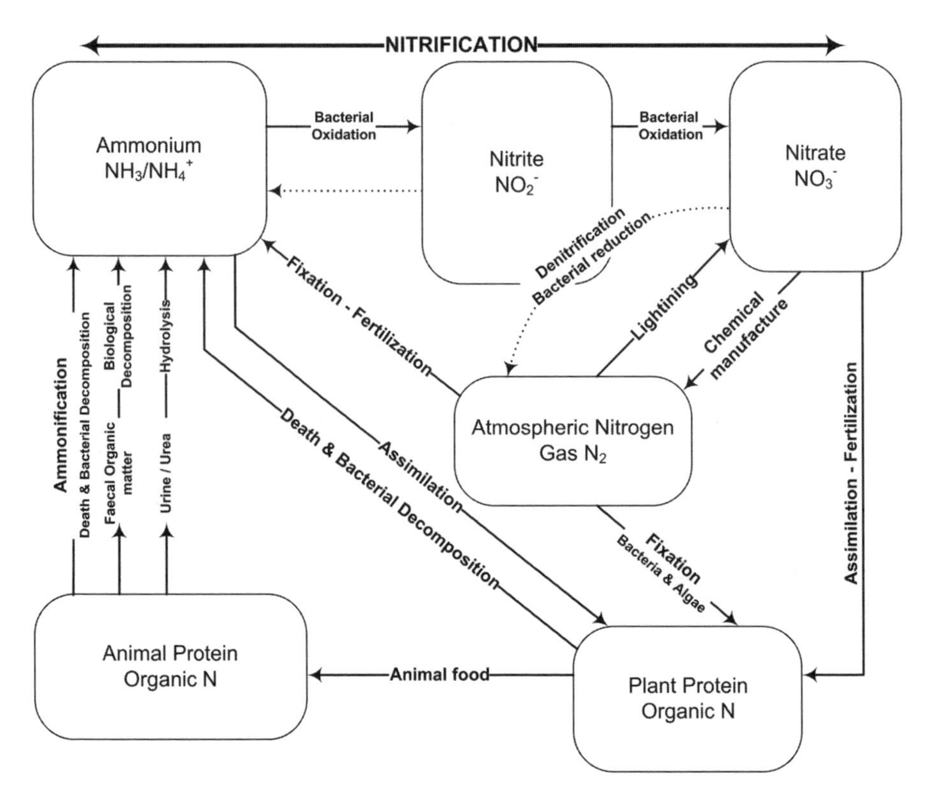

Figure 1. Details of the Nitrogen cycle (Environmental Protection agency, 1975).

$$NO_2^- + 0.5O_2 \xrightarrow{\text{Nitrobacter}} NO_3^- + \text{energy} \qquad (65-90\,\text{kJ}) \qquad (2)$$

The energy released in these reactions is used by the nitrifying organisms in synthesizing their organic requirements from inorganic carbon sources such as carbon dioxide, bicarbonate and carbonate. Nitrification is therefore closely associated with the growth of the nitrifying bacteria. In fact the major use of the energy gained from their oxidations of inorganic nitrogen is in the reduction of carbon dioxide to organic forms. Hagopian and Riley (1998) reported that the specific growth rate of nitrifying bacteria is unusually slow and that the energy produced has a low yield. He also added that a doubling time of 7–8 hrs is possible under ideal conditions (Hagopian and Riley, 1998). Other investigations that are more likely to resemble the reality resulted in higher doubling time data. For instance, doubling time was about 26 hrs for *Nitrosomonas* and 60 hrs for *Nitrobacter* (Hagopian and Riley, 1998, Belser, 1984). High doubling times like these is the key factor for the occurrence of washout of biomass in nitrification continuous processes using free cells. In addition, reaction (1) above clearly shows that as nitrification proceeds, hydrogen ions are produced, thus lowering the pH. Nitrification processes have been widely reported to be optimum at pH ranges between 7 and 8, with nitrification occurring satisfactorily at pH ranges 6.5 to 9 (Ruiz et al, 2003). Outside these pH ranges, inhibition has been noted. Therefore some form of buffering to a pH above 6 is necessary in a nitrification treatment process. In addition to sub optimal pH inhibition, other inhibitory factors and toxic substances have been reported to affect nitrification processes: low dissolved oxygen, presence of heavy metals, strong light (Schoberl and Engel, 1964; Painter, 1977; Johnstone and Jones, 1988).

A potentially attractive method to retain nitrifying biomass and offer some form of protection against adverse environment effects to nitrifying biomass is cell immobilization in suitable porous matrices. This topic has been the subject of numerous investigations are reported in the review of

Lozinsky and Plieva, 1998 and article of Sung-Koo Kim et al, 2000. The nature of the supporting matrices vary from natural to synthetic gels capable of encapsulating microorganism cells and allowing transport of nutrients and biochemical reaction products. Most natural gels like alginates and κ-carrageenan have been shown to be suitable immobilizing materials in terms of biocompatibility but suffer major drawbacks in terms of mechanical stability (Benyahia and Embaby, 2005a). In fact in academic studies, such gels disintegrate within a few days of preparation in aqueous media. On the other hand, synthetic gels offer better prospects even though most could have toxic properties towards the biomass (Willaert and Baron, 1996).

Amongst the most promising synthetic gels, Polyvinyl Alcohol (PVA) has been shown to offer the greatest potential for industrial applications. However, before it can be used for cell immobilization, it has to be cross-linked in order to create the three dimensional matrix that would entrap biomass. There are several methods reported in the literature for cross-linking PVA, but the most widely described include the so-called "boric acid" and "freeze-thaw" methods. The first method was shown to be inadequate for extended applications (Benyahia and Embaby, 2005b). The "freeze-thaw" method was shown to give excellent results and imparted both structural and mechanical stability to the immobilizing matrix. Thus the "freeze-thaw" method of cross linking PVA was adopted in this investigation and will be described further in the experimental sections.

1.3 *Load and pH shock in wastewater treatment plants*

Wastewater treatment processes can be subject to malfunctions due to equipment failure or operator errors. When these occur, there may be load and/or pH shocks to the biomass responsible for biological treatment. This is particularly serious for nitrification processes. Indeed, such load or pH shocks can cause the nitrification process to cease to be effective for several days, and even weeks before recovery takes place. In this investigation, the prospects of reducing the impact of load and pH shocks on nitrification using cell immobilization, will be studied.

2 EXPERIMENTAL

2.1 *Cell immobilization*

PVA solutions of stock polymer with an average molecular weight of 72,000 (BDH laboratory supplies, UK) were prepared at 80°C and left to cool to room temperature. Solutions of 10, 15 and 20% w/w PVA were prepared for various tests. Then a commercial slurry of *Nitrosomonas* cells (Cleveland Biotech Ltd, UK) was blended with the PVA solutions and uniformly homogenized. The overall thick mixture was poured into moulds and stored in a freezer set at −20°C for 24 hours. The frozen material was slowly thawed in a refrigerator set at 4°C and the cycle repeated five times. Upon completion of the last freezing cycle, the gel block was cut into small cubes with dimensions $1 \times 1 \times 0.8$ cm and allowed to thaw in the refrigerator.

The nitrifying biomass entrapped in the PVA gel matrix was reactivated in an aerated bubble column reactor with a medium containing 100 ppm ammonium carbonate $(NH4)_2CO_3$ in the absence of growth nutrients. The dissolved oxygen content of the bubble column was around 7.5 mg/L. The reactivation period lasted for about 5 days. Nitrification activity was monitored by collecting small samples from the bubble column and analyzing the medium for nitrite using a standard colorimetric method (spectrophotometer UV Philips, PU 8620 at 520 nm) and cells (Pye Unicam, UV/visible, 10 mm F.O.).

2.2 *Load shock tolerance tests*

The load shock tolerance tests were conducted in a series of bubble columns and consisted of adding a concentrated ammonium carbonate solution to the reactor such that the medium would have a concentration of 500 ppm of $(NH4)_2CO_3$. The nitrite liberation (measure of bacterial activity) was measured frequently before and after the shock application.

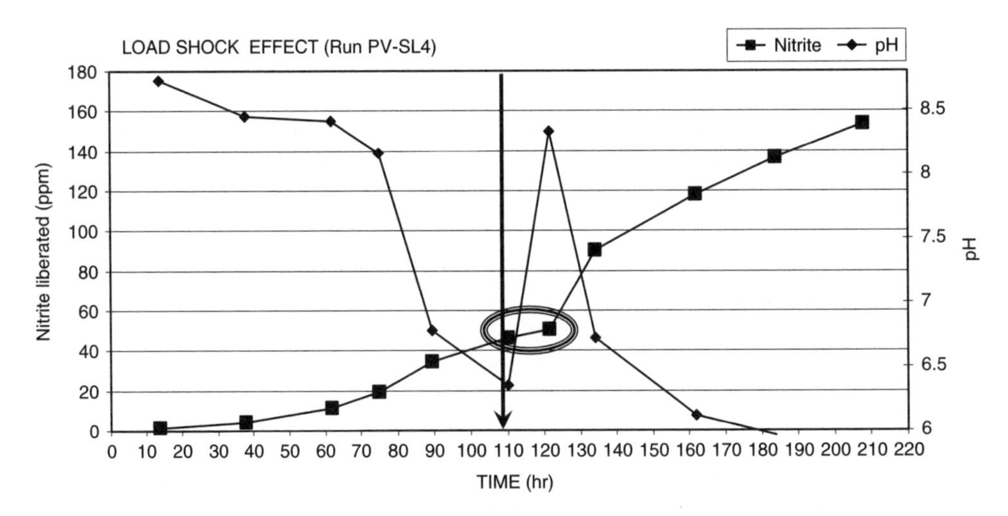

Figure 2. Response to load shock.

2.3 *pH shock tolerance tests*

The pH shock tolerance tests were conducted in a series of bubble columns and consisted of adding 0.1 M HCl solution to the reactor such that the medium would have a pH dropped to 5. The nitrite liberation (measure of bacterial activity) was measured frequently before and after the shock application.

3 RESULTS AND DISCUSSION

The objectives of the shock tolerance tests described above were to evaluate the merit of cell immobilization and anticipate its benefits to industrial applications. Thus, Figure 2 shows the response of the nitrification system using PVA immobilized *Nitrosomonas* to an ammonia load shock (from 100 ppm to 500 ppm as ammonium content). One can clearly see that the nitrite liberation was slowed down for about 10 hours and then recovered to a good level. The shock application was accompanied by a sharp increase in pH which came down following nitrite release. The response to the load shock is highlighted by the ellipse mark in Figure 2.

Figure 3 depicts the response of the immobilized nitrifiers to the pH shock. The pattern is similar to that of the load shock: a reduction in nitrite liberation for about 10 hours before nitrification activity resumed (shown as an ellipse highlight). The pH of the medium appear to have been slightly buffered by the carbonate in the solution. Nevertheless, the response clearly shows an activity declined followed by a remarkable recovery.

It appears that the PVA matrix provided some form of protection to the trapped cells and thus minimized the otherwise damaging effects of load and pH shock. The 10 hour recovery time compares quite favorably to the recovery period of several days, often extending to 2 weeks in traditional activated sludge processes with free cells.

4 CONCLUSION

In this work, Polyvinyl Alcohol (PVA) was shown to be a suitable immobilizing material for nitrifying cells. The "freeze-thaw" cross-linking technique imparted excellent mechanical stability to PVA. In order to evaluate the merits of cross linked PVA as a protective environment for nitrifying

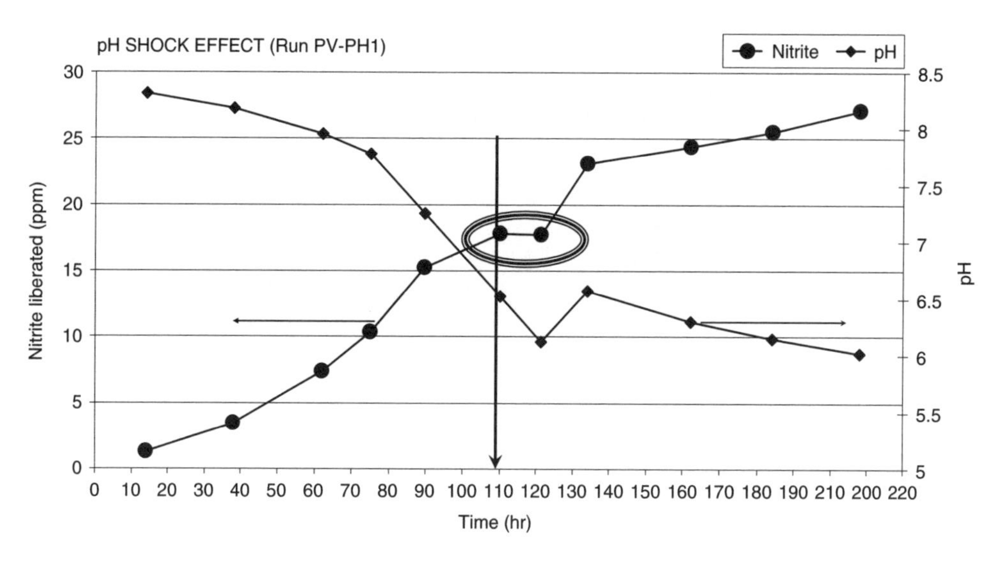

Figure 3. Response to pH shock.

biomass against shocks, a series of load and pH shock tolerance tests were successfully conducted and have shown that PVA does offer protection. The entrapped biomass recovered its nitrifying capacity in about 10 hours. The prospects of PVA as an immobilizing matrix for industrial applications are therefore good.

ACKNOWLEDGEMENTS

The authors would like to acknowledge the financial support of the UAE University research sector to this work under grant 02-11-7/17 and the award of a Master Scholarship to Ahmed Embaby by the Graduate Studies Deanship of the UAE University.

REFERENCES

Belser, L.W. 1984. Bicarbonate uptake by nitrifiers: effect of growth rate, pH, substrate concentration, and metabolic inhibitors. Appl. Env. Microbiol. Vol 48, pp1100–1104.

Benyahia, F. and Embaby, A. 2005a. Is immobilization a biomass retention technique only? *7th World Congress of Chemical Engineering,* Glasgow (UK). ISBN 0 85295 494 9.

Benyahia, F. and Embaby, A. 2005b. On the viability of PVA immobilized nitrifying bacteria for commercial applications. *7th World Congress of Chemical Engineering,* Glasgow (UK). ISBN 0 85295 494 9.

Environmental Protection Agency. 1975. Process Design Manual for Nitrogen Control. US. EPA, Washington DC.

Hagopian, D.S. and Riley, J.G. 1998. A closer look at the bacteriology of nitrification. Aquacultural Engineering. Vol 18, pp223–244.

Johnstone, B.H. and Jones, R.D. 1988. Effects of light and CO on the survival of a marine ammonium-oxidizing bacterium during energy source deprivation. Appl. Env. Microbiol. Vol 54, part 12, pp2890–2893.

Kim, S.K., Kong, I., Lee, B.H., Kang, L., Lee, M.G. and Suh, K.H. 2000. Removal of ammonium-N from a recirculation aquacultural system using an immobilized nitrifier, Aquacultural Engineering. Vol 21, pp139–150.

Lozinsky, V.I. and Plieva, F.M. 1998. Poly (vinyl alcohol) cryogels employed as matrices for cell immobilization. 3. Overview of recent research and developments. Enzyme and Microbial Technology. Vol 23, pp227–242.

Painter, H.A. 1970. A review of literature on inorganic nitrogen metabolism in microorganisms. Water Research, Vol 4, pp393–450.
Painter, H.A. 1977. Microbial transformations of inorganic nitrogen. Prog. Water Technol. Vol 8, pp3–29.
Ruiz, G., Jeison, D. and Chamy R. 2003. Nitrification with high nitrite accumulation for the treatment of wastewater with high ammonia concentration. Water Research, Vol 37, pp1371–1377.
Schoberl, P. and Engel, H. 1964. Behavior of Nitrifying Bacteria in the Presence of Dissolved Oxygen, Arch Mikrobiol. Vol 48: 393–400.
Willaert, R.G. and Baron, G.V. 1996. Gel entrapment and micro-encapsulation: methods, applications and engineering principles. Reviews in Chemical Engineering, vol 12, Freud Publishing House, London.

Reclaiming the Desert: Towards a Sustainable Environment in Arid Lands – Mohamed (ed.)
© 2006 Taylor & Francis Group, London, ISBN 0 415 41128 9

Author index